17⁵⁰/ee

Membranes of Mitochondria and Chloroplasts

Membranes of Mitochondria and Chloroplasts

ACS MONOGRAPH

Edited by EFRAIM RACKER
Albert Einstein Professor of Biochemistry and Molecular Biology —
Cornell University

VAN NOSTRAND REINHOLD COMPANY
New York   Cincinnati   Toronto   London   Melbourne

Van Nostrand Reinhold Company Regional Offices:
New York   Cincinnati   Chicago   Millbrae   Dallas

Van Nostrand Reinhold Company Foreign Offices:
London   Toronto   Melbourne

Copyright © 1970 by Reinhold Book Corporation
Library of Congress Catalog Card Number: 72–97168

Manufactured in the United States of America.
Published by Van Nostrand Reinhold Company
450 West 33rd Street, New York, N.Y. 10001

Published simultaneously in Canada by
D. Van Nostrand Company (Canada), Ltd.
15   14   13   12   11   10   9   8   7   6   5   4   3   2   1

GENERAL INTRODUCTION

American Chemical Society's Series of Chemical Monographs

By arrangement with the interallied Conference of Pure and Applied Chemistry, which met in London and Brussels in July, 1919, the American Chemical Society undertook the production and publication of Scientific and Technologic Monographs on chemical subjects. At the same time it was agreed that the National Research Council, in cooperation with the American Chemical Society and the American Physical Society, should undertake the production and publication of Critical Tables of Chemical and Physical Constants. The American Chemical Society and the National Research Council mutually agreed to care for these two fields of chemical progress.

The Council of the American Chemical Society, acting through its Committee on National Policy, appointed editors and associates (the present list of whom appears at the close of this sketch) to select authors of competent authority in their respective fields and to consider critically the manuscripts submitted. Since 1944 the Scientific and Technologic Monographs have been combined in the Series. The first Monograph appeared in 1921, and, up to 1967, 162 treatises have enriched the Series.

These Monographs are intended to serve two principal purposes: first, to make available to chemists a thorough treatment of a selected area in form usable by persons working in more or less unrelated fields to the end that they may correlate their own work with a larger area of physical science;

secondly, to stimulate further research in the specific field treated. To implement this purpose the authors of Monographs give extended references to the literature.

American Chemical Society

F. Marshall Beringer
Editor of Monographs

ASSOCIATES

Preface

The membranes of mitochondria and chloroplasts are the seat of the multi-enzyme systems which participate in the generation of biological energy. The precise role of the membrane in this process differs according to the two most popular current hypotheses of oxidative phosphorylation and photophosphorylation. According to the chemical hypothesis, the membrane functions as an organizer which is responsible for the efficient assembly of the proteins that catalyze the oxidation of substrates and the generation of ATP; in the chemiosmotic hypothesis the membrane must be vesicular, since it plays an essential role in the translocation of protons and separation of charges required for ATP formation. It appears therefore, that an understanding of the structure and function of the membranes is intimately linked to the elucidation of the mechanism of the generation of biological energy. Progress in the area of membrane-linked catalytic processes has been slow because of complexity of membrane organization which involves a network of numerous proteins and phospholipids. While soluble multi-enzyme systems have generally been rapidly elucidated once the individual catalysts had been separated, attempts to resolve and to reconstitute membrane-bound multi-enzyme systems have been only moderately successful. The major obstacles that have hampered progress are: (a) the strong forces of interaction that cement membrane components together; (b) the changes in the properties of resolved proteins (allotopic properties); and (c) the difficulties of devising assay systems for the individual catalytic components.

Nevertheless, recent years have witnessed important progress in some areas, particularly in the analysis of the ultrastructure of membranes and in the molecular biology of their phospholipid and protein components. A new concept which appears to be emerging from the work on mitochondrial membranes is that many, if not all, of the protein components play an important structural as well as catalytic role. Whether there is a major structural protein which has only an organizational function has not as yet been established.

The chapters in this book have been selected with emphasis on recent advances in our knowledge of membranes in general. Chapters on morphological aspects of membranes, on the physical chemistry of phospholipids and on model membranes deal with the more general problems in membranology. Current views of the mechanism of bioenergy production and of the role of the membrane in this process are discussed in a chapter on the inner membrane of mitochondria and chloroplasts. A chapter on the outer membrane on mitochondria deals with the composition of this membrane and its role in metabolism.

Ion transport has received considerable attention during the last decade, and numerous contributions bearing on the mechanism of the process have emerged from studies with mitochondria, chloroplasts and model membranes. A chapter on ion transport in mitochondria covers some of the recent developments in this rapidly expanding area of research. Finally, the assembly of mitochondrial membranes from their components has become a critical problem of interest to both biochemists and geneticists, and a chapter on the biogenesis of mitochondria therefore closes the book.

The assembly of this book has encountered difficulties not unlike those facing investigators who attempt to assemble a membrane. With time, patience, and the gracious cooperation of the authors, a book was assembled with some unity of purpose. Although limited to the structure and function of energy-generating membranes, it is hoped that the book will be useful to the investigators of membranes in general.

Ithaca, N.Y.   EFRAIM RACKER
August 1969

Contents

Experimental Phospholipid Model Membranes

T. E. Thompson and Fritz A. Henn
Department of Biochemistry
University of Virginia School of Medicine
Charlottesville, Virginia

I. INTRODUCTION

The object of this review is to discuss recent studies of several experimental models of biological membranes. Although many synthetic membranes have been investigated with the primary objective of understanding the biological system, this discussion will focus attention on physical models which have an organized phospholipid lamella of bimolecular dimension as a common feature.

The value of any experimental model depends on its relevance and its simplicity. It must accurately exhibit the salient features of the structure it is used to represent, and it must be simple enough to permit control of the operative parameters and, thus, allow analysis in greater detail than is possible with the referent structure. In essence, it should be a simple representative of the class of objects to which the referent structure belongs. Only if this condition is met, will the information obtained in model studies lead to useful and relevant conclusions. Phospholipid lamellar systems have many physical properties in common with biological membranes. Much of the available evidence suggests that they bear a close correspondence to the underlying structural element of natural membranes. In this review, a brief discussion of the central role played by membranes in biological systems and the development of current concepts of membrane structure will form the background, as well as a basis for the evaluation of the biological relevance, of studies on this type of model system.

A. Membranes as Organizational Elements in the Cell

An essential and unique characteristic of living systems is a spatial organization at dimensional levels which are supramolecular but subcellular. Many of the complex physical and chemical processes which the cell carries out depend upon the existence of assemblies of simple molecules and macromolecules. It is these organizational structures which form the basic elements of subcellular morphology. The membrane occupies a position of central importance among the organizational assemblies as the basic structural element of many subcellular organelles and the locus for much of the important biochemical activity associated with these organelles. Active transport phenomenon and the biochemical systems concerned with the energy transformations of oxidative phosphorylation and photosynthesis are among the most important of the biochemical functions localized in membrane structures.

In the case of oxidative phosphorylation, it is apparent that the complex of enzymes which forms the basis of this process requires the two dimensional structure of the membrane as a matrix in which the constituent reactions are spatially arranged. In principle at least, the organizational element of the membrane can play three roles in the coupled processes of electron transport and phosphorylation. First, it is possible that the spatial relationships between components of the systems are critical for both the basic activity and control of these processes. For example, it is not difficult to imagine that the rate of cyclic oxidation and reduction of two cytochromes adjacent to each other in the electron transport sequence depends critically on the fact that they be adjacent to each other spatially. Thus, the confinement of the cytochromes in a membrane matrix may be imposed to meet this spatial requirement. Second, because the membrane is a phase boundary, it may be that separation of reactant molecules by the boundary is an essential requirement, at one or more points, in the over-all process. Indeed, it is just this idea that forms the basis of the chemi-osmotic coupling hypothesis advanced by Mitchell.[112,113] In this scheme, the hydrogen and hydroxyl ions produced by electron transport are sequestered on opposite sides of the membrane. The electrochemical gradient thus generated drives an ATP hydrolysis reaction in the reverse direction through the agency of a membrane-bound asymmetric ATPase. Third, oxidative phosphorylation may constitute a heterogeneous reaction system utilizing not only the polar medium of the ambient aqueous phase, but also a nonpolar, water-poor medium as well. Organization of all or a portion of the mitochondrial lipid into a structure such as a bimolecular lamella in effect creates such a nonpolar, anhydrous phase.

Although there has been much speculation about the detailed mechanism of oxidative phosphorylation, there is no clear interpretation of the membrane-bound nature of this process. It is apparent, however, that knowledge of organization and structure-function relations in the mitochondrial membrane is essential to a complete understanding of this important process. In a similar sense, it is also evident that knowledge of the molecular organization of biological membranes is basic to an understanding of all other membrane localized phenomena including transport and photosynthesis.

In recent years, the problem of the molecular organization of biological membranes and the relations between this structure and its biologically important properties has been the focal point of intense research activity. Until very recently, the dominant hypothesis of molecular organization was that advanced over thirty years ago by Danielli and Davson.[38] In essence, the structure is postulated to consist of a bimolecular lamella of phospholipid arranged so that the long hydrocarbon chains of these molecules constitute its interior. Both faces of the lamella, formed from the polar moieties of the constituent molecules, are covered with a molecular layer of protein. The basic idea embodied in this hypothesis is that the structural form of biological membranes is to be attributed to the existence of a phospholipid lamella; a coherent lipid phase with a transverse dimension less than 100 Å. It is this concept which forms the bridge of relevance between biological membranes and the phospholipid lamellar models.

B. Current Concepts of Membrane Structure

The information on which the Danielli–Davson hypothesis is based, began to accumulate with Overton's study in 1895 of the permeability of the cell membrane.[129] In this investigation, he showed that lipid soluble substances readily penetrate the membrane, whereas polar substances do not. The suggestion that the cell membrane is a lipoid barrier resulted from this study. In 1925 Gorter and Grendel gave added support to this idea and refined it further.[55] They were able to demonstrate that the total lipid extracted from a single mammalian erythrocyte, when spread as a compressed monolayer at an air/water interface, occupied just twice the measured surface area of the cell. This result led them to propose that the lipid of the erythrocyte membrane is organized in a bimolecular lamella. Although it is now known that the Gorter and Grendel study suffered serious experimental shortcomings, their conclusion is supported by a recent redetermination of the basic data using modern monolayer techniques and methods of lipid extraction.[16] Additional information bearing on the structure of the cell membrane came from studies of its surface tension. In 1931 Harvey devised a method of measuring the surface tension of unfertilized eggs of *arbacia* by centrifuging them with just sufficient force to cause division of the egg into halves. From this experiment he calculated a surface tension of 0.2 dyne/cm for the outer surface of the egg.[72] The following year Cole obtained a value of 0.08 dyne/cm utilizing an independent method based on determination of the force necessary to flatten a spherical egg.[33] This low surface tension seemed incompatible with the properties of lipids known at the time. Furthermore, Danielli and Harvey were able to show that low surface tensions found on intracellular oil droplets were due to adsorbed protein.[39] This information lead Danielli and Davson in 1935 to propose the model for the plasma membrane shown in Figure 1-1.[38] The concept

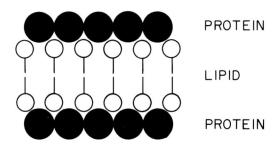

PROTEIN

LIPID

PROTEIN

FIGURE 1-1. Danielli model of membrane structure.

of the membrane as a lipid barrier was supported by studies on the optical birefringence of red cell ghosts carried out by Schmitt, Bear, and Ponder in 1936 [156] and by electrical impedance studies on cell suspensions reported by Cole and Curtis in 1938.[34] A membrane capacitance of about 1 $\mu F/cm^2$ obtained from the impedance work led directly to the conclusion that the membrane thickness must be less than 100 Å if the structure has a low dielectric constant, consistent with the known properties of bulk lipids.[34]

Strong independent support for the Danielli hypothesis was generated during the 1950's with the application of electron microscopy to biological systems. On the basis of extensive electron microscope studies, Robertson extended this hypothesis to include all membraneous elements of the cell.[142,143] This generalization, the unit membrane hypothesis, is based on the remarkable unity of structure displayed by biological membranes visualized in the electron microscope. The trilamellar structure seen after fixation and staining with a variety of agents is exhibited by membranes derived from almost any source. Although the over-all transverse dimension of the lamellar pattern is usually between 70 and 90 Å, there is some deviation from this range, and considerable variation in the widths of the three lamellar components.[47]

In recent years, important advances have been made in the identification and characterization of membrane lipids. It is now apparent that the bulk of these materials are amphipathic, that is, each molecule is comprised of distinctly polar and nonpolar moieties. In most animal cells these components are phospholipids[99,110]; whereas in plant or bacterial cells, sulfolipids[22] or glycolipids[23,100] may constitute a large proportion of these molecules. Although there is a great variety of naturally occurring amphipathic molecules, some uniform functional features exist within this structurally diverse family. The nonpolar region is always comprised of hydrocarbon chains of sufficient length to reduce water solubility to a negligible value. Furthermore, the chains always contain functional groups which promote liquid crystalline behavior in aggregates of these molecules in the biological temperature range. For example, the acyl groups of phospholipids found in animal cells are usually unsaturated, while amphipathic molecules of bacterial cells frequently contain, in addition, branched chains or cyclopropane groups.[100,179] A structural characteristic of the liquid crystalline state, assumed by phospholipids and related amphipaths in water, is an arrangement of the component molecules into concentric lamellae of bimolecular thickness. Thus, the propensity of these substances to form spontaneously membranous structures similar in transverse dimension to biological membranes constitutes strong support for the Danielli hypothesis.

Despite the evidence favoring this hypothesis, a number of important observations have recently led some investigators to question its generality. Benson has suggested that there are two classes of membranes which are both functionally and structurally dissimilar.[23] The first, which includes myelin, nuclear envelopes and the tonoplast, has essentially the Danielli structure. The second class, which includes such functionally complex units as mitochondrial and chloroplast membranes, is characterized by a structure formed by the fusion of lipoprotein subunits into a two-dimensional array. A consideration of the composition of several pure membrane preparations has led O'Brien to support this suggestion.[125]

Electron microscope studies, which originally formed the foundation for the unit membrane hypothesis, recently have afforded data of a contrary nature. For example, Sjöstrand has shown a marked difference in the thickness of mitochondrial membranes and plasma membranes of mouse kidney cells seen in the same thin section. He has suggested that this observation supports the idea that the membranes of the various subcellular organelles in a single cell are synthesized separately and do not have the common origin implicit in the unit membrane hypothesis.[162] Sjöstrand also has questioned the generality of the Danielli model on the basis of his ability to visualize globular subunits of mitochondrial and smooth endoplasmic reticular membranes in ultra-thin sections[163] in the electron microscope. Benedetti and Emmelot have observed globular subunits in the plasma membrane of rat liver cells using negative staining techniques. They have also reported the appearance of an hexagonal surface pattern on some membranes negatively stained at 37°C, which is absent when staining is carried out at 2°C. These observations have been used to support the suggestion that globular subunits exist in at least some membranes under certain conditions.[21] Electron microscope evidence for a subunit structure has also been presented by DiStefano.[45]

Green and co-workers have strongly advocated a similar view.[60,61] They have suggested that membranes are formed by the fusion of lipoprotein subunits into a two dimensional network.

Their model is based on two observations: (1) that mitochondrial membrane proteins bind hydrophobically to phospholipids with a precise stoichiometry, (2) that membranes can be fragmented by surface active agents into relatively homogeneous lipoprotein complexes. Similar complexes have been reported also when myelin is treated with lysolecithin,[53] and when bacterial membranes are treated with detergents.[150] These arguments and others have been brought together by Korn in a recent review in which he seriously questions the adequacy of the experimental evidence supporting the existence of a bimolecular lipid lamella in biological membranes.[95]

It is clear that two fundamentally different concepts of biological membrane structure are current today. One is based on the idea that the membrane is, in essence, a two dimensional array of macromolecular lipoprotein subunits in which each subunit acts as both a structural and a functional entity. The system properties of the membrane as well as its form are the summation of the characteristic properties of individual subunits which in turn derive from the detailed molecular architecture of each subunit. The central idea of the other concept is that a bimolecular lipid lamella constitutes the structural matrix of the membrane. In effect, this matrix is a continuous lipid phase which confers a limited set of physical properties on the membrane as a whole. The spectrum of physiologically important functional properties is gained by local modifications of the lipid matrix through its interaction with specific protein components. From either view point, the membrane is both a structural and a functional mosaic. The latter concept, which forms the basis of the Danielli hypothesis, is particularly amenable to examination through the use of experimental membrane models.

C. Phospholipid Membrane Models

Over the past twenty or thirty years a large number of investigations of the properties of nonbiological membrane systems have been carried out with the objective of understanding the physiologically important phenomena associated with cellular membranes. In general, two different types of model membranes have been chosen for these studies. On the one hand, membranes have been investigated which, though dissimilar to the referent biological system in both transverse dimensions and composition, are similar in a system property, such as permeability to a class of solutes, and, perhaps most important, are experimentally suitable for the study of the property. Although considerable insight into the biological problem has been derived from studies of this type, it is clear that relations between system properties and structure at the molecular level in the biological system can not be examined with this sort of model. On the other hand, membrane model systems have been investigated which are similar both in transverse dimension and composition to the biological system. Although these systems are, in principle, proper ones in which to examine the relations between properties and molecular structure, they have not always been suitable for the study of biologically relevant properties.

Among the earliest systems of this type to come under study as models of the biological membrane were monolayers at air/water and oil/water interfaces.[40,51] The relevance of this type of system is based on the Danielli concept of membrane structure, and, indeed, much of the information upon which the development of this idea is based has been derived from studies of interfacial systems. Detailed investigations of monolayers have led to conclusions which, when applied to the Danielli model, have set limits on the degree and type of molecular order to be found in the lipid phase of biological membranes. Important additional information bearing on the structure of biological membranes has been adduced from studies on water-poor smectic mesophases of phospholipids.[44,107,165,166] It has usually not proven possible, however, to examine in detail the mechanical, electrical, and permeability properties of the oriented lipid lamellae in these systems in terms which are directly relevant to structure–function problems in biological membranes.

In the past few years, however, a number of model membrane systems have been developed which are similar in both lipid composition and dimension to biological membranes, and, in addition, permit the experimental examination of biologically relevant structure–function relations. Tobias, Agin, and Powlowski in 1962 introduced a membrane formed from a millipore filter disc impregnated with phospholipid and cholesterol.[178] This system has been used extensively by Tobias and his co-workers to investigate the relations between lipid composition, divalent cation concentration, and membrane conductance which are markedly similar to those observed in electrically excitable biological membranes. A related system has been described by Saunders in which an insoluble phospholipid film is formed between an aqueous sol of the phospholipid and an aqueous electrolyte solution containing a divalent cation.[151,152] Schulman has developed an experimental system in which two aqueous phases are separated by a well-stirred, bulk, nonaqueous phase containing dissolved phospholipid.[146,147] Under certain conditions the rate limiting step for solute flux across the oil phase has been shown to be transport across the phospholipid layers at the two oil/water interfaces. This membrane system has been used to study the fluxes of ions and water across what is, in effect, a double monolayer of phospholipid.

Following the introduction of the Danielli hypothesis, a number of essentially unsuccessful attempts were made to generate in an aqueous phase a lipid membrane of bilayer dimension.[37,41,97] However, it was not until 1961 that the formation of a stable lipid bilayer in a form such that it could be utilized in experiments of direct biological relevance was first reported. The lipid bilayer membrane first described by Mueller, Rudin, Tien, and Westcott[121,122] has formed the basis of an extensive series of investigations carried out in a number of laboratories. Certain aspects of this work have been summarized in two excellent review articles.[177,148] More recently, Bangham and co-workers[13,14] have introduced the use of aqueous dispersions of phospholipid liquid crystals in order to examine the permeability properties of bimolecular phospholipid lamellae. This liquid crystal or "liposome" system is a potentially powerful model which in many respects is complementary to the Mueller–Rudin bilayer. The following discussion will be focused on these two phospholipid lamellar models: the Mueller–Rudin bilayer and the liposome. Attention is directed first to a description of the bilayer system, its formation, and basic physical properties. The discussion then turns to the relevance of these properties to the corresponding properties of natural membranes, and finally to a series of recent studies in which the basic properties of the bilayer model have been utilized to examine particular biological problems in detail. A similarly organized discussion of the liposome model follows that of the bilayer.

II. THE BILAYER SYSTEM

A. Formation

In 1961 Mueller and co-workers, described the formation of a thin lipid membrane separating two aqueous phases.[121,122] This planar membrane, generated from a complex mixture of brain phospholipids and other lipids dissolved in a chloroform–methanol solvent, formed by a process analogous to the generation in air of so-called black soap films.[124,128] Bilayer membranes have been generated in an aqueous phase from lipid sols and solutions encompassing a wide variety and number of components. In general, it has been found that the generating lipid solution must be composed of at least two components: one of which is an amphipathic molecule such as a phospholipid, the other a neutral lipid which is usually a straight chain paraffin.

Although several techniques of bilayer formation have been developed, the basic process of thin film generation can perhaps best be described with reference to a simple cell similar to that first described by Mueller and co-workers.[87] This cell, shown schematically in Figure 1-2, consists of two concentric chambers: the inner one a polyethylene cup 2 cm in diameter, and the outer one a 5 cm square lucite box. The wall of the inner chamber is pierced by a hole 1–2 mm in diameter in which the bilayer membrane is formed. Prior to membrane formation, the two chambers of the cell are filled with an aqueous electrolyte solution, and the cell and contents are brought to temperature equilibrium. Temperature control

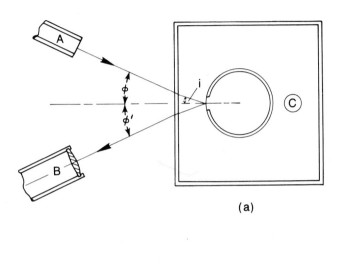

(a)

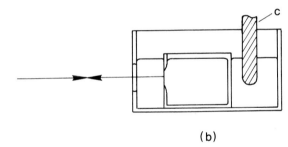

(b)

FIGURE 1-2. Diagrammatic representation of the cell in which membranes are formed; (a) plan, (b) elevation. A is a microscope illuminator, B is a 40× microscope, C is a thermistor element for temperature control. From Huang, C., Wheeldon, L., and Thompson, T. E. (1964). *J. Cell Biol.*, **8**, 151, Fig. 1. (Reproduced with permission)

is achieved by immersing the cell in a suitable thermostat. Isolation from mechanical vibrations, which frequently result in bilayer rupture, is gained by a vibration damping mounting for the thermostat.

To form the membrane, a small quantity of membrane-forming solution is applied with a fine sable brush to the aperture in the inner chamber wall. A film less than 100 μ in thickness, and frequently much thinner, can be formed across the aperture in this manner. This initial film is observed to thin continuously and spontaneously to a thickness of about 1000 Å. At this point, small regions of the film can be seen to undergo an abrupt transition to a much thinner structure, which has subsequently been shown to be 75 Å or less. The regions of extreme and limiting thinness rapidly increase in extent to fill the whole aperture. During this thinning process, the excess lipid is extruded to the circumference of the aperture to form a torus. The final state of the system is a bilayer membrane surrounded by a bulk phase lipid annulus contained within the aperture in the inner chamber wall.

The initial and final states of the system are illustrated diagrammatically in Figure 1-3. The sequence of events leading from the initial to final states

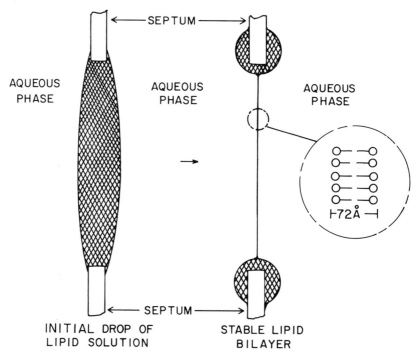

FIGURE 1-3. Schematic drawing illustrating the transition of a thick film to a bilayer.

can be followed in detail by observing the interference fringe systems generated by light reflected from the two film surfaces.[167] Thus, when the initial film is illuminated with white light, specular reflection is observed. As the gradually thinning film approaches a thickness of the order of the wave length of visible light, interference colors in the form of typical fizeau bands can be observed. The movements of these interference bands show that a considerable and sometimes violent flow is occurring in the thinning film. When a thickness of about 1000 Å is reached, specular reflection is again observed. It is in this film that the rapid and discontinuous transition to the terminal bilayer structure occurs. In order to observe the sequence of events leading to the formation of the bilayer, the system is illuminated with white light from an illuminator, A, (Figure 1-2), and the reflected light viewed with a low-power wide field microscope, B. The changes in the optical properties of the system occurring during the thinning process can be used to estimate quantitatively the thickness of the transition film and the terminal bilayer.

Much of the work on bilayers reported to date has utilized membranes formed by the brush technique. It has proven to be simple and, with practice, reliable. The principal drawback associated with the method is the difficulty in applying a reproducible quantity of membrane solution to the aperture. The size of the aperture together with the total amount of membrane solution introduced into it determines the relative sizes of bilayer area and torus. The relative sizes of these two phases are important for two reasons: (1) The mechanical fragility of the bilayer increases markedly as the size of the torus relative to the area of bilayer is decreased. (2) The question of whether a measured system property is to be associated with the bilayer, or torus, or both can be resolved only by deliberate variation of the relative sizes of the two phases. For example, in the case of electrical conductance or water permeability studies it is critically important to know the precise conduction or permeation routes across the system, since the properties of the bilayer and not of the torus are of interest.

Vreeman has introduced a method of bilayer formation which overcomes the disadvantage of the brush technique.[185] This method is based on a cuvette shown in Figure 1-4, similar to that used by Scheludko[155] and Duyvis[46] for soap film studies. The basic point of design is the connection of the aperture in which the membrane is formed to a syringe containing membrane solution. This is most conveniently accomplished through a channel within the wall of the septum separating the two aqueous compartments, as shown in Figure 1-4. A bilayer of controlled area is formed by the injection of a drop of membrane solution into the aperture with the syringe. Subsequent withdrawal of sufficient liquid gives a bilayer of the desired area. Using this cell with a 3 mm diameter aperture, Vreeman has been able to form bilayers with well-defined areas between 0.07–1.0 mm². The use of a cell based on the same design principle has recently been reported by Wobschall, Gordon, and Bolon[190] and by Howard and

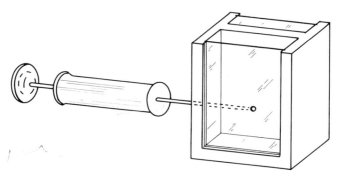

FIGURE 1-4. The teflon cell and syringe assembly used by Vreeman to form bilayers. From Vreeman, H. J. (1966). Konikl. Neded. Akad. van Wetenschappen (Amsterdam) Proc. **69-B,** 564, Fig. 2.

Burton.[83] Läuger, Lesslauer, Marti, and Richter have also described formation of bilayers with areas as large as 20 mm² using a basically similar injection technique.[98]

The small area of stable bilayer available for study has placed a serious limitation on certain types of investigations, particularly those concerned with permeability and optical properties. A number of successful attempts to make films of substantially larger areas have been reported. Van den Berg has described the formation of mechanically stable bilayers with areas up to about 50 mm².[182] His method consists of passing a septum containing an aperture through an interface between an aqueous phase and a membrane forming solution contained in a suitable cuvette. The design of the cell is such that, as the septum is passed vertically through the horizontal interface, it divides the aqueous phase into two compartments. Capillarity causes a layer of carbon tetrachloride placed under the aqueous phase to flow between the vertical edges of the septum and the cuvette walls to form a high resistance seal between the two aqueous compartments. Mechanical stability is achieved by allowing the top of the film to remain in contact with the floating membrane solution. A basically similar method of film formation has been utilized subsequently by Tien and Dawidowicz[175] to form bilayers of large area for optical studies. With their apparatus, the bilayer is formed on a polished teflon ring immersed in a continuous aqueous phase and, thus, cannot be used for permeability or electrical studies.

Generation of spherical bilayer membranes[83,130,131,123] has proven a very effective way of obtaining areas up to 100 mm² in a form suitable for the study of electrical and permeability properties. The spherical membrane system, which is very different in geometry from the planar membrane, in its final configuration consists of four phases: a spherical bilayer, a bulk lipid phase which takes the form of a lens covering less than 10% of the surface of the sphere, and two aqueous phases—one within the spherical volume encompassed by the bilayer, the other external to it.

A spherical bilayer is formed and immobilized for study in a density gradient in the following manner. A syringe microburette attached to a small diameter polyethylene tube is filled with an aqueous electrolyte solution. The end of the tube is coated with a membrane-forming solution and immersed in an aqueous density gradient. A drop of electrolyte solution coated with membrane solution is then discharged into the gradient from the syringe. The drop falls until it reaches its isopycnic position where it comes to rest. Figure 1-5 is a photograph of such a drop about 4 mm in diameter seen under white light illumination. The two spot reflections characteristic of a spherical surface can be seen. Within a few minutes the sphere shows a progressive asymmetry with increasing amounts of the bulk lipid phase migrating to the top of the sphere. The two spot reflections now show interference colors typical of films 2000–8000 Å thick. The drainage process is followed by an abrupt transition to a bilayer membrane surmounted by a lens of bulk phase lipid. This discontinuous transition has its direct analog in the planar bilayer when the excess lipid is extruded to form a torus around the circumference of the bilayer. Access to the interior of the sphere for electrical or permeability measurements is gained by the use of micropipettes. The properties of this type of bilayer have proven to be essentially identical to the corresponding properties of the planar membrane. In addition, the geometry of the system makes possible the determination of bilayer surface tension and direct isotopic determination of ion permeabilities.[132]

Tsofina, Liberman, and Babakov have described an entirely new method of bilayer formation which permits control of the surface pressure.[181] With this technique, the bilayer is formed by the apposition of two phospholipid monolayers, each at an oil/water surface, at least one of which is curved. The basic principle of the method is illustrated in Figure 1-6. Although this elegant procedure apparently permits control of film pressure, it may be that in permeability and electrical conductivity studies the existence of a very large bulk lipid phase affords an alternate conduction path between the two aqueous phases. This alternative route may dominate the system property and thus obscure the contribution of the bilayer itself.

In addition to cuvette designs which have been developed to solve problems of bilayer formation, a number of cells have been designed in order to study one or more specific bilayer property. Except in the cases of optical measurements,[101,175] fixation for electron microscope studies[77,123] and one bilayer composition investigation,[78] all cell designs permit formation of a bilayer which is interposed between two aqueous phases, as illustrated in Figure 1-2. For example, double chamber cuvettes designed specifically for the study of the osmotic permeability of bilayers to water

FIGURE 1-5. Spherical membrane during transition to a bilayer. The two spot reflections can be seen as well as two micro electrodes entering through the bulk phase cap.

have been described by a number of investigators.[27,62,86,138,139] Cells of the type which permit continuous sampling of one aqueous phase during permeability investigations have been designed in several laboratories.[2,189,191] Tien has recently reported the development of a special cell for interfacial free energy measurements.[173]

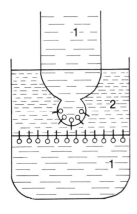

FIGURE 1-6. Principle of the formation of bilayers with controlled surface pressure. (1) aqueous phase, (2) phospholipid in heptane. From: Tsofina, L. M., Liberman, E. A. and Babkov, A. U. (1966). *Nature,* **212**, 682, Fig. 1.

B. Composition

Basic questions concerning the relation of the physical properties of the bilayer to its structure cannot be answered without knowledge of the chemical composition of the thin film. The composition problem has two aspects: First is the question of the number and type of components required in the membrane solution in order that bilayers may be formed. Second is the question of the actual composition of the bilayer itself. Answers to the first question have been obtained in a straightforward way by a number of investigators. On the other hand the second question is very much more difficult to answer. While it is quite clear that the composition of the membrane forming solution sets a limit on the chemical species which may be present in the bilayer, it is not necessary that the relative composition of the bilayer be the same as that of the parent bulk lipid phase.

Many reports have appeared in the past few years describing bilayer formation from lipid solutions of a variety of compositions. As a result of these investigations, it is apparent that the minimum restriction on the composition of the membrane-forming solution is that there be at least two components present, one of which is an amphipathic molecule, the other a liquid neutral lipid. The composition range of the membrane solution from which bilayers can be formed at a given temperature is usually small. The amount of the amphipathic component in the neutral lipid is commonly 1–5 weight per cent.

Phospholipids are the principal lipid component of most animal membranes and some plant membranes.[23] From the standpoint of biological relevance, therefore, this class of substances is the most interesting from which to form bilayers. A considerable number of phospholipids, both as pure compounds and complex mixtures derived from biological sources, have been used to generate these structures. Huang, Wheeldon, and Thompson[87] and Hanai, Haydon and Taylor[64,65] first reported the formation of bilayers using a purified phospholipid, egg phosphatidyl choline. Following this, a number of investigations have been described in which membrane solutions containing other purified phospholipids were utilized.[87,77,123,17,98,102,36,83] Bilayer membranes have been formed as well using mixtures of lipids derived from natural sources. In their pioneering work, Mueller and co-workers used a chloroform–methanol extract of bovine brain white matter as the source of amphipathic components.[121,122] Andreoli and co-workers have studied the properties of bilayers formed from the total lipid extract of sheep erythrocyte ghosts.[1,2] Mixed lipids derived from both human and rabbit red blood cells have been used by Wood and Morgan.[191] Recently, Howard and Burton[83] have described the formation of bilayer membranes from total lipid extracts prepared from a variety of natural membranes.

Amphipathic substances other than phospholipids have been used by several investigators.[123,168,175,174] Henn[76] has reported the formation and some physical properties of stable bilayer membranes using a purified mannolipid derived from the membranes of *Micrococcus lysodeikticus*. This type of compound, which is common in many plant membranes, particularly those of the chloroplast, may play a role analogous to that of the phospholipids in other types of membranes.[23]

Among the substances utilized as neutral lipid components are heptane,[101] n-tetradecane,[87] n-decane,[174,65,184,98] hexane,[185,101] octane,[118] n-dodecane,[173] α-tocopherol [121,122,181,159,42,2,189,17,27] and β-carotene.[101] In some systems the amphipathic molecule and neutral lipid have been dissolved in a mixed chloroform–methanol solvent.[87,121,122] Cholesterol has been present either in the lipids extracted from biological sources, or has been added to the membrane forming solution.

Knowledge of the chemical composition of the bilayer under a specified set of conditions is fundamental in understanding the relations between structure and properties. Although the prerequisite nature of the demand for chemical information about the bilayer has been recognized for some time, early attempts at composition analysis were unsatisfactory.[87,175] Recently, this problem has been attacked successfully by two independent methods with results which are generally in agreement. Henn and Thompson,[78] using radioactive components, directly determined the composition of bilayer fragments fixed with lanthanum nitrate and potassium permanganate; a procedure developed for the fixation of these membranes for electron microscopy.[77] Their study was carried out on bilayers formed from solutions of *E. coli B* phosphatidyl ethanolamine dissolved in decane, both with and without added cholesterol. The results clearly demonstrate the presence in the bilayer of all components of the bulk phase lipid solution. However, the molar ratio of decane to phosphatidyl ethanolamine in cholesterol-free membranes was found to be 11.8 ± 3.4, and in membranes containing cholesterol to be 10.5 ± 1.8. These figures are in marked contrast to the molar ratio of 700:1 for these components in the bulk lipid phase. In membranes containing cholesterol, 1.5 ± 0.2 was found for the molar ratio of cholesterol to phospholipid, a value closely similar to that in the bulk phase.

Redwood [139] and Cook *et al.*[35] have calculated bilayer composition from measurements of the interfacial tension of the membrane and the composition and interfacial tension of phospholipid monolayers at oil/water interfaces. These measurements, together with the reasonable assumption that the surface pressures of a lecithin monolayer at both the bulk and the bilayer interface are essentially equal, yielded a value of 1.41 for the molar ratio of decane to lecithin in the bilayer. A value of 0.40 was obtained for the molar ratio of cholesterol to lecithin in the bilayer surface[139] in a system containing lecithin, decane, and cholesterol.

The data both of Redwood and of Henn and Thompson show that there is considerable neutral lipid in bilayer membranes. Consideration of the calculated and measured surface areas together with electron micrographs of fixed membranes of the same type led Henn and Thompson to the conclusion that at least portions of the decane must be present as lenses of trapped hydrocarbon, while the remainder was unevenly distributed within the membrane to give a structure whose average thickness was 73 Å.[78] The considerably lower ratios for both decane to phospholipid and cholesterol to phospholipid obtained by Redwood are consistent with this interpretation; since these results are based exclusively on the surface properties of the bilayer and bulk phase, lenses of trapped decane would not contribute to the surface pressure. Taken together, the data strongly suggest that the bilayer membrane is a mosaic structure with portions of the area truly bimolecular in thickness, and other portions varying in thickness because of a trapped decane core. It is the average composition of the whole membrane that is obtained by the direct chemical method, while Redwood's technique yields the composition of the bimolecular regions only.

C. Stability

Haydon[73] has recently discussed the thermodynamic basis of bilayer formation following a general discussion of the thinning of liquid films presented by Mysels, Shinoda, and Frankel.[124] Drainage of the liquid contained between the two film faces through the action of gravity and capillary suction at the junction of film and torus, together with London-van der Waals forces and loss of liquid from the interior of the film all act to cause thinning. A stable film thickness can be obtained only when these forces are balanced by an opposing force which resists interpenetration of the molecules forming the opposing film faces. A consideration in quantitative terms of the relation between these forces and film thickness, has led Haydon to conclude that, in the range of thickness in which bilayer membranes occur, there exists a single stable energy minimum at a thickness of about twice the extended chain length of the neutral lipid molecules of the solvent. There is little detailed understanding, however, of the mechanism of bilayer formation and the general conditions for stability in particular systems.

D. Physical Properties

1. Bilayer Thickness.   One of the most interesting and biologically relevant structural features of the bilayer membranes is its extreme thinness. The very low optical reflectance of the structure indicates that the transverse dimension must be less than about 300 Å, and, hence, is in the size range of biological membranes. Considerable effort has been directed toward establishing its exact thickness. Three independent methods have been employed which yield results in substantial agreement.

The first critical determinations of bilayer thickness utilized an optical method based on quantitative measurements of the intensity of light reflected from the membrane combined with refractive index data derived from Brewster angle measurements.[85,87,172] The fundamental equations relating the thickness, d, of a thin film to the reflected light intensity, refractive indices, and angle of incidence are as follows:

$$I = 4I_0R^2 \sin^2 \frac{[2\pi n_1 d']}{[\lambda]} \qquad (1)$$

$$d' = d \cos r \qquad (2)$$

$$R^2 = \frac{[n_1 - n_0]^2}{[n_1 + n_0]} \qquad (3)$$

Here, I is the intensity of reflected light, I_0 the incident light intensity, λ the wave length of light, d' the optical path length through the film, and r the angle of incidence. n_1 and n_0 are, respectively, the refractive indices of the film and the medium surrounding it. It can be shown that there is a unique angle of incidence, the Brewster angle, at which the intensity will be zero for that component of the reflected light which has its electric vector in the plane of incidence. For the Brewster angle, β the following simple relation obtains:

$$n_1 = n_0 \tan \beta \qquad (4)$$

The value of n_1, the membrane refractive index, is calculated from experimental measurement of β and n_0 using Equation (4). By comparison of the optical parameters determined for both the bilayer and the transition film in which the bilayer forms, a thickness of 72 ± 10 Å was obtained for membranes derived from a solution of egg phosphatidyl choline and n-tetradecane in a chloroform–methanol solvent. This result strongly supports the suggestion that the thickness of the bilayer membrane is indeed close to twice the length of an extended phospholipid molecule. The same optical technique has been used by Tien and Dawidowicz to examine bilayers formed from glycerol distearate. These authors have reported a thickness of 50 ± 5 Å for films formed from this polyhydric alcohol in a hexane solvent,[175] a value about twice the extended length of this molecule.

A second approach to the estimation of bilayer thickness is based on measurement of the electrical capacitance of the structure. If the reasonable assumption is made that the membrane acts as a parallel plate condenser, then the following equation relates the experimentally determined capacitance C, to the dielectric constant ϵ, the area A, and the thickness d of the bilayer.

$$C = \epsilon A / 4\pi d \qquad (5)$$

Mueller and co-workers reported a capacitance of 0.7–1.3 $\mu F/cm^2$ for bilayers formed from bovine brain lipids, and found these values to be independent of frequency in the range 0 to 20 Kc, and voltages up to 50 mV.[123] Similar results have been reported from other laboratories.[9,65,66,158,189] Hanai, Haydon, and Taylor have determined the capacitance of membranes formed from solutions of lecithin in decane and showed it to be a linear function of bilayer area over a ten-fold range, and to be independent of frequency from 0–10^7 cps.[65,66,68] Using Equation (5) they calculated the thickness of the bilayer to be 48 ± 1 Å based on a measured capacitance of 0.38 ± 0.01 $\mu F/cm^2$ and the known dielectric constant of the bulk phase hydrocarbon. Although the dielectric constant in the bilayer is unmeasurable, it seems reasonable to use the value determined for the bulk phase. However, it is not known whether the two values are, in fact, identical. The difference in thickness obtained by the two methods is due in large measure to the fact that, whereas the hydrocarbon moiety of the constituent molecules contributes to the thickness determined by both methods, the polar moiety contributes only in the optical determination.[67,85]

A third, completely independent approach to the problem of thickness and structure is based on visualization of the bilayer in the electron microscope. Since the primary structural information about biological membranes comes from this source, it is particularly interesting to examine the bilayer under similar conditions. To date, two studies have been reported. Mueller and co-workers have presented a single electron micrograph of an $O_sO^-_4$ fixed bilayer membrane composed of mixed phospholipids derived from bovine brain white matter and additional neutral lipids.[123] This image, although about 90 Å in thickness, does not exhibit a trilamellar appearance. Recently, Henn, Decker, Greenawalt, and Thompson were able to obtain thin sections of a phosphatidyl ethanolamine-decane bilayer, fixed with $La(NO_3)_3 : KMnO_4$, which, when viewed in the electron microscope, revealed a continuous trilamellar image.[77] At high magnification (Figure 1-7) the trilamellar pattern of the synthetic bilayer is in complete agreement with the ubiquitous dark-light-dark pattern observed with cellular membranes. Microdensitometer tracings yield an average thickness of

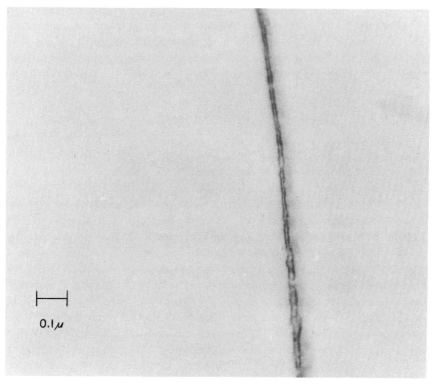

FIGURE 1-7. Electron micrograph of lanthanum nitrate, potassium permanganate fixed bilayer formed from phosphatidyl ethanolamine in n-decane. Taken by G. Decker and J. Greenawalt, Dept. of Physiological Chemistry, Johns Hopkins University, School of Medicine, Baltimore.

73 Å ± 22 Å, with a range of values from 37.5–116 Å. This result is in agreement with values obtained by optical and capacitance methods for similar bilayer membranes. In sections of the fixed bilayer, another image occasionally seen is a multilayer region similar in appearance to myelin figures (see Section III). These regions have a repeat distance of 38 Å and appear at the edge of the membrane. This repeat agrees well with values obtained for typical myelinic figures.[31] The larger thickness of the bilayer probably is due to the inclusion of decane in this structure. The trilamellar image also shows wide variation in spacing, whereas the multi-

ple-layered regions show a standard deviation in the repeat distance of less than 4 Å. Although the variation in thickness may be a fixation artifact, it is possible that the structure is, in fact, characterized by a varying thickness. If this is indeed the case, the liquid crystalline nature of this structure suggests that the variation in thickness seen in the electron microscope may reflect time-dependent fluctuations in thickness rather than a static structural characteristic. If the bilayer membrane exhibits fluctuations in thickness, then it is apparent that its local molecular organization must be time-dependent, and that the relations between the structure of this type of membrane and its physical properties must reflect this dynamic aspect of its organization.

2. Surface Tension.   The surface tension of bilayer membranes has been measured in several ways and found to be very low. The simplest procedure is to bow the membrane hydrostatically from a planar shape to a hemisphere of the radius R. The surface tension γ is related to the measured hydrostatic pressure, ΔP and to R by the following equation[170]:

$$\gamma = \Delta PR/4 \tag{6}$$

Values of about 1.0 dyne/cm have been obtained for bilayers formed from egg lecithin and n-tetradecane at 36°C.[87] Tien has used this method to measure the surface tension of membranes formed from oxidized cholesterol, cholesterol and dodecyl acid phosphate, or glycerol distearate as well as other surfactant molecules. In all cases, low values of the surface tension ($0.1 - 6.0$ dynes cm^{-1}) have been reported.[177] Pagano and Thompson, using spherical bilayers, have been able to determine a surface tension of 0.73 dyne/cm by measuring the radii of curvature of the bilayer and the two surfaces of the bulk phase cap (Figure 1-6). Redwood has also measured surface tensions, and reports similar low values for phospholipid bilayers.[139] The values obtained for these structures are in the range found for biological membranes.[170]

The composition, small transverse dimension, and low surface tension of the bilayer strongly support the view that the structure of the membrane is essentially that of a bimolecular lamella of amphipathic molecules arranged so that the polar ends of these molecules constitute each face of the lamella, and their hydrocarbon moieties its core. Within this core the hydrocarbon chains of the amphipathic component are in the liquid state, but exclude a volume roughly equivalent to their average extended chain length. Interspaced among these chains are molecules of the neutral lipid component. There may be regions containing variable amounts of neutral lipid which can give rise to minute lens-like structures of this component. If the amphipathic molecule is a phospholipid such as phosphatidyl choline, there is some evidence that the phosphate and trimethylammonium groups form a two-dimensional ionic lattice in the surface of the bilayer.[67,68] As will be seen, this picture of the molecular structure of the bilayer is supported by its electrical conductance, current–voltage characteristics, dielectric breakdown strength, and permeability properties.

3. Electrical Conduction Properties.   The resistance of simple lipid bilayer membranes unmodified by proteins or small molecules falls into two classes. Membranes with resistance $\geq 10^8\ \Omega\ cm^2$ are formed from a dispersion of most amphipaths in a neutral hydrocarbon such as decane. Lower resistances ($10^6 - 10^5\ \Omega\ cm^2$) result if the bilayer is formed from a chloroform-methanol solution of the amphipath and the neutral lipid.[123,85,65]

The principal problem in measuring bilayer resistance is that of defining the conduction path.[114] There are four possible paths through which current can flow: (1) surface conduction along the septum from one compartment to another, (2) between the bulk lipid in the torus and the hole in the support, (3) through the bulk lipid of the torus, and (4) through the membrane. The problem of surface conduction can be eliminated by careful sealing of the cell in which the membrane is formed. However, paths (2) and (3) can be ruled out only if it is possible to demonstrate that the resistance is strictly a function of membrane area. With low-resistance phospholipid bilayers, Miyamoto and Thompson have shown that the principal conduction route is through the membrane. The bilayer resistance of about $10^6\ \Omega\ cm^2$ is reproducible under carefully controlled conditions. It was also shown that the reciprocal of the resistance is a linear function of the concentration of NaCl in the aqueous phase in the range of 0.01–1.0 M.[114]

A number of investigators have reported difficulty in obtaining reproducible values[122,36,68] with high resistance bilayers. Haydon and co-workers studied the conductance as a function of area, varied over a 10-fold range, in this type of bilayer. These workers concluded that leaks around the circumference of the membrane probably form a significant conduction path. It has also been suggested that mechanical manipulations may be responsible for lowering the resistance of the membrane by causing a reversible alteration in structure.[1] The high resistance of this type of membrane is not markedly dependent on the salt concentration in the ambient aqueous phase. Tien and Dana[176] have reported the formation of a high resistance bilayer from a solution of cholesterol in dodecane with hexadecyltrimethylammonium bromide present in the aqueous phase. The resistance of this membrane is quite sensitive to the external ion concentration.

In many systems, if the lipid bilayer separates two aqueous electrolyte solutions of differing concentration, a diffusion potential develops. Potentials measured across membranes formed from lecithin and tetradecane in a chloroform–methanol solvent have been used to calculate transference numbers in two ion systems. Miyamoto and Thompson found the cation transference numbers listed in column 2, Table 1-1, for a series of monovalent cation chlorides.[114] Comparison with the bulk aqueous solution values listed in column 3 shows the transference number within the mem-

TABLE 1-1
CATIONIC TRANSFERENCE NUMBERS

1 Concentration[a] (moles/liter)	2 t+, Membrane @ 36°C	3 t+, Free Solution @ Temp. T	4 T (°C)
0.005 HCl[b]	0.365 ± 0.050	0.814	35
0.01 HCl[b]	0.502 ± 0.073	0.816	35
0.10 HCl[b]	0.293 ± 0.50	0.823	35
0.50 HCl[b]	0.657 ± 0.137	0.831	35
0.10 LiCl[b]	0.519 ± 0.132	0.317	25
0.10 NaCl[b]	0.698 ± 0.153	0.389	35
0.10 KCl[b]	0.719 ± 0.043	0.489	35
0.10 RbCl[c]	0.527 ± 0.066	0.494	18
0.10 CsCl[c]	0.706 ± 0.005	0.500	18

[a] Concentration of the aqueous phase in which the membrane was formed.
[b] Harned, H. S., and Owen, B. B. (1958). "Physical Chemistry of Electrolyte Solutions," Reinhold, New York.
[c] International Critical Tables, Vol. 6, p. 311, McGraw-Hill Book Co., New York, 1929.
From: Miyamoto, V. K., and Thompson, T. E. (1967). *J. Coll. Interface Science* **25**, 16.

brane to be quite different. Clearly the ionic environment in the membrane is different than bulk water. There is a similarity between the direction of the changes observed in the cation transference numbers in the bilayer and in bulk phase nonaqueous solvents as compared to the values obtained in aqueous solution. The transference number of Na+, K+, and Li+ are known to be larger in nonaqueous and in mixed solvent systems with dielectric constants lower than water.[70] In contrast to this behavior, the transference number of H+ has been shown to decrease as the weight fraction of dioxane is increased in a mixed dioxane–water solvent.[70] Since the large value of the transference number of the proton in aqueous solution depends upon the proton jump mechanism,* it is probable that the decrease observed in a dioxane–water system is due to a failure of this mechanism in the mixed solvent. On the bases of these observations it is probable that the membrane phase through which conduction occurs has a lower dielectric constant than bulk water, and its ability to participate in a proton jump conductance mechanism is much reduced.

Andreoli, Bangham, and Tosteson have determined cation transference numbers in membranes formed from the total lipid extracts of high potassium (HK) and low potassium (LK) sheep erythrocytes.[2] Since the relative permeabilities of HK and LK red cells to sodium and potassium are very different, it was of considerable interest to see if this selectivity difference

* For an explanation of this mechanism of protonic conduction in water see Robinson, R. A. and Stokes, R. H. (1959). "Electrolyte Solutions," Academic Press, Inc., New York, p. 121.

is exhibited by bilayer membranes formed from HK and LK lipids. The results indicate that the property of potassium selectivity does not in fact reside in the membrane lipids. Transference numbers for membranes made of either lipid extract are about 0.8 for both sodium and potassium. In contrast to this, however, are Na^+ and K^+ transferences of 0.55 and 0.59 reported by these authors for an egg phosphatidyl choline–decane bilayer. It is possible that the larger values obtained with the mixed lipid are due to the presence of acidic phosphatides in the total extract which confer a net negative charge on the bilayer. Henn, however, has examined the sodium transference numbers for bilayers formed from a variety of phospholipids in decane and found no dependence on the type of phospholipid.[76] This result suggests that the polar surface of the bilayer may not in itself be the major factor governing cation permeability.

Isotopic ion flux studies with planar bilayer membranes have thus far proven impossible because of the combination of the high specific resistance and small stable area.[114,185,34] The permeabilities of spherical bilayers to sodium and chloride have, however, recently been determined.[132] In membranes formed from egg phosphatidyl choline and n-tetradecane in a chloroform–methanol solvent, the chloride flux is 90.2 ± 0.8, and the sodium flux 0.39 ± 0.02 pmoles/cm²/sec at 30°C. Within the temperature range 10° to 30°C, a plot of the log [chloride flux] against the reciprocal of the absolute temperature is linear and yields a value of 10.7 ± 0.4 kcal/mole for the chloride permeation activation energy. The corresponding activation energy for sodium permeation is as yet undetermined. Following the treatment of Hodgkin,[80] values of less than 2.0 pmoles/cm²/sec for these fluxes can be calculated from the electrical conductance data for this system, and transference number data for a similar planar bilayer. The ratio of cationic to anionic fluxes is expected to be about 4.9 on the basis of the electrical transference data. The difference between the isotopic flux and the value calculated from the electrical parameters may be accounted for by assuming the existence of an auxiliary carrier mediated diffusion mechanism for chloride permeation. Large chloride permeabilities have also been reported for liposome preparations (Section III–B).

At low voltages most bilayer membranes show a constant resistance. However, if the impressed voltage exceeds about 30 mV, the resistance decreases progressively until dielectric breakdown occurs.[76,114] Mueller and co-workers found breakdown potentials ranging between 150 and 200 mV for bilayers formed from mixed brain lipids.[123] An average value of 198 ± 39 mV has been reported for low-resistance membranes.[114] In this system, breakdown is independent of the species of monovalent cation and the salt concentration in the aqueous phase.[114] Table 1-2 summarizes breakdown voltages for a number of additional bilayer systems.[76] It must be recognized that, because of the very small thickness of the bilayer, a transmembrane potential of 100 mV generates an electric field strength

TABLE 1-2

DIELECTRIC BREAKDOWN VOLTAGE AND SPECIFIC RESISTANCE OF SOME BILAYER MEMBRANES [76]

Membrane Composition	Specific Resistance (Ω cm^2)	Breakdown Voltage (mV)
P.E. (A)	2×10^5	180
P.C. (A)	8×10^5	200
Mannolipid (A)	1×10^7	>750
P.E. (B)	3×10^8	180
P.C. (B)	5×10^8	180
P.S. (B)	5×10^8	135

(A) = solution of $CHCl_3 - CH_3OH$ and tetradecane.
(B) = solution of cholesterol in decane.
P.E. = Phosphatidylethanolamine.
P.C. = Phosphatidylcholine.
P.S. = Phosphatidylserine.

within the membrane of $> 10^5$ V/cm. It is remarkable that the bilayer remains electrically intact in such a very large force field.

In general, there are two possible modes of current conduction in the bilayer membrane, electronic and electrolytic. None of the possible membrane components would be expected to participate in electronic conduction. However, because of the extremely small mass of the bilayer, it is possible that a compound capable of supporting electronic conduction, but present in only trace amounts in the initial membrane solution, is concentrated in the bilayer itself. On the other hand, the experimental evidence outlined above suggests that conduction is electrolytic with the ions of the aqueous phase acting as the principal current carriers. The magnitudes of the cation transference numbers argue that conduction is through a medium of relatively low-dielectric constant which is unable to sustain a proton jump conduction mechanism. An explanation of the nonohmic behavior of the bilayer might rest on the existence of ionic association within the conducting phase of the membrane as a result of its apparently low dielectric constant.[54,25] If this were the case, the degree of association would decrease, and hence, the conductance of the system would increase with increasing field strength at high values of the electric field.[126] Qualitatively, this is the behavior exhibited by the bilayer.

Conduction observed in liquid hydrocarbons has been postulated to be due to trace ionic and water impurities.[105,50] Upon careful purification, the resistivities were observed to increase. Current–voltage studies of purified n-hexane[105,192] showed an initially linear increase of current with potential. This ohmic conduction was postulated to be owing to impurities. Zaky *et al.*[192] observed a region of saturation in which no appreciable increase in current was observed with increasing voltage. This saturation region was followed by a nonlinear increase in current which persisted until dielectric breakdown occurred. The saturation region was not observed by Lewis.[105]

The breakdown mechanism of liquid dielectrics is not known, although it could be due to massive ionization of the dielectric caused by emission of electrons from the surface of the metallic electrode.[105] A similar mechanism could not be expected to operate in the bilayer membrane where, in effect, the "electrode" surfaces are the membrane/external aqueous phase boundaries. However, it might be possible that dielectric breakdown in the bilayer occurs because a critical concentration of ions in the membrane has been exceeded by invasion of ions from the external aqueous phase under the influence of the high field strength. It is interesting to note, in this respect, that in liquid dielectrics the field strength required for dielectric breakdown was observed to increase with purification of the liquid.[25]

On the other hand, it is possible that both dielectric breakdown and the nonlinearity of the current–voltage curves are due to a structural change in the membrane induced by the high fields. Such a change might be expected to have a time constant which might be large enough to give rise to a "dispersion" region in the breakdown voltage–frequency curve. No such region has been found in the interval from 20–20,000 cps.[158]

4. Permeability.   The permeability of bilayers to water has been determined by two methods. The first depends upon measurement of the flux of isotopically labeled water across the bilayer, and the second upon the measurement of the net volume flux of water moving across the structure in response to an osmotic gradient. Huang and Thompson, in a detailed study of the water permeability of lecithin–tetradecane membranes made from a chloroform–methanol solution of these materials, found the coefficient for tritiated water flux to be 4.4×10^{-4} cm/sec. For the osmotically driven flux, values ranging from $17.3–104 \times 10^{-4}$ cm/sec. were obtained depending on the lipid preparation used to form the bilayers.[86,171] Haydon and co-workers have reported values of 2.2×10^{-4} cm/sec and 19×10^{-4} cm/sec for the isotopically and osmotically determined coefficients, respectively, in lecithin–decane bilayers.[62,63] In all of these studies, the permeabilities of the bilayer were found to be within the range of values reported for biological membranes. Also, the osmotic value was found to be considerably larger than the isotopic coefficient, a situation identical to that obtaining in biological systems. The possibility that in the bilayer system the difference between the two types of coefficients is an isotope effect has been ruled out on the basis of the fact that the osmotic coefficient in pure D_2O is identical to that obtained in H_2O.[86,171,63] Although it initially seemed probable that the difference was an intrinsic bilayer property, recent experimental work by Cass and Finkelstein,[27] Haydon and co-workers[139,63] and Vreeman[185] has shown almost beyond doubt that it is due to the existence in the isotopic flux experiment of unstirred layers of aqueous phase adjacent to the bilayer. Similar layers are essentially absent in the osmotic experiment because of the relatively large concentration gradients generated adjacent to the bilayer by the flow of bulk water.

The identity of the isotopic and osmotic coefficients together with the liquid crystalline nature of the membrane and its high electrical resistance argue strongly against the existence in this structure of pores filled with water possessing bulk characteristics. On the other hand, the known solubility and diffusion coefficient of water in long chain hydrocarbons and the transverse dimension of the bilayer permit the calculation of permeability coefficients on the basis of a simple partition–diffusion model which are in good agreement with experimental values.[63,139,138,27] The magnitude of the temperature dependence of the osmotic coefficient also is consistent with this model. A value of 14.6 kcal/ml has been reported by Redwood [139] while 12.7 kcal/moles has been found by Price and Thompson for bilayers of slightly different composition.[138] If 4 kcal/mole is taken to be reasonable for the diffusion activation energy, about 8 kcal/mole for the partition process, the two processes in series would be expected to show an activation energy of about 12 kcal/mole. This is in reasonable agreement with the experimental values.

The permeability of the lipid bilayer to a number of other substances has also been examined. Vreeman measured the permeability of lecithin–decane membranes to ^{14}C-labeled urea, glycerol, and erythritol, and found tracer permeability coefficients of 4.2×10^{-6}, 4.6×10^{-6}, and 7.5×10^{-7} cm/sec, respectively.[185] These values all lie in the range found for the penetration of these substances through cell membranes. These data have been used by Vreeman[185] to argue against the existence of structured pores in the bilayer membrane. Wood and Morgan formed bilayers with red cell lipids extracted from human and rabbit erythrocytes, and examined their glucose permeability.[191] They found permeability coefficients for D-glucose of 6×10^{-8}, and 5×10^{-8} cm/sec for membranes formed from rabbit and human erythrocyte lipids, respectively. These results, when compared with the carrier-mediated glucose transport of human cells, forced the conclusion that the permeability of this simple model could not account for glucose transport in human red cells. However, the model membrane does account for the permeability of the rabbit red blood cell to glucose. The permeability of these synthetic films to several organic solutes has been investigated by Bean and Shepard,[19] who found that only un-ionized organic molecules have appreciable permeabilities. For example, indole has a permeability coefficient of 2.0×10^{-4} cm/sec, whereas tryptophan permeability is unmeasurable ($< 10^{-7}$ cm/sec).

Thus, not only are the electrical conduction properties consistent with a bimolecular lamellar structure for the bilayer, but it is also apparent that the permeabilities to water and other small molecules strongly support this view.

E. The Bilayer as a Model for Biological Membranes

If the Danielli hypothesis is correct and bimolecular lipid lamellae are structural components of biological membranes, then it is reasonable to expect that the physical properties of these membranes should display at least some of the physical properties of the component lamellae. A comparison is made in Table 1-3 between the characteristic bilayer properties, discussed in the preceding section, and the corresponding properties of natural membranes. It is apparent that striking similarities exist between the two sets of properties. In general, these similarities constitute strong presumptive evidence for the existence of bilayer regions in biological membranes. It is clear, however, that if this is indeed the case, the simple bimolecular lipid lamella can be the origin of only a very few of the physiologically important properties of biological membranes. The majority of these properties must be the products of the interactions of the lipid components with specific membrane proteins or other substances. The discussion that follows is focused on biologically relevant modifications in the basic properties of the lipid bilayer that are the result of the interaction of this structure with specific molecules.

The point of greatest discrepancy evident in Table 1-3 is electrical resistance. Characteristically, bilayer resistance is three to five orders of magnitude larger than the maximum value found in biological membranes. Thus, while the lipid barrier of the bilayer is truly a dielectric, such is not, in a strict sense, the case for the biological system. In addition, diffusion potential measurements on bilayer membranes separating aqueous phases containing alkali metal chlorides show variance on two points with respect to natural membranes. First the fraction of current carried by the cation in the bilayer is about twice the fraction carried by the chloride ion. Second, the bilayer permeabilities for the alkali metal ions are all about the same.[114,2] It is well known that in many biological membranes chloride is much more permeable than either sodium or potassium, and there exist, in general, marked differences among alkali metal ion permeabilities.[188,157,81] On the other hand, there is a close correspondence between the bilayer and natural membranes with respect to the electrical parameters of capacitance and dielectric breakdown strength.

Since the bilayer membrane is a high resistance barrier possessing essentially no capacity to discriminate between ions, the interesting question arises as to whether this structure can be modified in some manner so that it will display the ion permeability characteristics associated with the functional properties of one or another type of biological membrane. The answer to this question is yes. The conductance of the bilayer can be

TABLE 1-3
COMPARISON OF SOME PROPERTIES OF BILAYERS AND BIOLOGICAL MEMBRANES

Property	Biological Membranes (20–25°C)	reference	Bilayer (36°C)
1. Electron microscope image	trilaminar	a	trilaminar
2. Thickness, Å	60–100	a	60–75
3. Capacitance, μmf/cm^2	0.5–1.3	b	0.38–1.0
4. Resistance, Ω cm^2	10^2–10^5	c	10^6–10^9
5. Dielectric breakdown, mV	100	d	150–200
6. Surface tension, dynes/cm	0.03–1	e	0.5–2
7. Water permeability, μ/sec.	0.37–400	f	31.7*
8. Activation energy for water permeation, kcal/mole	9.6*	g	12.7*
9. Urea permeability μ/sec $\times$ 10^2	0.015–280	h	4.2†
10. Glycerol permeability μ/sec $\times$ 10^2	0.003–27	h	4.6†
11. Erythritol permeability μ/sec $\times$ 10^2	0.007–5	h	0.75†

† 20°C
* 25°C

[a] Elbers, P. F. (1964). "Recent Progress in Surface Science," Vol. 2, p. 443, eds. Danielli, J. F., Pankhurst, K. G. A. and Riddiford, A. C., Academic Press, New York.
[b] Pauly, H., and Packer, L. (1960). *J. Biophys. Biochem. Cytol.* **7**, 603.
[c] Cole, K. C. (1959). "Proceedings of the First National Biophysics Conference," ed. Quastler, H. and Morowitz, H. Yale University Press, New Haven.
[d] Shanes, A. M. (1958). *Pharmacol. Rev.* **10**, 59.
[e] Ackerman, E. (1962). "Biophysical Science," pp. 236–239. Prentice-Hall, Englewood Cliffs, N.J.
[f] Dick, D. A. T. (1959). *Int. Rev. Cytol.* **8**, 387.
[g] Hampling, J. (1960). *Gen. Physiol.* **44**, 365.
[h] Vreeman, H. J. (1966). *Proc. Koninkl. Ned. Akad. Wetenschap* **69B**, 542.

brought into the biological range, and ion selectivity can be conferred on the system by the addition of a variety of substances to either the aqueous phase or the membrane solution.

A comparison of the specific resistances of several different types of simple bilayers shows that considerable variance exists among systems. For example, there is a difference of about two orders of magnitude between the resistance of a lecithin–decane bilayer[65] and one formed from lecithin and tetradecane dissolved in a chloroform–methanol solvent.[114] The lower resistance of the latter bilayer is shown to be due to the presence of methanol in the membrane by the fact that the addition of low concentrations of methanol to the aqueous phase of a lecithin–decane membrane causes the resistance to fall to about 1×10^6 Ω cm^2. The resistance, however, is unaffected by a similar addition of chloroform.[120] Thus, the presence of a small uncharged organic molecule of relatively high dielectric constant lowers bilayer resistance. A number of other small organic molecules such as dipicrylamine, tetraphenylboron, uranylacetate,[120] and dinitrophenol [24] have been shown to lower the resistance, and, in some cases, confer ion selectivity on the system. Similar effects on resistance have also been reported for a number of surfactants.[159]

The role of 2,4-dinitrophenol as a mitochondrial uncoupler has been examined in light of the Mitchell hypothesis by studying the effect of this substance on the resistance of bilayer membranes. The chemiosmotic coupling hypothesis for oxidative phosphorylation proposed by Mitchell postulates that electron transport generates a transmembrane pH gradient which drives the enzymatic synthesis of ATP from ADP and inorganic phosphate through the reversal of a membrane-bound asymmetric ATPase.[112,113] On the basis of this scheme, Mitchell has proposed that uncouplers such as dinitrophenol are not inhibitors of specific enzymatic reactions as such, but uncouple because they act as transmembrane proton carriers and thus prevent the generation of the transmembrane pH gradient necessary for ATP synthesis. Using a lecithin–decane bilayer, Bielawski, Thompson, and Lehninger have shown that dinitrophenol lowers the electrical resistance of this structure by an amount that would be sufficient to account for the uncoupling effect of DNP in mitochondria.[24] Subsequent studies by Hopfer, Lehninger, and Thompson[82] and by Skulachev and co-workers[164] have demonstrated that bilayer resistance is lowered by a variety of uncouplers. On the basis of diffusion potential measurements, these studies have shown in addition that the increased conductivity is due exclusively to hydrogen and/or hydroxyl ion conduction.

Certain ions have been shown to alter the resistance of bilayer membranes. For example, Miyamoto and Thompson observed that Fe^{+++} lowered the resistance of a lecithin–tetradecane bilayer by a factor of at least 10^3 at concentrations below 10^{-5} M.[114] Concomitant with this change, the bilayer becomes permselective for anions. Although Fe^{+++} is known to promote auto-oxidation of unsaturated fatty acids, studies on bilayers formed above 50°C from synthetic lecithins with saturated acyl groups have shown that the effect of iron cannot be the result of auto-oxidation. These results may be related to the effects of iron on some biological membranes. It is interesting that, although Fe^{+++} produces a precipitous fall in bilayer resistance, Cd^{++}, Mn^{++}, and Cu^{++} all appear to increase resistance of phosphatidylcholine membranes when the concentration of these ions in the aqueous phase is below 10^{-3} M.[114] Läuger, Lesslauer, Marti, and Richter, investigating bilayers similar to those studied by Hanai and co-workers, found that I^- lowers the resistance over 10^3 fold.[98]

Although interaction with the various substances described above lowers the resistance of bilayers, and in some cases the ion selectivity is altered, in none of these instances has cation discrimination of the type seen in biological membranes been observed. It is of considerable interest, however, that both low resistance and cation specificity remarkably similar to that observed in biological membranes can be induced in bilayer membranes by a number of macrocyclic peptides. Bangham, Standish, and Watkins[13] and Chappell and Crofts[31] observed that valinomycin, a cyclic dodecadepsipeptide, produced a markedly larger increase in potassium than sodium permeability in liposome preparations (see Section III). A similar effect on bilayer systems was first reported by Mueller and Rudin[117] for this peptide, as well as the cyclohexadepsipeptides, enniatins A and B, the tetralactones, monactin and dinactin and gramicidin A, B, and C. Studies of the effects of valinomycin on bilayer membranes of several different compositions and under a variety of conditions of pH and ionic strength have been carried out subsequently by Lev and Buzhinski[104] and by Andreoli, Tieffenberg, and Tosteson.[5]

The addition of valinomycin to the aqueous phase of a lipid bilayer at a concentration of 10^{-7} g/ml causes a drop in resistance of about five orders of magnitude. Concentrations as low as 10^{-10} g/ml produce a measurable resistance drop in bilayers formed from a number of purified phospholipids. Bilayers formed from mixtures of these compounds, as well as from oxidized cholesterol derivatives, respond in a similar manner to low concentrations of this peptide. The low-resistance state induced by valinomycin is in the range of 10^3 Ω cm^2, a value well within the range observed for biological systems. Perhaps the most remarkable effect, however, is the cation selectivity of the modified bilayer. The permeability to K$^+$ is two to three hundred times larger than the sodium permeability. The permeabilities of the alkali metals and H$^+$ have the following order, a series that is the same as that observed in most natural membranes:

$$P_{H^+} > P_{Rb^+} > P_{K^+} > P_{Cs^+} > P_{Na^+} > P_{Li^+}$$

These same cyclic antibiotics have been shown to affect cation permeabilities in several biological systems. Valinomycin, for example, has been shown to uncouple oxidative phosphorylation and stimulate respiration in mitochondria. These effects, which are associated with an uptake of K$^+$ and a concerted extrusion of H$^+$ by the mitochondria[30,58,115,136,108,160,161] have been shown to be sensitive to relatively minor changes in the structure of the peptide.[136] It is of considerable interest that the effectiveness of compounds of this type in the mitochondrial system is faithfully reflected in the alterations of ion permeability produced by them in the bilayer system.[117] Recently Tosteson and co-workers have reported that valinomycin produced a marked increase in permeability of sheep erythrocytes to K$^+$ but had little effect on either the passive or active moment of Na$^+$.[180]

The mechanism of action of these macrocyclic compounds in either the bilayer or the biological system is unknown. It seems certain, however, that since the effects on ion permeability in the bilayer are dependent on details of the macrocyclic structure, but not on the composition of the lipid of the bilayer, the effect is most probably due to a primary inter-action between the cation and the peptide ring rather than the result of a rearrangement of the bilayer lipids induced by the peptide.[117] It is tempting to speculate that the peptide ring acts as a hydrophobic cage for the ion, and thus permits it to pass with relative ease through the hydrocarbon portion of the bilayer.[5,117] It is interesting to note that another class of antibiotics including nigericin and dianemycin, which has been shown to induce ion permeability changes in mitochondria, has no effect on the lipid bilayer system.[137] Pressman and co-workers have suggested that this class of antibiotics, which possess an ionogenic group, induce permea-bility changes in the natural membrane by formation of an electrically neutral complex with cations.[137] Since a neutral complex would not con-tribute to current conduction, no effect on bilayer resistance would be ex-pected. It would be interesting to examine this hypothesis by measuring the cation flux through nigericin treated bilayers isotopically.

Recently, Andreoli and Monohan have found that the two polyene anti-biotics amphotericin-B and nystatin cause a marked reduction in the re-sistance of bilayers containing cholesterol, but not in cholesterol-free sys-tems.[3,4] Bilayer membranes formed from sheep erythrocyte lipids contain-ing cholesterol and n-decane characteristically have a resistance of about 10^8 Ω cm^2 and show cation selectivity ($t_{K^+} = 0.85$, $t_{Cl^-} = 0.15$). If con-centrations of nystatin less than 2×10^{-5} M or amphotericin-B less than 2×10^{-7} M are added to the ambient aqueous phase, the resistance falls to about 10^2 Ω cm^2. Under these conditions $t_{Cl^-} = 0.92$. The related anti-biotic, filipin has no effect on membrane resistance.

In earlier studies, van Deenen and co-workers had shown that the polyenes interacted preferentially with lipid monolayers[43] and phospholipid bilayers[184,94] only if these structures contained cholesterol or related sterols. In the bilayer studies, the result of interaction was observed to be a marked mechanical instability. Weissman and Sessa have recently examined the interaction of these antibiotics with phospholipid liposomes (Section III).[187] Using the leakage rates of glucose, phosphate, and chromate as a measure of liposome integrity, these investigators have concluded that the presence of cholesterol is a requirement for amphotericin-B and nystatin interaction, but not for the interaction of filipin.

The polyene antibiotics are cyclic lactones with a number of conjugated double bonds in the ring.[186,127] Amphotericin-B and nystatin both contain carboxyl and amino groups, while filipin bears no ionogenic groups.[28,127] The cellular site of biological activity of this class of antibiotic is apparently the plasma membrane where alterations in permeability as well as mechani-cal properties are apparent.[32,26,71,90,91,92,93] However, biological activity is only obtained with cell types in which the sterol content of the plasma membrane is appreciable. The growth of fungi, yeast, and certain protozoa which have considerable membrane-bound sterol is inhibited by these antibiotics. In contrast, bacteria which do not have sterols in their membranes are not affected by the agents.[96] The similarity between the effects of these antibiotics on cell membranes and their effects on the properties of sterol-containing phospholipid lamellar systems is striking. The relative simplicity of the model system offers the hope that reasonable suggestions about the mode of action of these materials *in vivo* may be made on the basis of studies in lamellar lipid systems.

Returning to the modifications of the electrical properties of bilayers, without question the most remarkable system is that first described in a series of papers by Mueller, Rudin, and co-workers,[116,118,120,121,122,123] and extended by Bean and co-workers.[17,18,19,20] These investigators have described the production of unusual electrical properties in a mixed lipid bilayer through interaction with an unidentified proteinaceous material (EIM) derived from cultures of *Enterobacter cloacae*. Recently they have produced very similar effects in the bilayer using alamethicin, a cyclic peptide antibiotic containing nineteen amino acids.[119] These properties are similar in many respects to the corresponding properties of nerve axons, muscle, and electrically excitable algae. The presence of EIM in the aqueous phase causes bilayer resistance to decrease about five fold when the applied potential exceeds a threshold value in the range of 15–50 mV.[32] With alamethicin, the resistance decreases continuously and nonlinearly by a factor of 10^4 in the range 0–100 mV. No ion selectivity is apparent. Bilayers capable of interaction with these materials can be formed from a variety of mixtures. The best results have been obtained using membrane forming solutions consisting of mixed brain lipids in α-tocopherol, lecithin in decane, sphingomyelin in α-tocopherol, or oxidized cholesterol in octane.[120]

The electrical properties of the bilayer modified with EIM or alamethicin have been utilized in a system in which protamine present in the ambient aqueous phase and a transmembrane potential generated by a KCl gradient produce characteristics of negative resistance, delayed rectification and bistable potentials analogous to the characteristics of electrically active biological membranes.[118,120,119] Under appropriate conditions, the system shows a rhythmic oscillation between two states which is reminiscent of action potential activity. Mueller and Rudin have suggested that EIM develops cation selective channels in the bilayer. The addition of protamine converts a fraction of these channels to anion selective. Under these conditions the transmembrane voltage can be made, in effect, to open or close either set of channels. If a critical set of conditions is maintained, a regenerative effect causes the system to oscillate between two conductive states with a resultant oscillation in transmembrane voltage. The molecular details of the mechanism and structure, which are interpreted in terms of the field-dependent phenomenon, are as yet unknown. It seems possible, however, that the relative simplicity of the system together with complete knowledge of its composition will enable these questions to be answered in molecular terms.

In addition to EIM and the cyclic peptides, the interaction of lipid bilayer membranes with other types of protein has been the subject of several studies. Hanai, Haydon, and Taylor examined the capacitance of lecithin–decane membranes in the presence of insulin, bovine serum albumin, and egg albumin and found no effect on the membrane capacitance.[69] Tsofina, Liberman, and Babakov, however, employing the method of generating bilayers with controllable surface pressure described in Section II–A, found that, at low pressure, albumin did interact to produce a 100-fold drop in electrical resistance and an increase in capacitance.[181]

Maddy, Huang, and Thompson using the total protein derived from beef erythrocytes were able to show that the surface tension of the membranes is decreased in the presence of this material and that the optical reflectance markedly increased.[109] This protein, however, produced no change in any of the electrical properties of the bilayer. In contrast to this observation, Van den Berg, using the same protein, reported that resistance of the membranes is lowered 200–500 fold by this material.[183]

A remarkable but as yet little understood phenomenon is the transient impedance change occurring in a bilayer system when an enzyme and its substrate interact, or an antigen–antibody interaction occurs. This evanescent effect first described by del Castillo and co-workers,[42,179] has been reported subsequently by two other laboratories.[190,83]

The modifications of bilayers discussed above have generated system properties similar in some cases to the properties of biological membranes. The most striking effects are those produced through the interaction with macrocyclic peptides. Indeed, the degree and type of cation selectivity exhibited by a valinomycin-treated bilayer is not found in any other nonbiological system of such small transverse dimension. Intensive efforts are underway in a number of laboratories to reconstruct in a bilayer system the more complex physiological and biochemical functions displayed by biological membranes *in vivo*. The success which Mueller and co-workers have had in simulating the electrical phenomena associated with action potential activity offers the hope that these efforts will be fruitful.

Although it is important that a complex activity analogous to that found in a biological membrane can be generated in a bilayer system, the full significance of the achievement is realized only if the model system can be utilized to understand its biological counterpart. It is clear that any bilayer system has two important limitations which restrict a full analysis of its properties in molecular terms. First is the question of composition of the modified bilayer. For example, it is essential to know the amount of valinomycin in the bilayer, if ion permeability is to be understood. This information, however, is almost impossible to obtain in a straightforward way. Second, the form of the bilayer prohibits the use of optical methods such as absorption, rotatory dispersion, and circular dichroism to answer questions about the conformations of proteins which have interacted with the bilayer. It is just these questions, as well as others, however, which can be answered with phospholipid liquid crystal dispersions. The so-called liposome is, in many important respects, complementary to the bilayer as a model for biological membranes. The discussion in the next section turns to a consideration of this interesting system.

III. LIPOSOME SYSTEMS

A. Formation and Structure

With the exception of lysophosphatides and some acidic phosphatides, naturally occurring phospholipids disperse in water and aqueous salt solutions to form tubules and spherulites ranging in size from fractions of a millimeter down to several hundred angstroms. These aggregates, called myelinics, or more recently liposomes, constitute a disperse smectic mesophase of phospholipid. Within each tubule, or spherulite, the lipid is organized into concentric bimolecular lamellae, each separated from its neighbor by an interspersed lamella of water.[10,44] It is quite probable, on both experimental and theoretical grounds, that whatever the size of the liposome, each bimolecular lamella is a continuous, completely closed surface.[74,13] Thus, liposomes consist of an aqueous phase entrapped within a system of phospholipid lamellae, each one of which is a replica of the bilayer postulated by Danielli to be the structural form of the lipid component of biological membranes.

In 1964, Bangham, Standish, and Watkins[13] utilized a liposome preparation to study the diffusion of univalent ions across phospholipid lamellae, an investigation prompted by the Danielli hypothesis. Following this work, a number of studies have been reported in which the molecular organization of the liposome has been exploited in order to investigate questions relevant to problems posed by biological membranes. The discussion in this section is concerned with the formation, structure, and properties of liposomes and with the question of the relevance of these properties to the corresponding properties of natural membranes.

Dry solid phospholipids are very hygroscopic. For example, at 25°C egg phosphatidyl choline will absorb about 44% of its own weight in water.[44] This hydration process, which can readily be observed by adding a small piece of lecithin to a volume of water, results in the spontaneous formation of microscopic cylindrical structures which appear to grow out like hairs from the insoluble lecithin. If the system is shaken, the hydrated lecithin is dispersed in the form of tubules and spherulites which appear as birefringent structures in the polarizing light microscope. A typical liposome is seen in the electron microscope to consist of a large number of concentric lamellae of phospholipid, each separated from the other by a water filled gap; the thickness of each lipid lamella is about 50 Å, and thus, each sheet must be a bimolecular layer of phospholipid molecules. Most naturally occurring phospholipids and mixtures of these materials, as well as some synthetic compounds, form disperse liquid crystalline phases in aqueous medium consisting of closed bimolecular lamellae. However, the formation, as well as the details of the size and shape of liposomes in dispersion are governed by several parameters. The most important of these are temperature, mode of formation, and the species of phospholipid.

Liposomes can be formed only above the critical temperature at which melting of the hydrocarbon chains of the phospholipid occurs.[44] In the pure, anhydrous phospholipid, this transition is identified by line broadening of the infrared absorption band associated with the CH rocking frequency and by changes in the x-ray diffraction pattern characteristic of the liquid state of hydrocarbons. It occurs well below the true melting temperature, and gives rise to the liquid-crystalline state, characterized by the coexistence of crystalline order in the polar regions and a degree of disorder usually associated with the liquid state, in the hydrocarbon chains of the phospholipid.[29] Liposomes are thus a liquid crystalline mesophase and have frequently been referred to as liquid crystals.

Liquid crystal transition temperatures for most naturally occurring phospholipids and mixtures of these compounds usually are within the range of temperatures encountered in biological systems. In general, the transition temperature increases as the chain length increases and as the degree of unsaturation and branching of the hydrocarbon chain decreases. Thus, the relatively low transition temperatures of natural phospholipids can be attributed to the unsaturated or branched chain acyl groups present in these molecules.

Although aqueous suspensions of liposomes can be formed at room temperature with most naturally occurring phospholipids, a lecithin such as distearoyl phosphatidyl choline with fully saturated, unbranched chains has a transition temperature of about 83°C [29] and does not form dispersions much below this temperature.[44] In general lecithins with both acyl groups longer than fourteen carbons and unbranched will not form stable dispersions in water at room temperature.[154] It is interesting that the condition of liquid crystallinity apparently is also necessary for the formation of the Mueller–Rudin bilayers. Although it has been reported that these structures can not be formed at room temperature from synthetic distearoyl or dipalmitoyl phosphatidyl choline,[87] above 50°C stable membranes can be generated from these materials. This is consistent with the expectation that the transition temperature is probably lower in the multicomponent bilayer system than the value determined for the pure anhydrous phospholipid.

The size and shape of the liposome, and thus, the stability and optical clarity of the dispersion, are governed in large measure by the degree of mechanical agitation used in its preparation. For example, if small amounts of egg lecithin are hand shaken in a volume of water, most of the phospholipid is found in forms ranging in size from millimeters to microns with many of spherical shape. Each crystal consists of a large number of concentric bilayers. The dispersion is very opalescent and will settle on standing for a short period. On the other hand, if the dispersion is prepared by prolonged sonication under a nitrogen atmosphere, a remarkably homogeneous dispersion is produced in which all of the liquid crystals are between 200–300 Å in diameter.[84] Each liposome consists of a single continuous bimolecular lamella. Figure 1-8, an electron micrograph, shows the form of the liquid crystals in a preparation of egg lecithin sonicated for $2\frac{1}{2}$ hr under nitrogen at 0°C, then negatively stained with ammonium molybdate.

Although gross dispersions of myelinic figures in water have been known for many years, the first serious effort to produce a clear, stable dispersion of the pure well-characterized phospholipid, and to study the size of the dispersed particles was that reported by Robinson in 1960.[145] Using a chromatographically purified egg lecithin preparation, he was able to produce an optically clear dispersion by the addition to water of a concentrated solution of the lecithin in ether followed by evaporation of the ether with nitrogen. Light scattering studies on this preparation led him to conclude that the molecular weight of the liquid crystal was about 20×10^6, and that its shape was that of a cylinder with a diameter of 910 Å, assuming the thickness to be equivalent to a bimolecular leaflet (69 Å).

Ultrasonic dispersion is perhaps the most widely used method of forming stable aqueous phospholipid dispersions of uniform liposome size.[6,7,8,52,48,154,153,134,149,103] The first serious study of the size and shape of the liquid crystals produced by this method was carried out by Saunders, Perrin, and Gammack.[154] These investigators used viscosity, light scattering, ultracentrifugation, and free-diffusion measurements to characterize liposomes formed from egg lecithin. On the bases of these data, they concluded that the liposome could be an oblate ellipsoid with either a molecular weight of 2×10^6 and a thickness equal to one bilayer, or a molecular weight of 16×10^6 and a thickness equal to two bilayers. Saunders and his co-workers have followed this initial investigation with studies on liposome size and shape in ultrasonically dispersed phospholipids of ox brain,[52] and a group of synthetic phospholipids.[8] In addition, the effects of various salts in the aqueous dispersion medium and the time of sonication on liposome size and shape have been investigated by this group.[6,7] They have concluded that, with egg lecithin, little reduction in size is obtained with sonication times in excess of 90 min.[6] Recently, Huang, using electron microscopy and a variety of hydrodynamic methods, has examined the size and shape of liposomes of egg lecithin prepared in aqueous salt solutions by sonication under nitrogen for 2–3 hr followed by gel filtration. Under these conditions, the variation of liposome size can be reduced to less than 1%.[84]

Unlike sonication, the method of dispersion employed by Bangham and

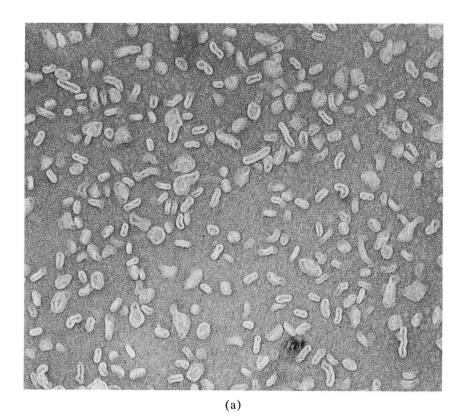

(a)

FIGURE 1-8. Electron micrograph of egg phosphatidyl choline liposomes negatively stained with ammonium molybdate. The liposomes were prepared by sonication in a Branson instrument for 2½ hours under nitrogen at 0°C. (A) 130,000×, (B) 400,000×. Taken by G. Decker and J. Greenawalt, Dept. of Physiological Chemistry, Johns Hopkins University, School of Medicine, Baltimore.

co-workers utilizing gentle mechanical shaking, gives lecithin dispersions showing great heterogeneity in liposome size and shape.[13,14,15,133] These preparations, although relatively stable, are very opalescent. Fleischer and Klouwen, however, have achieved clear dispersions by the dialysis against water of a number of phospholipids solubilized in an aqueous butanolcholate solvent.[49] An electron microscope study of their dispersions showed them to be composed of spherical particles 100–200 Å in average diameter.[59]

Recently, Papahadjopoulos and Miller have examined the form and

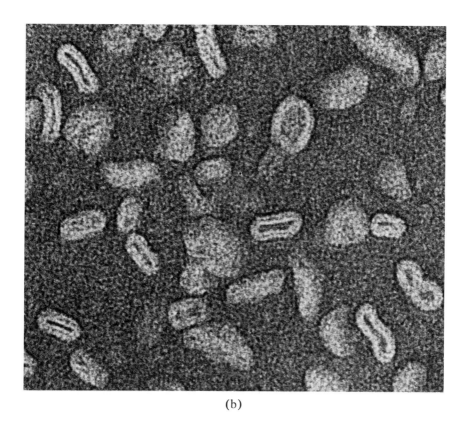

(b)

FIGURE 1-8. (Continued)

structure of liposomes in dispersion prepared by several methods from a variety of purified natural phospholipids and mixtures of these compounds.[134] These investigators used x-ray diffraction, electron microscopy, and optical birefringence measurements to examine in detail the thickness of both lipid lamellae and interspaced aqueous layers as functions of the pH and ionic strength of the aqueous phase. As might be expected, the thickness of the water compartment between lipid lamella was found to be independent of pH and ionic strength in lecithin dispersions. However, the thickness of this compartment showed a marked dependence on these variables in liposomes formed from acidic phospholipids such as phosphatidyl serine, phosphatidic acid, and phosphatidyl inositol. Only egg phosphitidyl ethanolamine gave liposome dispersions which tended to aggregate and settle out rapidly. Although lamellar in structure, the gross form of liposomes obtained with this material was found to be noticeably different from liposomes derived from the other phospholipids.[134]

It seems certain that the phospholipid in aqueous dispersions of the type described above is present in bimolecular lamellae which entrap an aqueous phase. The individual liquid crystals in the dispersion may be either aggregates consisting of many concentric alternating lipid and water lamellae, which in size cover a large range, or small vesicles with an overall dimension of several hundred angstroms bounded by a single bilayer. In either case, the form of the system, while permitting study of bilayer permeability as a mass transfer process, excludes determination of the electrical conductance of the bilayer. The small size and homogeneity of the liposomes prepared by prolonged sonication make these preparations particularly attractive for the study with the conventional techniques used in macromolecular physical chemistry of interactions between proteins and phospholipid bilayers.

B. Liposomes as Membrane Models

1. Permeability to Ions.   In 1965, Bangham, Standish, and Watkins described a study of the diffusion of univalent ions out of liquid crystals of lecithin in aqueous dispersion. The impetus for this investigation was the realization that the lamellar form of the liquid crystal, with its entrapped aqueous phase, is an experimentally suitable model for the examination of the permeability properties of a Danielli bilayer of phospholipid.[13] The basis of this study was the determination of the rate of efflux of an ionic species trapped during preparation within the aqueous compartments of a lecithin liquid crystal dispersion. In outline the procedure was as follows: lecithin was dispersed by hand shaking in an aqueous medium of specified composition. The external aqueous phase was then adjusted to the desired composition by rapid dialysis or passage of the dispersion through a Sephadex G-75 column. Appearance in the external phase of the ion under study was determined as a function of time. Thus, in essence the method is similar to that widely used for the determination of solute efflux rates from cell suspensions. In this study, liquid crystal dispersions were prepared from purified egg phosphatidyl choline and from this material containing a small quantity of a long chain ion such as dicetyl phosphoric acid or stearyl amine. Incorporation of this ion into the phospholipid lamella creates a net surface charge density on the otherwise electrically neutral surface. With a net charge on the faces of the lipid lamellae, it might be expected that the thickness of the interspersed aqueous layers would be a function of the ionic strength, and that the bilayer would show permselective properties dependent on the sign and magnitude of the surface charge density. Both expectations are realized in liquid crystals to which long chain ions have been added. However, it is difficult to account for the effects of the charge on the basis of a simple electrostatic model.[13]

Perhaps the most interesting result of this investigation is the observation that anion permeabilities are several order of magnitude greater than cation permeabilities.[13] This is in direct contrast to the situation in Mueller–Rudin bilayers in which the ratio of cation to anion permeabilities determined from electrical transference number measurements is about 5.[114,2] Pagano and Thompson, however, have recently determined permeabilities both from electrical and isotopic tracer measurements in a spherical bilayer (Section II–A,D). In this system, the permeability ratio for the two ions based on electrical parameters is about 4.9, in agreement with previous work on planar bilayers. The isotopically determined ratio is about 0.004 in agreement with the situation obtaining in the liquid crystal system.[132] This result strongly suggests the presence of a chloride carrier system.

In contrast to the situation in natural membranes, the liposome system shows little difference in the permeabilities of Li^+, Na^+, K^+, and Rb^+. As might be expected, however, the permeabilities to these monovalent cations are increased as the surface charge of the liquid crystal is made negative by the incorporation of a long chain anion; a decrease is observed, however, if the bilayer is charged with the opposite sign by incorporation of stearyl amine. On the other hand, the exchange diffusion rates of a series of monovalent anions, are significantly different. In a pure lecithin dispersion, anion permeabilities are ordered as follows:

$$Cl^- > I^- > F^- > NO_3^- > SO_4^= > Phosphate^{[13]}$$

The diffusion of both cations and anions is strongly dependent upon the species of phospholipid in the dispersion, and, as discussed above, upon the amount and type of long chain ion incorporated into the liquid crystal. In general, the diffusion of chloride through bimolecular lamella of egg or beef phosphatidyl choline, phosphatidyl serine, and mixtures of phosphatidyl ethanolamine and cholesterol is greater than the diffusion of sodium or potassium. On the other hand, these cations diffuse more rapidly than chloride through liquid crystals of phosphatidic acid and phosphatidyl inositol.[133,135]

In addition to the dissimilarity between anion and cation diffusion rates, the temperature dependence of each process is quite different. Chloride permeation through neutral phospholipids has an activation energy of 4–5 kcal/mole. This value increases to about 8 kcal/mole in negatively charged systems (e.g., phosphatidic acid. In either case, the plot of log permeability versus temp^{-1} is linear in the range studied (22–60°C). Potassium diffusion, on the other hand shows a discontinuity in the Arrhenius plot at about 37°C in phosphatidyl choline and phosphatidyl serine systems. Above this temperature, the activation energy is 15–17 kcal/mole, while below the value is similar to that obtained for chloride.[133,135]

In order that permeability measurements of this type be related to permeation of the bimolecular phospholipid lamella of the liquid crystal, it is essential that each lamella be a continuous, closed surface without gross leaks. Perhaps the strongest evidence in support of the integrity of the lamellar system is the great difference in the anion and cation diffusion rates. In the event of gross leakage, the efflux rates would be expected to be the same. Also, the fact that the cation efflux increases if the surface of the bilayer is made negative, or decreases if it is made positive by the incorporation of the appropriate long chain ion, argues against gross leakage due to mechanical rupture of the lamellae.[13]

Recently Papahadjopoulos and Watkins have examined the question of leakage in some detail.[135] On the basis of the ratio of anion to cation efflux rates as a function of liquid crystal size, these investigators conclude that, while some gross leakage may occur in large multilamellar liquid crystal systems, leakage diminishes as the size of the liquid crystal is decreased. The largest values for the permeability ratios were obtained for sonic dispersions. Among a number of naturally occurring phospholipids, only phosphatidyl ethanolamine liquid crystals showed evidence of gross leaks. Furthermore, these investigators calculate the electrical resistance of a single bimolecular lamella of lecithin to be about 1×10^7 Ω cm^2 on the basis of chloride efflux data and indirect estimates of the surface area of lamellae involved in permeation. This value compares favorably with the values measured in the planar bilayer system (Section II–D). It seems probable, then, that diffusion in liposome dispersions of naturally occurring phospholipids is translamellar.

The basic ionic permeability properties outlined above can be modified by the addition of a number of substances to the system. Bangham, Standish, and Weissman have demonstrated that the permeability to Na$^+$ of liposomes formed from mixtures of lecithin, cholesterol, and dicetylphosphoric acid is greatly increased by the addition of one of a number of steroids to the aqueous phase.[14] Moreover, they have observed that the ability of a specific steroid to enhance Na$^+$ permeability in a liquid crystal dispersion is closely correlated with the ability of the steroid to promote lysis of lysosomes isolated from rabbit liver. Cortisol, cortisone, and cortisone acetate as well as chloroquine all have been shown to protect lysosomes against lysis by steroids. It is interesting that each of these materials act as an antagonist to the steroids which promote increased Na$^+$ permeability in the liquid crystal system.[14,15] A number of local anesthetics cause effects on cation permeability in liquid crystals which correlate with the biological potency of these agents.[12,15]

2. Permeability to Water and Nonelectrolytes. Bangham, Standish, and Watkins have examined the diffusion of tritiated water from lecithin liquid crystals by the method described in the previous section.[13] They concluded that the permeability must be at least as great, if not greater than chloride permeability. These investigators noted in the same report that liposome lysis occurred under certain conditions in hypotonic media, an observation that also suggests that the water permeability of the lipid lamella is large.

Recently Bangham, De Gier, and Greville have examined the osmotically driven net volume flux of water across the lamellae of liposomes formed from mixtures of egg phosphatidyl choline and the phosphatidic acid derived enzymatically from it.[11] Water fluxes were determined under various osmotic conditions from changes in the optical extinction of the dispersions at 450 mμ as a function of time. The validity of this procedure was confirmed by volume determinations in centrifuged pellets of the liquid crystals. The total surface area of phospholipid involved in water permeation was obtained from measurements of the difference in the amount of UO_2^{++} required to produce a given surface potential in a monolayer of phospholipid both with and without liquid crystals dispersed in the subphase, and knowledge of the area of phospholipid per absorbed ion of UO_2^{++}.

The negatively charged liposomes examined in this study behaved as ideal osmometers when alkali-metal salts, glucose, sucrose, and mannitol were employed as solutes. Using water fluxes induced by gradients of these solutes, and estimates of the area of lipid lamellae involved in water permeation, permeability coefficients were obtained ranging from $(0.8-16) \times 10^4$ cm sec^{-1} at 20°C. These results compare favorably with the range of values reported both for Mueller–Rudin type bilayers and for natural membranes (Section II–E).

Nonideal osmotic behavior was observed with a number of solutes. The time rate of change of optical extinction was used to measure the relative permeabilities of these materials. The following series of decreasing permeabilities was obtained for negatively charged liposomes: ethylurea, methylurea, ethylene glycol, ammonium acetate, propionamide, glycerol urea malonamide, and erythritol. It is interesting that this series is similar to that reported for erythrocytes[75] and rat liver mitochondria.[169]

Rendi has reported an unusual process in which water is extruded from liposomes of phospholipids prepared from either soy beans or rat liver mitochondria when both serum albumin and a divalent metal chelate of EDTA are added together to the system.[140] The apparent loss of water has been followed by optical extinction changes and direct gravimetric determination. Since no information is available on the form of the liquid crystals in this system, it is difficult to speculate on the origin of the process. It seems clear, however, that it is not osmotically driven. Rendi has noted that water extrusion in the liposome system is analogous in detail to the loss of water from rat liver mitochondria, nuclei, microsomes, and spinach chloroplasts under similar conditions.[111,141]

3. Interactions of Liposomes with Proteins and Other Substances. It seems clear that the Danielli hypothesis must be considered a first approximation to the structure of biological membranes. The many biochemical functions associated with cellular membranes must of necessity depend upon the existence of more complicated relationships between the proteins and lipids of the system than of those expressed by this model. The central issue in the problem of understanding the structure–function relations in biological membranes is the detailed nature of the interactions of proteins with ordered phospholipid arrays in aqueous media. Phospholipid dispersions, particularly those prepared by prolonged sonication which exhibit a remarkable homogeneity in both size and shape, offer an almost ideal system for the investigation of these interactions. Interactions between sonically dispersed liposomes and a specific protein may be studied by the usual array of transport and equilibrium methods developed for the study of macromolecular interactions. These include ultracentrifugation, light scattering, and molecular sieve chromatography. In addition, the liposome is well suited for use with optical methods depending upon light absorption and optical activity.

Litman and Thompson have reported a preliminary study of the interaction of whale metmyoglobin with sonic dispersions of egg phosphatidyl choline.[106] The interaction was studied spectrophotometrically by observing changes in the Soret band of the protein as a function of the ratio of phosphatidyl choline to metmyoglobin. Under conditions of complex formation, the Soret band was observed to shift from 409 mμ–422 mμ. This shift is quite different from that observed in acid or guanidine hydrochloride denaturation of metmyoglobin. It is, however, similar to spectral changes observed upon ligand substitution of the iron in the heme. This suggests that the heme group may be involved in complex formation with the liposome. The ratio of phospholipid to protein in the complex was calculated to be about 200. Examination of the complex by electron microscopy indicated that no change had occurred in the basic structure of the liposome as a result of complex formation.

Papahadjopoulos and Miller have examined the liposomes formed from phosphatidyl serine in the presence of cytochrome-c and concluded that a complex is formed under some conditions.[134] The presence of this protein in the aqueous phase appears to promote formation of single lamella vesicles without sonication in phosphatidyl choline systems. No effect on either cation or anion permeabilities was observed in these interactant systems.[135]

There are many examples in the literature of the use of so-called "solubilized" lipid preparations to study the dependence of enzymatic systems on this type of material. In most cases, the solubilized lipid is uncharacterized physically, and apart from its effect on the enzymatic system, no attempt has been made to examine the details of complex formation in terms of the physical state of the lipid. In some instances, however, physical studies have been carried out on the lipid dispersion. As an example, Gotterer in a recent study has used well-characterized sonic dispersion of various phospholipids to examine the lipid requirements of a partially purified β-hydroxybutyrate dehydrogenase prepared from rat liver mitochondria.[56,57]

IV. CONCLUSION

The macroscopic configurations of both the Mueller-Rudin bilayer and the liposome enable one to ask two important questions about these systems. First, what is their molecular architecture? Second, what are their physical properties? The information presented in the preceding discussions leaves little doubt that in both systems the molecules are arranged in the form of a bimolecular lipid lamella. The extensive physical studies carried out on these systems have established within relatively narrow limits the range of permeability, mechanical and electrical properties which will be displayed by such a bimolecular lamella of lipid. In addition, considerable information is available concerning the dependence of these properties on temperature, and the composition of both the ambient aqueous phase and the lamella itself.

With regard to the problem of the structure of biological membranes, it is clear that the arrangement of component molecules in both the bilayer and the liposome closely approximates the lipid lamella of the Danielli hypothesis. Since this is the case, it is apparent that the basic physical properties of biological membranes which should derive from the Danielli lamella, if such a structure is indeed present in natural membranes, are the properties of the experimental liposome and bilayer systems. A comparison of the established properties of a wide variety of natural membranes reveals a marked similarity to the corresponding physical properties of the synthetic systems. This correspondence constitutes strong circumstantial evidence for the existence of a bimolecular lipid lamella in biological membranes. This viewpoint is supported by the analogous effects on the system properties in both biological and bilayer membranes caused by a number of substances such as the macrocyclic and polyene antibiotics, dinitrophenol, and heavy metals. At the present time, all of these diverse data are most simply explained by the existence of a bimolecular lipid lamella in biological membranes. If this viewpoint is adopted, however, an important question remains unanswered; How much of the area of a specific biological membrane is occupied by the lipid bilayer? The available data have little to offer in answer to this question.

It is clear, however, that if the lipid bilayer is the structural element in biological membranes, all membrane area cannot be underlain by a bilayer. Many of the physiological properties associated with natural membranes are not found to be the properties of a simple bilayer. Among these are such functionally important properties as oxidative phosphorylation, photosynthesis, and active transport. The functional specificity represented by these biochemical systems can only be generated by local modifications of the lipid bilayer through interaction with specific proteins. If this viewpoint is operationally correct, then it is apparent that the cardinal direction for research with phospholipid model membranes is investigation of the interactions in aqueous media between membrane proteins and organized lamellar phospholipid systems. Both the Mueller–Rudin bilayer and liposome systems are well suited for this type of work.

REFERENCES

1. Andreoli, T. E. (1966), *Science* **154**, 195.
2. Andreoli, T. E., Bangham, J. A., and Tosteson, D. C. (1967), *J. Gen. Physiol.* **50**, 1729.
3. Andreoli, T. E., and Monahan, M. S. (1968), Biophys. Soc. Meeting Abstr. MC-4.
4. Andreoli, T. E., Monahan, M. S. (1968), *J. Gen. Physiol.* **52**, 300.
5. Andreoli, T. E., Tieffenberg, M. and Tosteson, D. C. (1967), *J. Gen. Physiol.* **50**, 2527.
6. Attwood, D., and Saunders, L. (1965), *Biochim. Biophys. Acta* **98**, 344.
7. Attwood, D., and Saunders, L. (1965), *Biochim. Biophys, Acta* **116**, 108.
8. Attwood, D., Saunders, L., Gammack, D. B., de Hass, G. H. and van Deenen, L. L. M. (1965), *Biochim. Biophys. Acta* **102**, 302.
9. Babkov, A. N., Ermishkin, L. N. and Liberman, E. A. (1966), *Nature* **210**, 953.
10. Bangham, A. D. (1963), "Advances in Lipid Research," Vol. 1, p. 65, ed. Paoletti, R. and Kritchevsky, D., Academic Press, New York.
11. Bangham, A. D., de Gier, J., and Greville, G. D. (1967), *Chem. Phys. Lipids* **1**, 225.
12. Bangham, A. D., Standish, M. M., and Miller, N. (1965), *Nature* **208**, 1295.
13. Bangham, A. D., Standish, M. M., and Watkins, J. C. (1965), *J. Mol. Biol.* **13**, 238.
14. Bangham, A. D., Standish, M. M., and Weissman, G. (1965), *J. Mol. Biol.* **13**, 253.
15. Bangham, A. D., Standish, M. M., Watkins, J. C., and Weissman, G. (1967), "Symposium on Biophysics and Physiology of Biological Transport." p. 183, ed. Bolis, L., Caprara, V., Porter, K. R., and Robertson, J. D., Springer, New York.
16. Bar, R. S., Deamer, D. W., and Cornwell, D. G. (1966), *Science* **153**, 1010.
17. Bean, R. C. (1966), *Aeronutronic Publication* No. U–3494.
18. Bean, R. C., and Shepard, W. C. (1965), Abstr. 37, p. 19c 150th Amer. Chem. Soc. Meeting.

19. Bean, R. C., Shepard, W. C. and Chan, H. (1968), *J. Gen. Physiol.*, **52**, 495.
20. Bean, R. C., Shepard, W. C., D'Agostino, C., and Smith, L. (1966), *Fed. Proc.* **25**, 2641.
21. Benedetti, E. L., and Emmelot, P. J. (1965), *J. Cell. Biol.* **26**, 299.
22. Benson, A. A. (1963), *Adv. Lipid Res.* **1**, 387.
23. Benson, A. A. (1964), *Ann. Rev. Plant Physiol.* **15**, 1.
24. Bielawski, J., Thompson, T. E., and Lehninger, A. L. (1966), *Biochem. Biophys. Res. Commun.* **24**, 948.
25. Bjerrum, N. (1926), *Biol. Med. Mat. Fys. Med.* **7, No. 9.**
26. Butler, W. T., Alling, D. W., and Cotlove, E. (1965), *Proc. Soc. Exp. Biol. Med.* **118**, 297.
27. Cass, A., and Finkelstein, A. (1967), *J. Gen. Physiol.* **50**, 1765.
28. Ceder, O., and Ryhage, R. (1964), *Acta Chem. Scand.* **18**, 558.
29. Chapman, D. "The Structure of Lipids" (1965), p. 119, John Wiley & Sons, New York.
30. Chappell, J. B., and Crofts, A. R. (1965), *Biochem. J.* **95**, 393.
31. Chappell, J. B., and Crofts, A. R. (1966), "Regulation of Metabolic Processes in Mitochondria," Vol. 7, p. 293, ed. by Tager, J. M., Papa, S., Quagliariello E., and Slater, E. C., B.B.A., Library Elsevier: Amsterdam.
32. Cirillo, V. P., Harsch, M., and Lampen, J. O. (1964), *J. Gen. Microbiol.* **35**, 245.
33. Cole, K. S. (1932), *J. Cell. Comp. Physiol.* **1**, 1.
34. Cole, K. S., and Curtis, H. J. (1938), *J. Gen. Physiol.* **21**, 591.
35. Cook, G. M. W., Redwood, W. R., Taylor, A. R., and Haydon, D. A., (1968) *Kolloid. F. F. Polyn.,* **227**, 28.
36. D'Agostino, C., and Smith, L. (1965), "Biophysics and Cybernetic Systems, Symposium," p. 184, eds. Mayfield, M., Callahen, A., and Fogel, L. S., Spartan Books, Inc., Washington, D.C.
37. Danielli, J. F. (1936), *J. Cell. Comp. Physiol.* **7**, 393.
38. Danielli, J. F., and Davson, H. (1935), *J. Cell. Comp. Physiol.* **5**, 495.
39. Danielli, J. F., and Harvey, E. N. (1934), *J. Cell. Comp. Physiol.* **5**, 483.
40. Davies, J. T., and Rideal, E. K. (1961), "Interfacial Phenomena," Academic Press, New York.
41. Dean, R. B., Curtis, H. J., and Cole, K. S. (1940), *Science* **91**, 50.
42. del Castillo, J., Rodriquez, A., Romero, C. A., and Sanchez, V. (1966), *Science* **153**, 185.
43. Demel, R. A., Kinsky, S. C., and van Deenen, L. L. M. (1965), *J. Biol. Chem.* **240**, 2749.
44. Dervichian, D. G. (1964), "Progress in Biophysics," Vol. 14, p. 263, ed. Butler, J. A. V., and Huxley, H. E., Macmillan Co., New York.
45. DiStefano, H. S. (1966), *Z. Zellforsch.* **70**, 322.
46. Duyvis, E. M. (1962), Ph.D. Thesis, Utrecht.
47. Elbers, P. F. (1964), in "Recent Progress in Surface Science," Vol. 2, p. 433, eds. Danielli, J. F., Pankhurst, K. G. A., and Riddiford, A. C., Academic Press, New York.
48. Fleischer, S., and Brierley, G. P. (1961), *Biochem. Biophys. Res. Commun.* **5**, 367.

49. Fleischer, S., and Klouwen, H. (1961), *Biochem. Biophys. Res. Commun.* **5**, 378.
50. Forester, E. O. (1964), *J. Chem. Phys.* **40**, 92.
51. Gains, G. L. (1966), "Insoluble Monolayers at Liquid Gas Interface." Interscience, New York.
52. Gammack, D. B., Perrin, J. H., and Saunders, L. (1964), *Biochim. Biophys. Acta* **84**, 576.
53. Gent, W. L. G., Gregson, N. A., Gammack, D. B., and Raper, J. G. (1964), *Nature* **204**, 553.
54. Goldman, D. E. (1943), *J. Gen. Physiol.* **27**, 37.
55. Gorter, E., and Grendel, P. (1925), *J. Exp. Med.* **41**, 439.
56. Gotterer, G. S. (1967), *Biochem.* **6**, 2139.
57. Gotterer, G. S. (1967), *Biochem.* **6**, 2147.
58. Graven, S. N., Lardy, H. A., Johnson, D., and Rutter, A. (1966), *Biochem.* **5**, 1729.
59. Green, D. E., and Fleischer, S. (1963), *Biochim. Biophys. Acta* **70**, 554.
60. Green, D. E., and Perdue, J. F. (1966), *Proc. Nat. Acad. Sci.* **55**, 1294.
61. Green, D. E., and Tzagoloff, A. (1966), *J. Lipid Res.* **7**, 587.
62. Hanai, T. and Haydon, D. A. (1966), *J. Theoret. Biol.* **11**, 370.
63. Hanai, T., Haydon, D. A., and Redwood, W. R. (1966), *Ann. N.Y. Acad. Sci.* **137**, 731.
64. Hanai, T., Haydon, D. A., and Taylor, J. (1964), *J. Koll. Z.* **195**, 41.
65. Hanai, T., Haydon D. A., and Taylor, J. *Proc. Roy. Soc. (Lond)* **A. 281**, 377.
66. Hanai, T., Haydon, D. A., and Taylor, J. (1965), *J. Gen. Physiol.* **48**, 59.
67. Hanai, T., Haydon, D. A., and Taylor, J. (1965), *J. Theoret. Biol.* **9**, 278.
68. Hanai, T., Haydon, D. A., and Taylor, J. (1965), *J. Theoret. Biol.* **9**, 422.
69. Hanai, T. Haydon, D. A., and Taylor, J. (1965), *J. Theoret. Biol.* **9**, 433.
70. Harned, H. S., and Owen, B. B. (1958), "Physical Chemistry of Electrolyte Solutions," pp. 217–243, Reinhold, New York.
71. Harsch, M., and Lampen, J. O. (1963), *Biochem. Pharmac.* **12**, 875.
72. Harvey, E. N. (1931), *Biol. Bull.* **60**, 67.
73. Haydon, D. A. (1968), *J. Amer. Oil Chem. Soc.* **45**, 230.
74. Haydon, D. A., and Taylor, J. (1963), *J. Theoret. Biol.* **4**, 281.
75. Hedin, S. G. (1897), *Arch. Gen. Physiol.* **68**, 229.
76. Henn, F. A. (1967), Ph.D. thesis, Johns Hopkins University.
77. Henn, F. A., Decker, G. L., Greenawalt, J. W., and Thompson, T. E. (1967), *J. Mol. Biol.* **24**, 51.
78. Henn, F. A., and Thompson, T. E. (1968), *J. Mol. Biol.* **31**, 227.
79. Hildebrand, J. G., and Law, J. H. (1964), *Biochem.* **3**, 1304.
80. Hodgkin, A. L. (1951), *Biol. Rev.* **26**, 399.
81. Hodgkin, A. L., (1964), "The Conduction of the Nervous Impulse," Charles C. Thomas, Springfield.
82. Hopfer, U., Lehninger, A. L., and Thompson, T. E. (1968), *Proc. Natl. Acad. Sci.* (U.S.) **59**, 484.

83. Howard, R. E., and Burton, R. M. (1968). *J. Amer. Oil Chem. Soc.* **45,** 202.
84. Huang, C. (1969), *Biochem.* **8,** 344.
85. Huang, C., and Thompson, T. E. (1965), *J. Mol. Biol.* **13,** 183.
86. Huang, C., and Thompson, T. E. (1966), *J. Mol. Biol.* **15,** 539.
87. Huang, C., Wheeldon, L., and Thompson, T. E. (1964), *J. Mol. Biol.* **8,** 148.
88. International Critical Tables. (1929), Vol. **6,** p. 311, McGraw-Hill, New York.
89. Katz, M., and Thompson, T. E., Unpublished results.
90. Kinsky, S. C. (1961), *J. Bacteriol.* **82,** 889.
91. Kinsky, S. C. (1962), *J. Bacteriol.* **83,** 351.
92. Kinsky, S. C. (1963), *Arch. Biochem, Biophys.* **102,** 180.
93. Kinsky, S. C., Auruch, J., Permeett, M., Rogers, H. B., and Shonder, A. A. (1962), *Biochem. Biophys. Res. Commun.* **9,** 503.
94. Kinsky, S. C., Luse, S. A., van Zutphen, and van Deenen, L. L. M. (1967), *Fed. Proc.* **26,** 3394.
95. Korn, E. D. (1966), *Science* **153,** 1491.
96. Lampen, J. O. (1966), *Symp. Soc. Gen. Microbiol.* **16,** 111.
97. Langmuir, I., and Waugh, D. F. (1938), *J. Gen. Physicol.* **21,** 745.
98. Läuger, P., Lesslauer, W., Marti, E. and Richter, J. (1967), *Biochim. Biophys. Acta* **135,** 20.
99. Lehninger, A. L., Wadkins, C. L., Cooper, C., Devlin, T. M., and Gamble, J. L. (1958), *Science* **128,** 450.
100. Lennarz, W. J., and Talmo, B. (1966), *J. Biol. Chem.* **241,** 2707.
101. Leslie, R. B., and Chapman, D. (1967), *Chem. Phys. Lipids* **1,** 143.
102. Lesslauer, W., Richter, J., and Läuger, P. (1967), *Nature* **213,** 1224.
103. Lester, R. L., and Smith, A. L. (1961), *Biochim. Biophys. Acta* **47,** 475.
104. Lev, A. A., and Buzhinski, E. P. (1967), *Cytology* (Russ.) **9,** 102.
105. Lewis, T. J. (1964), Advancement of Science, **20,** 501.
106. Litman, B. J., and Thompson, T. E. (1967), *Fed. Proc.* **26,** 834.
107. Luzzati, V., and Husson, F. (1962), *J. Cell. Biol.* **12,** 207.
108. Lynn, W. S., and Brown, R. H. (1966), *Arch. Biochem. Biophys.* **114,** 271.
109. Maddy, A. H., Huang, C., and Thompson, T. E. (1966), *Fed. Proc.,* **25,** 933.
110. Marinetti, G. V., Erbland, J., and Stotz, E. (1958), *J. Biol. Chem.* **233,** 562.
111. McCaig, N., and Rendi, R. (1964), *Biochim. Biophys. Acta* **79,** 416.
112. Mitchell, P. (1961), *Nature* **191,** 114.
113. Mitchell, P. (1966), "Chemiosmotic Coupling in Oxidative and Photosynthetic Phosphorylation," Glynn Research Laboratories, Bodmin, Cornwall, England.
114. Miyamoto, V. K., and Thompson, T. E. (1967), *J. Coll. & Interface Sci.* **25,** 16.
115. Moore, C., and Pressman, B. C. (1964), *Biochem. Biophys. Res. Commun.* **15,** 562.
116. Mueller, P., and Rudin, D. O. (1963), *J. Theoret. Biol.* **4,** 268.

117. Mueller, P., and Rudin, D. O. (1967), *Biochem. Biophys. Res. Commun.* **26,** 398.
118. Mueller, P., and Rudin, D. O. (1967), *Nature* **213,** 603.
119. Mueller, P., and Rudin, D. O. (1968), *Nature* **217,** 713.
120. Mueller, P., and Rudin, D. O. (1968), *J. Theoret. Biol.,* **18,** 222.
121. Mueller, P., Rudin, D. O., Tien, H. T., and Wescott, W. C. (1962), *Circulation* **26,** 1167.
122. Mueller, P., Rudin, D. O., Tien, H. T., and Wescott, W. C. (1962), *Nature* **194,** 979.
123. Mueller, P., Rudin, D. O., Tien, H. T., and Wescott, W. C., ed. Danielli, J. F., Pankhurst, K. G. A., Riddeford, A. C. (1964), "Recent Progress in Surface Science," Vol. 1, p. 379. Academic Press, New York.
124. Mysels, K. J., Shinoda, K., and Frankel, S. (1959), "Soap Films," Pergammon Press, London.
125. O'Brien, J. S. (1967), *J. Theoret. Biol.* **15,** 307.
126. Onsager, L. (1934), *J. Chem. Phys.* **12,** 599.
127. Oroshnik, W., and Mebane, A. D. (1963), in "Fortschritte der Chemie Organischer Naturstoffe," ed. Zechmeister, L. Springer-Verlag, Vienna.
128. Overbeek, J. Th. G. (1960), *J. Phys. Chem.* **64,** 1178.
129. Overton, E. (1895), *Vjschr. Naturf. Ges.* (Zurich) **40,** 159.
130. Pagano, R., and Thompson, T. E. (1967), 11th Ann. Biophysical Soc. Meeting, Abstr. TD-1.
131. Pagano, R., and Thompson, T. E. (1967), *Biochim. Biophys. Acta,* **144,** 666.
132. Pagano, R., and Thompson, T. E. (1968), *J. Mol. Biol.* **38,** 41.
133. Papahadjopoulos, D., and Bangham, A. D. (1966), *Biochim. Biophys. Acta* **126,** 185.
134. Papahadjopoulos, D., and Miller, N. (1967), *Biochim. Biophys. Acta* **135,** 624.
135. Papahadjopoulos, D., and Wadkins, J. C. (1967), *Biochim. Biophys. Acta* **135,** 639.
136. Pressman, B. C. (1965), *Proc. Nat. Acad. Sci.* (U.S.), **53,** 1076.
137. Pressman, B. C., Harris, E. J., Jagger, W. S., and Johnson, J. H. (1967), *Proc. Nat. Scad. Sci.* (U.S.) **58,** 1949.
138. Price, H. D., and Thompson, T. E. (1968), 12th Biophysical Soc. Meeting, Abstr. MC-5.
139. Redwood, W. (1967), Ph.D. Thesis, Cambridge University.
140. Rendi, R. (1964), *Biochim. Biophys. Acta* **84,** 694.
141. Rendi, R., and McCaig, N. (1964), *Arch. Biochem. Biophys.* **104,** 267.
142. Robertson, J. D. (1958), *Biophys. Biochem. Cytol.* **4,** 349.
143. Robertson, J. D. (1959), *Biochem. Soc. Symp.* **16,** 3.
144. Robertson, J. D. (1964), in "Cellular Membranes in Development," ed. Locke, M., p. 1. Academic Press, New York.
145. Robinson, N. (1960), *Trans. Farad. Soc.* **56,** 1260.
146. Rosano, H. L., Duby, P., and Schulman, J. H. (1961), *J. Phys. Chem.* **65,** 1704.
147. Rosano, H. L., Schulman, J. H., and Weisbuch, J. B. (1961), *Ann. N.Y. Acad. Sci.* **92,** 457.

148. Rothfield, L., and Finkelstein, A. (1968), *Ann. Rev. Biochem.* **37**, 4–5.
149. Rouser, G. (1958), *Amer. J. Clin. Nutr.* **6**, 681.
150. Salton, M. R. J., and Netschey, A. (1965), *Biochim. Biophys. Acta.* **107**, 539.
151. Saunders, L. (1956), *Proc. 2nd. Int. Congr. of Surface Activity London,* **2**, 56.
152. Saunders, L. (1960), *J. Pharm. Pharmacol.* **12** (suppl.), 253T.
153. Saunders, L. (1963), *J. Pharm. Pharmacol.* **15**, 155.
154. Saunders, L., Perrin, J. H., and Gammack, D. (1962), *J. Pharm. Pharmacol.* **14**, 567.
155. Scheludko, A. (1959), *Qzvest. Khim. Inst. Bulgar. Nauk.* **7**, 123.
156. Schmitt, F. O., Bear, R. S., and Ponder, E. (1936), *J. Cell. Comp. Physiol.* **9**, 89.
157. Schoffeniels, E. (1967), "Cellular Aspects of Membrane Permeability," p. 81, Academic Press, New York.
158. Schwan, H. P., Huang, C., and Thompson, T. E. (1966), Abstract, Biophysical Society Meeting.
159. Seufert, W. D. (1965), *Nature* **207**, 174.
160. Shemyakin, N. N., Ovchinnikov, Yu. A., Ivanov, V. T., Kiryushkin, A. A., Zhdanov, G. L., Ryabova, I. D. (1963), *Experientia* **19**, 566.
161. Shemyakin, M. M., Vinogradova, E. I., Feigina, M. Yu., Aldanova, N. A., Loginova, N. F., Ryabova, I. D., and Pavlenko, I. A. (1965), *Experientia* **21**, 548.
162. Sjöstrand, F. S. (1963), *J. Ultrastruct. Res.* **9**, 340.
163. Sjöstrand, F. S. (1964), "Intracellular Membraneous Structures," p. 103 (eds. Seno, S. and Cowdry, E. V.), Chungoku Press Ltd., Okayama.
164. Skulachev, V. P. (1967), Abstr. of the Intern. Cong. of Biochem. Tokyo, Japan, Symp. VI-2, 7.
165. Small, D. M. (1968), *J. Amer. Oil Chem. Soc.* **45**, 108.
166. Stoeckenius, W. (1962), *J. Cell. Biol.* **12**, 221.
167. Strong, J. (1958), "Concepts of Classical Optics," p. 222, W. H. Freeman & Co., San Francisco.
168. Taylor, J., and Haydon, D. A. (1966), *Disc. Farad. Soc.* **42**, 51.
169. Tedeschi, H., and Harris, D. L. (1955), *Arch. Biochem. Biophys.* **58**, 52.
170. Thompson, T. E. (1964), "Cellular Membranes in Development," p. 83, ed. M. Locke, Academic Press, New York.
171. Thompson, T. E., and Huang, C. (1966), *Ann. N.Y. Acad. Sci.* **137**, 740.
172. Thompson, T. E., and Huang, C. (1966), *J. Mol. Biol.* **16**, 576.
173. Tien, H. T. (1967), *J. Phys. Chem.* **71**, 3395.
174. Tien, H. T., Carbone, S., and Dawidowicz, E. A. (1965), *Nature* **212**, 718.
175. Tien, H. T., and Dawidowicz, E. A. (1966), *J. Coll. & Interface Sci.* **22**, 438.
176. Tien, H. T., and Diana, A. L. (1967), Preprints 41st National Colloid Symposium, Buffalo, New York.
177. Tien, H. T., and Diana, A. L. (1968), *Chem. Phys. Lipids* **2**, 55.
178. Tobias, J. M., Agin, D. P. and Powlowski, R. (1962), *J. Gen. Physiol.* **45**, 989.

179. Toro-Goyco, E., Rodriguez, A., and del Costillo, J. (1966), *Biochem. Biophys. Res. Commun.* **23,** 341.
180. Tosteson, D. C., Cook, P., Andreoli, T., and Tieffenberg, M. (1967), *J. Gen. Physiol.* **50,** 2513.
181. Tsofina, L. M., Liberman, E. A., and Babkov, A. U. (1966), *Nature* **212,** 618.
182. van den Berg, H. J. (1965), *J. Mol. Biol.* **12,** 290.
183. van den Berg, H. J. (1967), Abstr. **53,** 153rd Meeting Amer. Chem. Soc., April.
184. Van Zutphen, L. L., Van Deenen, M., and Kinsky, S. C. (1966), *Biochem. Biophys. Res. Com.* **22,** 393.
185. Vreeman, H. J. (1966), *Proc Koninkl. Nederl. Akademie van Wetenschappen* (Amsterdam) **B, 69,** 542.
186. Waksman, S. A., and Lechevalier, H. A. (1962), "The Actinomycetes," Vol. 3, William & Wilkins, Baltimore.
187. Weissman, G., and Sessa, G. (1967), *J. Biol. Chem.* **242,** 616.
188. Whittam, R. (1964), "Transport and Diffusion in Red Blood Cells," p. 76 Williams & Wilkins, Baltimore.
189. Wirth, F. P., Morgan, H. E., and Park, C. R. (1965), *Fed. Proc.* **24,** 588.
190. Wobschall, D., Gordon, D., and Bolon, R. (1967), 11th Ann. Biophysical Soc. Meeting, Abstr. TD-5.
191. Wood, R. E., and Morgan, H. E. (1966), 11th Ann Biophys. Soc. Meeting, Abstr. TD-3.
192. Zaky, A. A., Tropper, H., and House, H. (1963), *Brit. J. Appl. Phys.* **14,** 651.

2

Electron Microscopy of Mitochondrial and Model Membranes

Walther Stoeckenius
Cardiovascular Research Institute
University of California
San Francisco Medical Center

I. INTRODUCTION

The high resolution of the electron microscope could only be used for a study of cell fine structure after preparation techniques had been developed that preserved the structure at a level far below that required for light microscopy. This was achieved around 1950. Then, for the first time, it became possible to resolve optically the cell membrane. As expected, every cell was found to be bounded by a membrane usually somewhat less than 100 Å thick. However, the observation that almost all cell organelles were also bounded and, in some cases subdivided by very similar membranes, was unexpected. Two of the most striking examples were found in the chloroplasts and mitochondria. Since then, the analysis of ultra-

Note added in proof: Work on this manuscript was finished in March of 1968. Since then a more detailed and critical review on general membrane structure and the models has been prepared. It may serve to clarify aspects of these topics, which are treated only cursorily here. (W. Stoeckenius and D. M. Engelman, 1969, *J. Cell Biol.*, **42**, 613.

structure has progressed much further for the membranes of these two organelles than for any other membrane, and the results have profoundly affected our ideas about the structure of biological membranes in general.

Since the discovery of the ultrastructural features of mitochondria,[59,78,19] numerous reviews of the subject have appeared.[63,65,39,40] Most modern textbooks of histology and cytology also contain satisfactory descriptions of these observations. Therefore, no comprehensive treatment of the subject will be given here. Instead, a critical evaluation of some more recent observations and their interpretation will be attempted.

II. MEMBRANE MODELS AND MODEL MEMBRANES

A. The Danielli Model

As discussed in more detail elsewhere,[88,89] mitochondrial membranes show the same basic structural and functional characteristics as other cellular membranes, and the general arguments for a common basic structure of all cellular membranes also apply to mitochondrial membranes. The most successful attempt so far to define the basic structure of membranes has been the Danielli model for the cell membrane and its extension to all other cellular membranes in the unit-membrane theory of Robertson.[75] The most distinctive feature of the Danielli model is a central continuous bilayer of lipid that forms the structural backbone of the membrane. While no single argument for this model is compelling, the number of independent observations and conclusions that are at least compatible with it let it appear rather well supported. They can briefly be summarized as follows:

1. Chemical analyses of isolated membrane fractions show that lipids and proteins are the main constituents. Usually a major part of the lipids consist of phospholipids, but other amphiphile lipids may take their place. The lipids and proteins are present in amounts compatible with a Danielli-type structure.

2. Impedance measurements on suspensions of cells and membrane-bounded cell organelles are compatible with the assumption that they are bounded by a continuous hydrocarbon layer of the same thickness found in lipid bilayers.

3. The myelin sheath consists of concentric, closely packed layers derived from the Schwann cell membrane. When its molecular structure, which is well supported by x-ray diffraction and polarized light data from living nerve fibers, is extrapolated to the Schwann cell membrane, it is found to be identical to the Danielli model.

4. Bilayers of isolated membrane lipids and lipoprotein complexes in water form spontaneously, and are found to be the most stable form of these compounds under conditions approaching the physiological milieu of cells.

5. When such artificial membrane lipid and lipoprotein bilayers are pre-

pared for electron microscopy by the procedures used for cells and tissues, they retain their lamellar structure and appear as unit membranes in the electron microscope.

6. When these membrane lipoprotein preparations are finely dispersed in water, they form vesicular structures of variable size, indistinguishable morphologically from dispersed cellular membrane fractions.

7. The electrical and permeability properties observed in model systems of thin lipid and lipoprotein films are compatible with the assumption that a similar structure is present in cellular membranes.

8. Freeze-etching experiments indicate that a preferential cleavage plane parallel to the surface is located in the center plane of cellular membranes and that lipid bilayers under similar conditions also split preferentially along this plane.[6,14]

Since the first description of the Danielli model, a number of variations have been introduced, and it appears necessary to describe our present concept of this model in more detail.

The central continuous layer of lipid molecules is usually described as a bimolecular leaflet, i.e., a layer of lipid two molecules thick with the hydrophilic groups of the lipid molecules facing outward, and the hydrophobic part of the molecules—mostly hydrocarbon chains of fatty acids—facing inward. However, this is neither a sufficiently detailed nor an entirely correct description. We have good reason to assume that the fatty acid chains in the hydrophobic interior of the membrane are in a rather disordered state approaching that in a liquid hydrocarbon.[44] In the spontaneously forming liquid-crystalline phases of membrane lipids, the thickness of such layers under conditions approaching those in living cells is always found to be less than twice the length of two fully extended molecules. Consequently, the surface area occupied by one molecule is considerably larger than in a compressed surface film or in the "classical" bimolecular layer with more nearly straight and parallel hydrocarbon chains, which was originally derived from studies of compressed lipid monolayers at an air–water interface. These general conclusions from observations of a variety of membrane lipids and model systems have recently been confirmed specifically for mitochondrial lipids through x-ray diffraction studies in Luzzati's laboratory. An average area per molecule of 65 Å^2 was found in the lamellar liquid crystalline phase at physiological temperatures as compared to 41 Å^2 per molecule in a closely packed array.[23] Birefringence studies on other lipids which directly measure the order and orientation in the hydrocarbon chain layer lead to the same result.[76]

The lipid backbone of the Danielli model should, therefore, be described as a lipid leaflet less than twice the length of two fully extended lipid molecules wide with a disordered central hydrocarbon layer and two surface layers of hydrophilic groups not closely packed. When the degree of disorder in the hydrocarbon layer is high and/or extensive interdigitation exists between the hydrocarbon chains of molecules facing in opposite directions, the total thickness of the "bilayer" may approach the length of one fully extended lipid molecule. X-ray analysis indicates that such a bilayer may exist in chloroplast membranes.[48,35,36]

The conformation of the protein part of the membrane has been depicted as a monolayer of fully extended protein molecules, or as a layer of globular protein molecules, or as a combination of both. There is little, if any, direct evidence to support one or the other of these views. Consequently, the structure of the protein layers cannot be specified. That only a layer of fully extended polypeptide chains is present can, of course, be excluded. This is incompatible with the function of membranes as carriers of enzymes and with the isolation of native membrane proteins which are irreversibly denatured when they are fully unfolded. However, it has been shown that proteins adsorbed to lipid surfaces may be irreversibly denatured through unfolding.[16] The question then arises, could there be a basal layer of unfolded protein to which additional protein in a globular form is bound? Evidence for such a multiple structure has been found in model systems.[16] It is not known, however, if membranes *in vivo* are formed in a similar way—that is, by adsorption of a protein from a solution of low concentration to a free lipid surface, and therefore, the model experiments are no compelling argument for assuming such a structure. That some conformational changes in proteins can take place when they are bound to a membrane is to be expected, and can be inferred from the observed differences in the activity and specificity of enzymes in the free and bound state. So far there is no reason to assume the existence of a general structural principle for the conformation of the protein layers in most or all cellular membranes. It will be necessary to explore the structure of the protein in many different membranes in detail before we can draw any conclusions about the existence of such a common basic structure for the protein layers. At present, few, if any, reliable data are available for the protein structure of any membrane.

The binding of protein will also influence the structure of the lipid layer and will probably tend to increase the area of the layer. That, at least, has been observed when protein is adsorbed to a monolayer of lipid at an air–water interface, even at high film pressures.[15,45,17] This would lead to an increase in the area per molecule for the lipid and to a decrease of the thickness of the lipid layer.

Some evidence for this effect can be obtained from model systems. Figure 2-1 shows the lamellar liquid-crystalline phase of a total lipid extract from beef brain. The lamellar structure is not as clearly visible here as it is in the electron micrographs of brain phospholipids because the average number of double bonds per hydrocarbon chain is lower in the total brain lipid preparation, and less heavy metal accumulates per unit area in the planes of hydrophilic groups. Therefore, the dense lines representing these

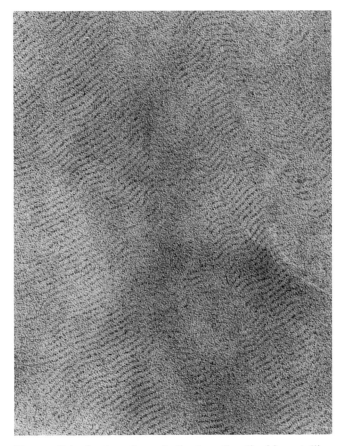

FIGURE 2-1. Section through the lamellar liquid-crystalline phase of a total lipid extract from beef brain. Magnification 320,000×. (307/62)

planes are not very conspicuous. X-ray diffraction data, however, show that the order of the long spacings is as high as in the phospholipid liquid-crystalline phase. The thickness of the lipid layers in this system is 52 Å as compared to 45 Å for the phospholipid system. Under appropriate conditions, addition of excess cytochrome c to a suspension of the total brain lipid will result in the formation of a precipitate that contains cytochrome c and all the lipid of the suspension. It forms a lamellar liquid-crystalline phase. The thickness of the cytochrome c-lipid layers in this phase is found to be 68 Å. If one assumes that the thickness of the lipid layer is unchanged, this would allow for a layer of protein only 8 Å thick on both surfaces of the lipid, i.e., one fully extended polypeptide chain. This is incompatible with the observations that native cytochrome c can be reisolated from such complexes, with the enzymatic activity of cytochrome c-lipid complexes[13] and with ORD studies on protein conformation in these complexes.[96] One must, therefore, conclude that the thickness of the lipid layers has been reduced, owing to the binding of protein. This is borne out by the electron

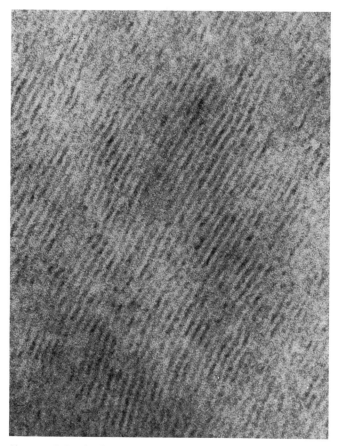

FIGURE 2-2. Section through the lamellar liquid-crystalline phase formed by cytochrome c and a total lipid extract from beef brain. Magnification 320,000×. (442/62)

micrographs which show a considerable increase in the width of the dense bands representing the bound protein and a reduction in the width of the light bands which represent the hydrocarbon layer (Figure 2-2).

Obviously these data must be regarded as preliminary. At least an electron density distribution across the layer will have to be obtained, and the conformation of the hydrophilic groups of the lipids must be explored so that a more precise model of the structure can be drawn. The important point, however, is that in bimolecular lipid leaflets the area per molecule is considerably greater than in a close-packed film, and that this area may further increase during reaction with protein. Obviously, this conclusion has to be taken into consideration when the question is raised if the amount of lipid found in a membrane fraction is sufficient to cover the surface of the cells or subcellular particles with a continuous bimolecular layer, and it is also of importance for a discussion of the possible types of binding between lipid and protein in a Danielli-type membrane.

Since the discovery of "structural protein," its role in the structure of mitochondrial membranes has been a serious problem which still appears far from being solved. When prepared according to Richardson, Hultin, and Fleischer,[74] it is apparently not a uniform protein but a mixture of many components.[24] A structural role for it has never been directly demonstrated, but only inferred from its insolubility under physiological conditions. It may well contain soluble proteins irreversibly altered through the solvent extraction procedure. Its discovery does not preclude that other, especially enzyme, proteins also have a structural function. It has to be discussed here because of the often stated conclusion that the largely hydrophobic interaction of "structural protein" with lipid is incompatible with the Danielli model. Once it is realized that the area per molecule of lipid in a bilayer is considerably larger than in a close-packed film, it becomes obvious that between the hydrophilic groups at the bilayer surface enough hydrocarbon interface can be exposed to allow hydrophobic bonding between lipid and protein without extensive penetration of protein between the hydrocarbon chains.

These conclusions, too, derive support from electron microscopical observations on model systems. Finely dispersed mitochondrial lipids or other phospholipids can bind cytochrome c to form vesicles bounded by a membrane that is identical in appearance to the typical unit-membrane (Figure 2-3). If "structural protein" is used instead of cytochrome c, the result is essentially the same. Again, vesicles are formed which are bounded by a membrane identical in appearance to the cytochrome c-lipid membrane. If "structural protein" prepared according to Richardson, Hultin, and Fleischer[74] is used, usually masses of unreacted insoluble protein(s) persist which adhere to the vesicles, and few free vesicles are seen. Density gradient centrifugation can at least partly separate these from the unreacted protein and their sedimentation behavior shows that the vesicles are not free lipid but lipoprotein (Figure 2-4). If the better soluble and less heterogeneous F_4[99] is used instead, no significant amount of unreacted protein is observed, and the preparation is found to consist only of vesicles bounded by a "unit-membrane" (Figure 2-5).

In the case of the cytochrome c-lipid complex, the structure is generally assumed to be that of the Danielli model, and this is supported by the X-ray

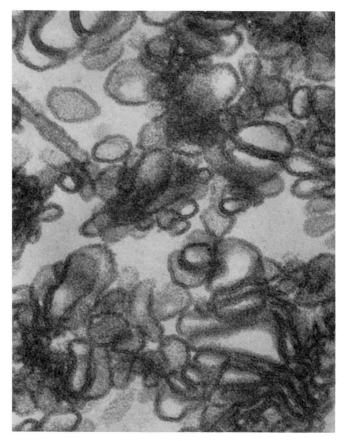

FIGURE 2-3. Section through vesicles formed from highly dispersed mitochondrial lipid and cytochrome *c*. Magnification 160,000×. (H 242/65)

diffraction evidence on the liquid-crystalline phase of the cytochrome c-brain lipid complex. Obviously, a more detailed study of both model membranes is called for. However, it would be very surprising if the "structural protein" or F_4-lipid membranes had a structure entirely different from the practically identical appearing cytochrome c-lipid membrane.

The cytochrome c-lipid and F_4-lipid complexes have so far mainly been used as structural models for mitochondrial membranes. Obviously, they could also be used for functional studies; but so far, little work on this aspect of the model systems has been done. More data are available on functional aspects of two other model systems, one developed by Bangham[3] and the other by Mueller and Rudin[52] and their collaborators.

Bangham uses a rather coarse dispersion of phospholipids and related

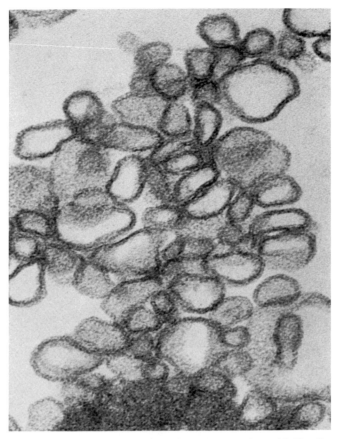

FIGURE 2-4. Section through vesicles formed from highly dispersed mitochondrial lipid and mitochondrial "structural protein." Note mass of unreacted protein at bottom of picture. Magnification 192,000×. (H 994/66)

compounds in water. Under the conditions chosen, the lipids form small spherules consisting of concentric lipid bilayers separated by water layers, thus forming a series of concentric aqueous compartments. The rather complex geometry makes a quantitative interpretation of permeability studies difficult.[4] Qualitatively, however, it can be shown that the response to compounds that modify the permeability is very similar in natural membranes and in the model system.[12] Although the molecular structure of the system has not been explored very carefully, there is little doubt that the structure of the lipid bilayers is essentially the same as that found in the lamellar phase in the liquid-crystalline state, i.e., very similar to the structure postulated for the bilayer in the Danielli model.[60,61]

Much better data can be obtained for the permeability and electrical properties in the Mueller–Rudin system. Unfortunately, the structure and composition of this model membrane is less well defined than Bangham's. It consists of a thin lipid layer, typically ~1 mm² in area, that separates

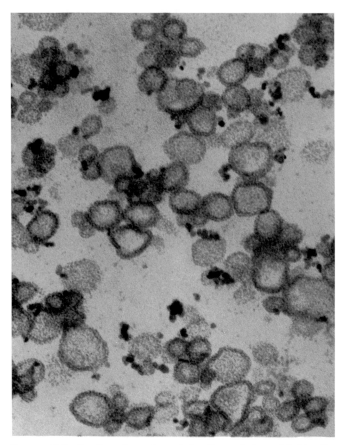

FIGURE 2-5. Section through vesicles formed from highly dispersed phospholipid and F_4. The dense particles are colloidal thorium dioxide which has been added as a marker of the extravesicular space. Magnification 192,000×. (H 108/66)

two aqueous compartments which are both accessible during the course of the experiment. The membrane is formed from a solution of phospholipids in chloroform, decane, tetradecane, or similar solvents, which remains present as a bulk phase around the edge of the membrane. The exact composition of the thin membrane is not known (see also Chapter

1). Recent studies indicate that the thickness of the membrane originally assumed to be ~ 70 Å—which is already too high for a phospholipid bilayer—may still have been underestimated, and that the actual value may be twice as high.[25] Proper sampling of the thin membrane is technically very difficult, but it may contain as much as 10 moles of decane per mole of phospholipid.[26] There is little doubt that hydrophilic groups of the phospholipid molecules are concentrated at the oil–water interface in this model membrane and that the general orientation of the molecules is the same as in the lamellar liquid-crystalline phase; however, it seems that the interior hydrophobic phase may be considerably thicker than in a typical bilayer. Nevertheless, many properties of these thin lipid films, for instance their electrical resistance and water permeability,[10] are amazingly similar to those found in natural membranes, and agents that influence the permeability of natural membranes show very similar effects in the model system. Uncouplers of oxidative phosphorylation in concentrations comparable to those effective in mitochondria selectively increase the permeability for protons.[27,82] Selective ion permeability, resting and action potentials, rhythmic activity, and similar phenomena known from natural membranes can also be induced in this model by the addition of natural components like valinomycin[70,1,54] or the cyclopeptide alamethicin and proteins.[53] These additives are assumed to form ion specific channels which transport unidirectionally but can be reversed. A sharp distinction between an actual channel and a carrier mechanism cannot be made. Recently, Mueller and Rudin[55] have shown that such a mechanism could also explain the coupling of electron transport to phosphorylation.

While neither the structural nor the functional aspects of these model membranes have yet been explored in sufficient detail, what is known so far tends to support the Danielli model rather than any other model of membrane structure.

B. The Subunit Model

The structural models for membranes that can be grouped together as subunit models differ rather widely from each other, and only a very general description can be given here. The specific case of the mitochondrial inner membrane will be discussed in detail later. Mitochondria and chloroplasts contain the two types of membranes that have contributed most of the data for a subunit model of membrane structure, and in both cases an *a priori* argument for a subunit structure can be made (see p. 65). Before we consider the evidence and the arguments in detail, it appears necessary to define the term subunit. In the cases of viruses and protein molecules, the definition is relatively simple and has been discussed by Caspar and Klug[9]; for the case of the more complicated membrane structure, this remains to be done.

Subunits could be units either of structure or of function.* Structural subunits would be particles of similar size, shape, and composition directly bound to each other and determining the morphological character of the membrane. Functional subunits would have to contain the minimum number of all components necessary to carry out one function of the membrane. A requirement for the identification of structural subunits is that they can be demonstrated in the intact membrane by morphological techniques such as electron microscopy or x-ray diffraction, and that they can be isolated. This implies that the bonds between subunits are different from the bonds that hold the components of the subunit together. It should also be possible to show that the components of structural subunits exist in constant proportion in the membrane. Isolation is not required for the identification of functional subunits; it is only necessary to show that the components are bound to the membrane, and that no soluble components are required for their function. However, since complex functions can be resolved into sequences of reactions, the choice which sequence to call a functional subunit seems often rather arbitrary. It might, therefore, be useful to restrict the general term subunit to cases where structural and functional subunits coincide, or where it can be shown that the components of a functional subunit are grouped together on the membrane. One would expect that, at least in the case of a functionally highly specialized membrane like the mitochondrial membrane, the bulk of the membrane material consists of one or a few classes of such subunits.

Membrane subunits are usually depicted as globular lipoprotein particles. There is little justification for this; a small segment of a Danielli-type membrane would fit the term as well. In a Danielli-type membrane, one could also envisage the subunits as restricted to the protein layers without seriously violating the concept. The lipid bilayer on which the protein subunits assemble would serve as the less specific structural backbone and general diffusion barrier with a low degree of order in the plane of the membrane. In this form the Danielli and the subunit model are not incompatible. The latter rather appears as a further development of the former.

It is often assumed that the subunit model implies a self-assembly of membranes from the subunits either *in vivo, in vitro,* or both. This, of course, is not necessarily so, but given our current trend of thinking it would simplify matters considerably if it were found to be true. The most frequently presented evidence for a subunit structure is the disaggregation of a membrane by chemical or mechanical means into small particles that appear as a more or less homogenous population with one or another technique of observation, usually analytical ultracentrifugation or electron microscopy. In some cases, reaggregation into membrane-like structures has

---

* There could also exist units of assembly or growth. However, so little is known about the biogenesis of membranes that a discussion of this possibility appears premature. The available data, however, do not support a subunit model.

been claimed. To establish a subunit structure it is necessary, however, to show that the presumed isolated subunits are present as recognizable entities in the intact membrane. With the possible exception of the photoreceptor membranes in the retina[5] and in chloroplasts,[62,48,37] this has not been achieved for any membrane so far. In fact, reinvestigation of presumed isolated subunits from PPLO membranes[18] and halobacterium membranes[90] has failed to substantiate the original claims. It remains doubtful if any other membrane subunit preparation would withstand a similar closer scrutiny.

It should be pointed out that one cannot necessarily expect that the integrated function of a subunit can be demonstrated in the isolated state. A slight rearrangement of its components might prevent that, or a difference in environment on two sides of the subunit which can only be maintained when the subunit is part of an intact membrane might be required for function.

If we accept these criteria for the recognition of subunits, it becomes obvious that the respiratory chain together with the mechanism for oxidative phosphorylation would be a likely candidate for a functional subunit of the inner mitochondrial membrane. It remains to be seen if its arrangement in the membrane can be established, if it can be isolated as a structural subunit as well, or if a different structural subunit exists. None of the postulated subunits in the mitochondrial membrane, including Green's tripartite particle (see page 72) and Sjöstrand's lipoprotein globules (see p. 67), are big enough to accommodate all the known components of the respiratory chain and oxidative phosphorylation mechanism.

In contrast to the Danielli model for which a number of model systems have been developed, no model membranes have been found so far for which a subunit structure could be claimed with any degree of confidence.

III. MITOCHONDRIAL MEMBRANES

All available evidence clearly indicates that the components of the respiratory chain and oxidative phosphorylation are bound to the inner mitochondrial membrane and that they must be spatially arranged in ordered arrays so that they can interact in the proper sequence necessary for the conservation of energy. It has been the aim of morphological studies to obtain direct evidence for this implied structural organization. Differences in structure of mitochondria and mitochondrial membranes in different functional states are very likely to exist, but direct morphological evidence for this is still largely lacking. We will be concerned here only with the components found on isolated mitochondrial membranes.

The idea of a repeating ordered respiratory assembly is largely due to the observations that the flavoproteins and cytochromes occur in constant proportions,[11] and that respiratory activity and content of respiratory enzymes are proportional to the amount of inner membrane.[7] Recently it has been shown that the soluble enzymes of the Krebs cycle also occur in constant proportions.[69] This may indicate that the supramolecular order also extends to these components. Ordered complexes of such enzymes, usually found in the soluble fraction of mitochondrial homogenates, have been isolated.[73] However, their structural relation to the membranes has not been established.

Information on the structure of mitochondrial membranes has been obtained nearly exclusively through electron microscopy. The electron optical resolution of today's microscopes should be more than sufficient to reveal order at the molecular level. The main problem arises in the preservation of the structure at this level and in the introduction of sufficient contrast to distinguish biological molecules. Mainly, sectioning and negative staining techniques have been used. While no major discrepancies in the results obtained with these two techniques have been encountered, it is still not possible to reconcile some of the findings satisfactorily. This mainly concerns the exact location and configuration of inner membrane particles,** and will be discussed in detail later. No systematic study of mitochondria with the freeze-etching technique has been published so far. This is regrettable because freeze-etching would very likely throw some light on the problem of the inner membrane particles. The incidental observations with this technique that have been published [8,51] seem to agree reasonably well with the gross morphological picture of mitochondria obtained from sectioned material, but structures corresponding to the inner membrane particles have not been clearly identified.

So far, x-ray diffraction techniques have added little to our knowledge of mitochondrial membrane structure. The only relevant study published [92,93] seems to confirm the structure obtained from electron microscopy of sections, and the high angle pattern observed is compatible with the existence of a liquid hydrocarbon phase similar to that observed in the liquid-crystalline model systems. The mitochondrial membranes are much too complex to allow a correlation of O.R.D. or C.D. spectra with a specific membrane model.[85,97] Impedance measurements indicate a capacitance similar to that of other membranes, thus providing an argument for the existence of a continuous lipid layer in the membrane.[67,68] However, one

---

** These particles had been termed "elementary particles" when they were first discovered by Fernandez-Moran, Green, and their collaborators.[20] The use of this word disappeared with the realization that the particles did not contain all the components of the electron transport chain as originally assumed. The most widely used terms now are "subunits," "inner membrane subunits," or "projecting subunits." There is little justification for this, since it appears most likely that the 85 Å spherical "head piece" is the mitochondrial ATPase, and nothing is known about the "stalk." Obviously, the term subunit would lose all meaning if it were applied to every enzyme that is isolated from, or seen on, a membrane. I prefer to retain the term "inner membrane particle." "Inner membrane sphere," proposed by Racker, would serve as well except that its abbreviation I.M.S. will be read as "inner membrane subunit" by almost everyone.

would like to see these measurements repeated in a simpler and more homogeneous system, e.g., isolated outer and inner membrane fractions.

A. Intact Mitochondrial Membranes

The images obtained from sections through intact mitochondria with different types of fixation and embedding techniques have been described in detail elsewhere.[88,91] Mitochondrial membranes do not differ significantly from other cellular membranes in their general properties and in their reaction with fixatives and embedding media.[89] The unit membrane structure is more difficult to demonstrate in mitochondria than in the membranes of most other cell organelles. This is apparently due to differences in the nature of the hydrophilic groups of the lipids, and possibly also the proteins, and can be overcome by the use of suitable fixation and staining techniques.[88,91]

When the unit-membrane is visible, the dense lines may either appear as rows of granules (Figure 2-6) or as continuous dense bands (Figure 2-7). The rows of granules may reflect the distribution of stainable groups on the surface of the membrane, but have also been interpreted as evidence for a subunit structure.[79,80] Unfortunately, it cannot be excluded that they are an artifact, due to an aggregation under the electron beam of the heavy metals used as fixatives and/or stains.[88] Therefore, they do not constitute a convincing argument for a subunit model. Cross-bridges between the two dense layers have also been described, but it is easy to see that, owing to the finite thickness of the section, any plane of sectioning that is close to but not exactly at 90° to the plane of the membrane, would give rise to such an image when the dense lines consist of rows of granules. The typical unit membrane appearance can, of course, be considered an argument for a Danielli-type membrane.

There is no distinct difference in the appearance of the inner and outer mitochondrial membrane in sections. One sometimes has the impression that the unit membrane structure is less frequently seen in the outer membrane, and that the two dense lines appear thinner than in the inner membrane. This, however, is difficult to verify, owing to the variability in the appearance of both membranes and the difficulties in determining the angle of sectioning. At any rate, it may only reflect differences in the distribution and number of stainable groups which cannot be evaluated at present.

Negative staining of isolated intact mitochondria does not reveal much detail. Usually, only the outer compartment, i.e., the space between outer and inner membrane, is penetrated by the stain and a whole mitochondrion is too thick an object to allow a resolution of fine detail. Frequently, however, disrupted mitochondria are also found in the preparations. Their number may be considerably increased by previous swelling in dilute salt solutions or distilled water by a brief sonication, or by spreading of the material at an air–water interface.

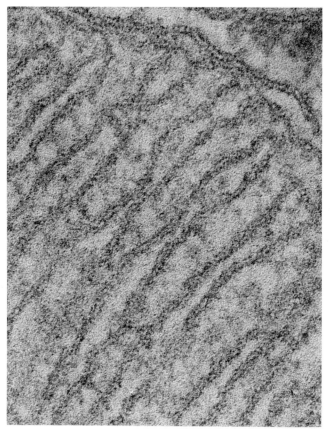

FIGURE 2-6. Part of a mitochondrion from a tissue section with weak staining in the dense bands of the membranes. They appear discontinuous and granular. Magnification 352,000×. (1212/62)

In disrupted mitochondria, the inner and outer membrane are clearly distinguishable in negatively stained preparations.[87] The outer membrane appears thin and smooth, the inner thicker and studded with the inner membrane particles. The outer membrane of some plant mitochondria shows a surface structure that appears as a semiregular array of closely packed dense spots ~30 Å in diameter surrounded by a lighter rim. These may represent pits in the surface. This structure has so far only been found in mitochondria from beans[64] and Neurospora (Stoeckenius, unpublished). A structure consisting of a close-packed array of hollow cylinders 60 Å in diameter has been described for the outer membrane of rat and beef liver mitochondria.[63] These cylinders are not consistently observed; it has

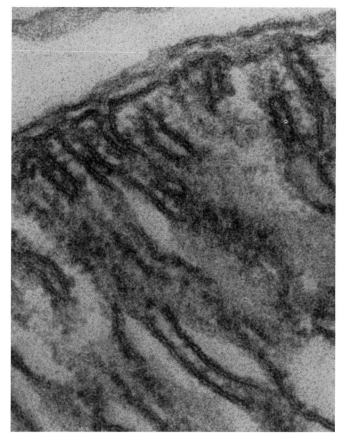

FIGURE 2-7. Part of a mitochondrion from a section heavily stained. The dense layers of the membranes appear continuous. Compare to Figure 2-6 but note the difference in magnification. Magnification 192,000×. (H 756/64)

not been conclusively shown that they are indeed derived from the outer membrane, and their significance remains obscure.

The inner membrane particles are so well known that it seems hardly necessary to describe them here, but we shall examine the evidence for their location on the inner membrane because this has recently been questioned.[46] The problems arise from the fact that the particles can only be clearly identified in negatively stained preparations of disrupted mitochondria where the original topographical relationship of the membranes is usually lost. The evidence that they are located on the inner membrane was originally deduced from pictures of mitochondria where the outer membrane was still mostly intact and could be seen to envelop the inner membrane, and where the cristae could still be at least tentatively identified. Additional evidence came from the high proportion of membrane studded with particles seen in heart mitochondria which contain a high proportion of inner to outer membrane when compared to the lower proportion of particle-studded membrane to smooth membrane observed in liver mitochondria which have fewer cristae. More recent observations both by sectioning and negative staining on inner and outer membrane fractions characterized by their enzyme content confirm this interpretation.[64,83,84] So there is little room for any doubt that these particles are characteristic components of the inner membrane only.

The question then arises, where on or in the inner membrane the particles are located. Again the images of partially disrupted mitochondria, where the relationship of outer and inner membrane and the cristae apparently can still be recognized, strongly suggest that it is the inner surface, i.e., the side facing the matrix, that carries the particles. Additional evidence for this is derived from the observation that when only the outer compartment of an intact mitochondrion is penetrated by the negative stain the inner membrane particles are not seen. The objections to these conclusions are that artifacts of unknown nature may be introduced by the drying of a suspension of partially disrupted mitochondria in phosphotungstate, and that the mechanism of negative staining and the three-dimensional configuration of membranes and similar objects embedded in a thin film of negative stain are not really understood. It is not surprising, therefore, that a dissenting view has been presented that places the particles on the outer surface of the inner membrane.[46] However, to me this view seems to lack sufficient support to be seriously considered at present.

Functional data seem to indicate location of the particles on the matrix side of the inner membrane. These arguments are derived from a comparison of intact mitochondria with fragmented mitochondria, usually called submitochondrial particles, and prepared by sonication. These consist of pieces of mitochondrial membrane which close to form vesicles, a general property of biological membranes. "Submitochondrial vesicles," *** therefore, would be a better term, and will be used here. The vesicles, which are able to carry out oxidative phosphorylation, behave in many respects as if the membrane had been turned inside out; this includes the transloca-

*** Obviously, the term "submitochondrial vesicles" should only be used when it can be demonstrated that the membrane fragments actually form closed vesicles. This usually seems to be the case when mitochondria are mechanically disrupted, e.g., by sonication or passage through a French press.[21] Membrane pieces that fail to close up into vesicles may also be obtained, especially when detergents are used to disrupt mitochondria. They might be called "submitochondrial fragments." A clear distinction between these two types of submitochondrial membrane preparations should be made, because functional differences between preparations might be explainable on the basis of this difference in structure. The importance of this distinction in any discussion of the chemiosmotic mechanism is obvious.

tion of K^+ and H^+ and adenine nucleotides through the membrane, the accessibility of substrate, and the sensitivity to atractyloside.[34,49,50,98,38] When negatively stained, these vesicles are seen to carry the inner membrane particles on the outside, indicating that the matrix side of the membrane is now facing the medium. These results, therefore, seem to be consistent with the morphological argument, and satisfactorily account for many of the observed facts. It should be noted that the so-called digitonin-particles are usually assumed to have retained the original orientation of the membrane. Because digitonin preferentially reacts with cholesterol, and cholesterol is only found in the outer mitochondrial membrane,[41,66] one might expect that digitonin treatment would leave the inner membrane essentially intact.

This rather satisfactory and consistent picture has, unfortunately, been seriously challenged by recent work from Racker's laboratory.[42,43] Not only was it shown by electron microscopy that both digitonin treatment as well as sonication can produce submitochondrial vesicles that carry the inner membrane particles on their outer surface, but these particles also actively accumulate Ca^{++}. Oxidative phosphorylation in these particles can be inhibited by an antibody against F_1, again indicating that the inner membrane particles are on the outside of the actively transporting particles. While these results do not necessarily force a reevaluation of our position on the location of the inner membrane particles in intact mitochondria, they may require reevaluation of Mitchell's chemiosmotic hypothesis. Further investigation of this phenomenon seems indicated.

The work of Racker and his collaborators[30,31,32,72] has established that the "inner membrane particle" is identical with coupling factor F_1. It has been isolated and shown to have ATPase activity,[70a] to be a protein molecule of 280,000 molecular weight,[68a] and to be a component of the oxidative phosphorylation system. It can be bound by submitochondrial vesicle membranes from which it has been artificially removed, and oxidative phosphorylation activity is then restored. Its removal from the membrane has apparently no influence on the activity of the electron transport chain. Results confirming the identity of the inner membrane particle with the mitochondrial ATPase have later been obtained by Tzagoloff and Oda.[57,95] Indications for an additional "structural" role of the particle have recently been reported. This role is deduced from the fact that, even when the particle is enzymatically inactive, it can still restore high levels of oxidative phosphorylation in membranes which have residual F_1 activity. These findings clearly rule out that the particle is an artifact of the electron microscopical preparation technique[2,81] or an essential component of the electron transport chain, possibilities that have been debated for some time.

These findings apply to the spherical 85 Å particle only. The stalk seems to be left on the membrane when the particle is removed. No convincing pictures of isolated particles with stalks have been obtained, and the role of the stalk in the function of the respiratory chain and/or oxidative phosphorylation remains obscure.

When it became evident that the inner membrane particle could not be the morphological substrate of the electron transport chain as originally claimed, an attempt was made to save the concept of the "elementary particle" or "inner membrane subunit" by including the stalk and the piece of underlying membrane to form a new entity that was termed the "tripartite particle" consisting of the head piece ($=$ inner membrane particle), the stalk, and the base piece. While this appears as a reasonable concept, such a particle has not been isolated. Sometimes in negatively stained preparations, the inner membrane seems to show a periodic variation in thickness equal to the center-to-center distances of the attached inner membrane particles, and this has been taken as evidence for the existence of the tripartite particle in the membrane. However, these observations are very rare and usually do not show the presumed periodicity clearly enough to constitute a convincing argument.

While the existence of inner membrane particles *in vivo* can hardly be doubted, they must not necessarily exist in the configuration and relation to the membrane seen in negatively stained preparations. The particles can be shown to be labile to the commonly used fixatives. Other procedures avoiding chemical fixation, such as freeze-drying or freeze-substitution, render the matrix very dense, and these effects may explain the failure to observe the particles in sectioned material. Consistent with this view, is the observation that submitochondrial vesicles before and after removal of the particles show no obvious differences in membrane structure when observed in sections.[88] Nevertheless, that the particles can only be convincingly demonstrated in negatively stained preparations remains a disturbing fact. Moreover, in sectioned mitochondria from heart muscle or insect flight muscle, the cristae appear so densely packed that it seems questionable if there is enough space left between them to accommodate the particles in the configuration observed on the cristae of the same mitochondria when they are disrupted and negatively stained. While the dense packing of the cristae may itself be an artifact of fixation and embedding, this remains to be proven. As things stand now, the question of the configuration of the inner membrane particle and its structural relationship with the membrane in intact mitochondria cannot be considered closed, and more work on the problem seems highly desirable.

B. Dissection and Reconstitution of Mitochondrial Membranes

One approach to an understanding of the respiratory chain and oxidative phosphorylation has been the attempt to isolate the components of these multienzyme systems and to reconstitute functional complexes from the components. Thus far, these efforts have been only partially successful, but the results obtained are of great interest from the structural point of view and offer, at present, perhaps the most promising approach to an interpretation of membrane structure at the molecular level. We shall consider here only results of combined biochemical and morphological observations, which means mainly the work from Racker's and Green's laboratories. For a review of the functional aspects of the reconstitution work see King.[33]

A separation of the mitochondrial ATPase from the membrane, and the electron transport chain and reconstitution of phosphorylating vesicles have already been described when the inner membrane particles were discussed (see p. 69). Structural protein and F_4 with regard to their combination with lipid to form membrane-like structures have been described as model membranes (see p. 59). CF_0, a factor that confers oligomycin sensitivity and cold stability to the mitochondrial ATPase, remains to be discussed.[30] It is prepared from submitochondrial vesicles that have been stripped of inner membrane particles by treatment with trypsin and urea. These TU-particles can be further fractionated in the presence of cholate, and a protein fraction is obtained that has less than 10% of the original dehydrogenase activity, and contains only traces of respiratory pigments and $\sim 5\%$ residual lipid. It binds F_1 and lipids, and when this complex is investigated it is found that the oligomycin sensitivity and cold stability of the ATPase, which are lost when F_1 is removed from the membrane, are now restored. In negatively stained or sectioned preparations, CF_0 appears as thin flakes of varying size (Figure 2-8). These could be interpreted as being derived from small pieces of unit membrane that have split along the central light layer, or as the two dense layers of the unit membrane that are closely apposed after removal of the central light layer. The latter possibility is emphasized by the observation that often a narrow central gap can be observed in cross sections through the flakes. In negatively stained preparations, the complex of CF_0 and F_1 shows the inner membrane particles (F_1) attached to the flakes (CF_0). They are characteristically bound only to one side of any flake, and separated from it by a small gap bridged by a stalk corresponding to the appearance of the particles on intact inner membranes (Figure 2-9). This indicates that an asymmetry across the flakes exists which may be the same asymmetry that existed in the original membrane and has been preserved through the fractionation procedure. When lipid is also added, vesicles are formed, bounded by a unit membrane when seen in sections. They are found to carry particles on the outside when observed with negative staining techniques. They are, thus, morphologically indistinguishable from the original submitochondrial vesicles.

The work of Green and his collaborators is somewhat more difficult to review because details in the interpretations of their data have changed considerably over the last few years, and the following discussion is based mainly on some recent papers[47,94] and the latest review article that has come to my attention.[22] According to Green, the electron transport chain, located on the inner mitochondrial membranes, can be resolved into four complexes which are lipoprotein particles. Under proper conditions they can recombine, forming a membrane that will catalyze the oxidation of

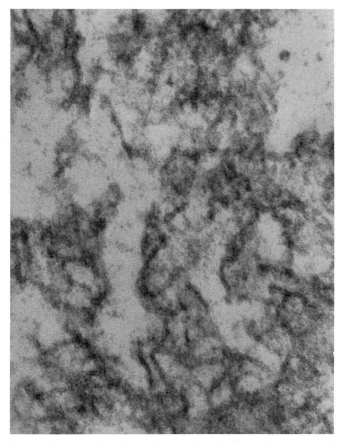

FIGURE 2-8. Section through a pellet of CF_0. Thin sheets of protein in short pieces are the main morphological element recognizable in these preparations. Magnification 192,000×. (H 696/65)

DPNH and succinate by oxygen through the same sequence of reactions found in intact inner membranes. The four complexes are assumed to be located in the base pieces of the tripartite particles discussed earlier (see page 72). Complex I contains the segment of the chain from DPNH to coenzyme Q, Complex II the segment from succinate to coenzyme Q, Complex III the segment from coenzyme Q to cytochrome c, and Complex IV the segment from cytochrome c to oxygen. Each complex, as an isolated lipoprotein particle, carries one set of the components necessary for the

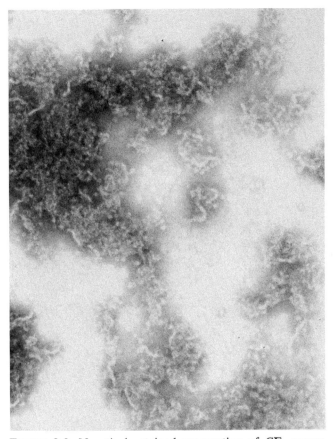

FIGURE 2-9. Negatively stained preparation of CF_0 recombined with F_1. The material is heavily aggregated which makes it difficult to clearly observe its morphology. Where the thin flakes of CF_0 are recognizable, F_1 particles can typically be seen attached to only one side of the flake by a short stalk. Magnification 160,000×. (555/65)

function of the corresponding segment of the chain. Cytochrome c and coenzyme Q are envisaged as mobile carriers that can shuttle back and forth within the membrane, making close contact between the complexes for electron transport activity unnecessary. The intact or reconstituted

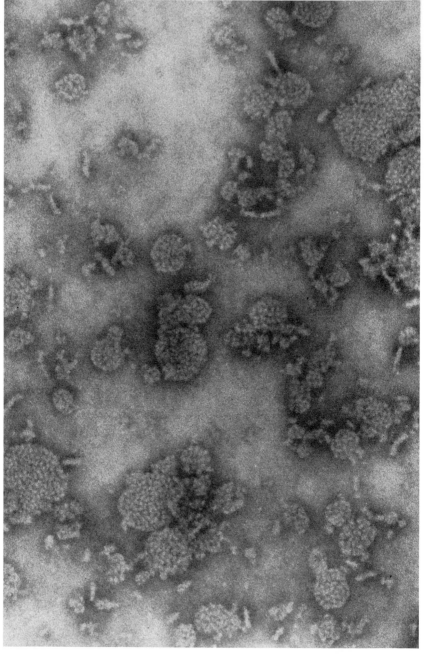

FIGURE 2-10. Negatively stained preparation of Complex IV (cytochrome oxidase) in the most highly dispersed form observed. Magnification 192,000×. (H 1462/66)

membranes are thought to consist of the complexes aggregated side by side into a layer one particle thick and bent to form closed structures. These would take the form of vesicles in disrupted mitochondria and in the reconstituted membranes or the more complex shapes of the inner membrane in intact mitochondria. The same structural principle has been postulated for the outer mitochondrial membrane and, in fact, for all other cellular membranes. However, data supporting this view have been published mainly for the inner mitochondrial membrane. For most of the other membranes assumed to have the same basic structure, the data are scarcer and more controversial.

This model for the inner mitochondrial membrane represents a true subunit structure for which all criteria, isolation of the subunit, demonstration of its composition and function, reconstitution of the membrane morphologically and functionally, and identification of the subunit in the intact membrane are claimed to be established. The model rests mainly on the following evidence: (1) Electron microscopy of negatively stained inner mitochondrial membranes in which the tripartite particle was supposedly recognized. I have dealt with this on page 72 and do not think the evidence is conclusive, a view shared by others.[71] (2) Evidence for a hydrophobic bonding between lipid and protein which supposedly rules out the existence of a bimolecular lipid layer and a Danielli-type structure. This also has been discussed, and cannot be considered conclusive (page 59). (3) Direct observation of the isolated complexes and reconstituted membranes in the electron microscope. In collaboration with Drs. B. and S. Fleischer, I have repeated these experiments for Complex IV, the first complex that was investigated in detail by Green and his collaborators, and for which the most extensive data have been published. The results will be briefly presented here.

Complex IV in its dispersed state, i.e., still containing the bile salts used in its preparation, has been reported to consist of essentially monodisperse 50–100 Å particles.[47] Our preparations show what are apparently the same particles with a diameter of $\sim$55Å. A few bigger particles of $\sim$100 Å diameter may also be observed. However, these particles always occur in groups varying in diameter between 250 and 1800 Å. Very few, if any, free particles are seen. The appearance is somewhat variable from one preparation to another. The given values are for the most highly dispersed forms (Figure 2-10); a less disperse preparation is shown in Figure 2-11. Within the groups, the average particle spacing is rather constant $\sim$70 Å, but no long-range order exists. Some particles appear fused into short, sometimes parallel chains with the same $\sim$70 Å distance between the chains. Fixation and embedding of the material reveals that the occurrence of particles in groups is apparently not a drying artifact of the negative staining technique. The bulk of the material seen in sections appears as short pieces of unit membrane which have the same size

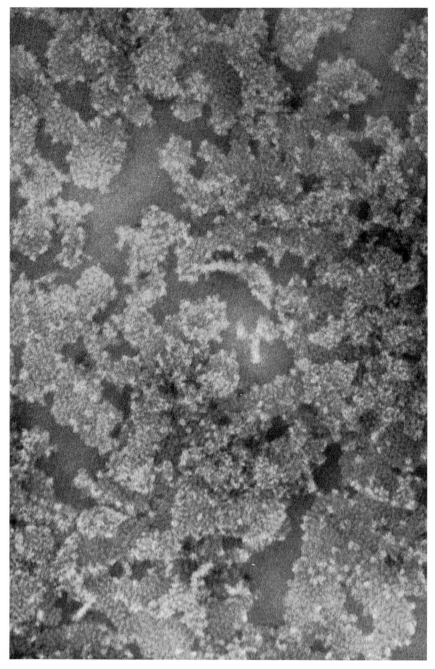

FIGURE 2-11. Same as Figure 2-10 but from another preparation that shows larger aggregates of particles and a few bigger ∼ 100 Å particles. Magnification 193,000×. (449/65)

distribution as the groups of particles found in the negatively stained preparations (Figure 2-12). When cross sections through pellets of centrifuged samples are examined, a small amount of amorphous material may be found on top of the pellets; the bulk of the material, however, has a well-defined membrane structure. The total thickness of these pieces of unit membrane is somewhat variable, but usually wider than that of the inner mitochondrial membrane with an average thickness of ~100 Å determined from several preparations. The central light band of the unit membrane appears denser than in intact mitochondrial membranes.

When the bile salt concentration is reduced through dialysis or dilution of the preparation, bigger aggregates are formed as reported by McConnell *et al.*[47] These take the form of vesicles which show particles on their surface that appear identical to those seen in the preparation before removal of the bile salt. We can confirm this but find that the vesicles typically are not completely closed, and often show large holes in their walls (Figure 2-13). In sections, the vesicles are found to be bounded by a typical unit membrane ~70 Å thick (Figure 2-14). The holes are not seen so frequently, as in the negatively stained preparation. This is to be expected from the randomness of the sections, but easily leads to the false impression that mainly closed vesicles are present.

We conclude that Complex IV consists of small pieces of inner mitochondrial membrane that carry ~55 Å particles on their surface. The increased thickness of the membrane and the higher density of the central layer are attributed to the uptake of bile salts preferentially into the lipid layer of the membrane, though a reorientation of membrane components cannot be excluded. When the bile salt concentration is reduced, the membrane pieces aggregate to form defective vesicles; the membrane is reduced in thickness, and the central layer appears lighter—in short, it becomes very similar to the appearance of the intact inner membrane, but with negative staining shows a surface structure consisting of ~55 Å particles. The inner membrane stripped of the inner membrane particles (e.g., Racker's TU- or SU-particles), however, appears smooth when observed with the negative staining technique. We attribute this to the fact that the components of the respiratory chain and, perhaps, additional unidentified proteins are so closely packed in the intact membrane that the negative stain cannot penetrate into this layer. In the cytochrome oxidase preparation, most of these components have been removed. The resulting gaps are now filled by the negative stain, and the remaining components thus rendered visible in negative contrast. The same appearance of particles on the surface of a membrane is found in the cytochrome oxidase preparation of Jacobs[28,29] prepared by different techniques. It seems possible that Oda and his collaborators[56,58,77] have succeeded in removing some or all of these particles from the membranes of Complex IV and identifying them as cytochrome a. Unfortunately, their published observations are

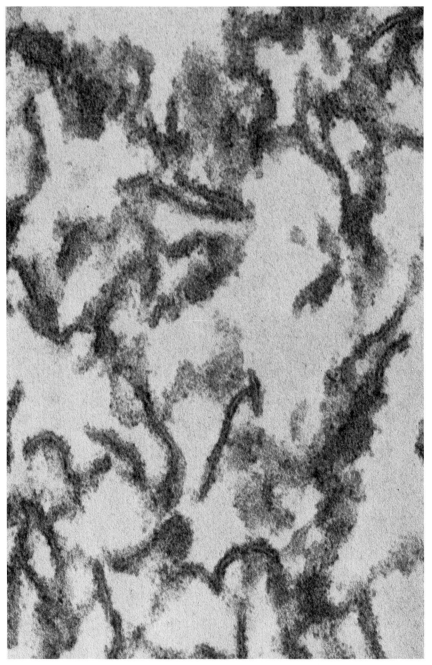

FIGURE 2-12. Section through a pellet of the same Complex IV (cytochrome oxidase) preparation shown in Figure 2-10. It apparently consists of short pieces of unit membrane. Magnification 191,500×. (812/67)

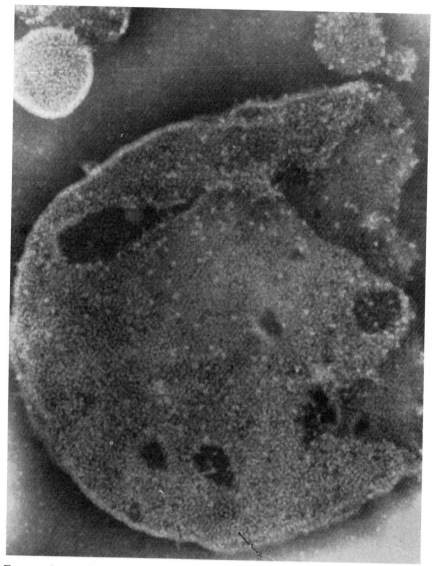

FIGURE 2-13. Negatively stained preparation of Complex IV (cytochrome oxidase) after reduction of the bile salt concentration through dilution. The membrane fragments aggregate into more or less complete vesicles. Magnification 160,000×. (H 1480/66)

not detailed enough to be absolutely convincing or to be discussed in detail. The same holds true for the resolution and reconstitution of the other components of the inner membrane reported by this group.

The interpretation presented here differs sharply from that of Green

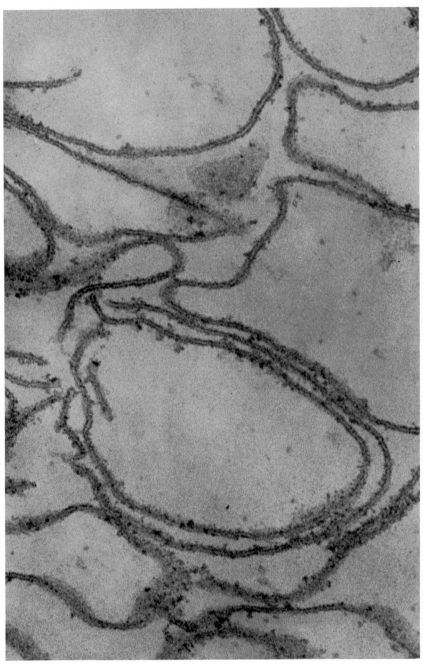

FIGURE 2-14. Same preparation as Figure 2-13 after fixation, embedding and sectioning. The vesicles are seen to be bounded by a typical unit membrane. Magnification 192,000×. (H 15/67)

and his collaborators. This, however, cannot be said for the data. Close inspection of their published electron micrographs shows that they are entirely consistent with the description given here, and not significantly different from our electron micrographs, even though they may not show the features stressed here as clearly. The published micrographs of the other complexes (I, II, and III) also raise doubts about their nonmembranous particulate nature.

It is possible, of course, to argue that the particles have been a monodisperse suspension of lipoprotein particles—"molecularized" or "monomerized," in Green's terminology,—and that some reaggregation to the membranous form occurred during negative staining and fixation. We do

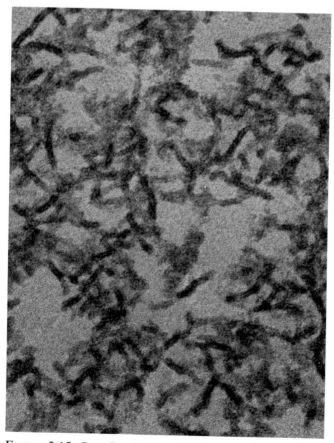

FIGURE 2-15. Complex IV (cytochrome oxidase) after lipid extraction with acetone-water. The sectioned material shows that the unit membrane appearance of the membrane fragments is preserved. Magnification 191,500×. (633/67)

not think this is likely because it would be rather surprising if such different techniques as negative staining and fixation with formaldehyde or OsO_4 should give rise to aggregates of the same size and configuration. The holes found in the vesicles formed after removal of bile salt could also be used as an argument that they are assembled from irregular membrane pieces. Finally, acetone–water extraction of the lipids from Complex IV does not lead to a disappearance of the unit-membrane structure (Figure 2-15). This persistence of unit-membrane structure after lipid extraction is a distinctive feature of the inner mitochondrial membrane[21] (see also page 85) and has not been found in any reconstituted membrane so far (Stoeckenius, unpublished). In any case, the starting material for the preparation of the complex was a membrane. All stages in the preparation that have been observed also show membrane structure. What has to be proven, therefore, is that at some stage in the process the bulk of the material was not in the form of a membrane. This has not been convincingly done so far.****

IV. CONCLUSIONS

The fine structure of mitochondrial membranes is apparently still far from being solved. The subunit model as proposed by Green is unsatisfactory, because it lacks experimental support and does not take into account many of the observed properties of mitochondrial membranes such as their electrical impedance and permeability properties, their appearance as unit membranes in the sections, and their behavior upon lipid extraction (see below). The only evidence for the identification and isolation of subunits comes from electron microscopy, and has been shown to be inconclusive at best. The originally claimed stoichiometry for the recombination of the four complexes to form a functional electron transport chain would have been a strong argument for "monomerized" subunits, but it has recently been withdrawn.[94] Nevertheless, the subunit model is attractive because it emphasizes the main functional aspects of mitochondrial membrane structure and, even though the inner membrane particles and surface structure seen in cytochrome oxidase preparations cannot be properly called subunits, they may still be taken as indicators for the existence of a subunit structure in the inner membrane. If the observed low degree of order in the arrangement of these particles is indicative of the

---

**** As long as it is probable that even only one of the four complexes has not been completely disaggregated and lost all membrane structure, no reconstitution of a functional membrane from nonmembranous components can be claimed. That membrane lipids and proteins can recombine to form membrane-like structures has been known for some time.[80] The crucial question is: Is it possible to form a functional membrane from molecularly dispersed membrane components in a self-assembly process without participation of an existing membrane pattern? This has not been answered by the work of Green and his collaborators, even aside from the fact that full reconstitution of the inner mitochondrial membrane would also require reconstitution of oxidative phosphorylation.

degree of order existing in the packing of the hypothetical subunits, it is not surprising that they have not been detected thus far by x-ray diffraction.

The Danielli model is supported by the morphological observations on sectioned and freeze-etched mitochondria, the impedance measurements and the comparison with model systems presumably containing lipid bilayers. It is compatible with all other observations except one: Virtually complete lipid extraction of mitochondria or submitochondrial vesicles does not lead to a collapse or dissociation of the protein layers in the membrane. After extraction, the unit-membrane structure is found to be essentially unchanged. While this is in good aggreement with the existence of two continuous protein layers on the surface of the membrane, it requires the introduction of a minor modification—i.e., cross-links stable to lipid extraction between the two protein leaflets of the Danielli model.[21] A serious shortcoming of the Danielli model is its failure to fulfill one of the main purposes of a model, i.e., to explain the specific functions of the biological structure.

According to what we presently know about the properties of lipid bilayers, they could serve as the structural backbone and the general permeability barrier of a membrane. Special sites and/or carrier mechanisms with a local modification of the structure would have to be introduced to approach the permeability characteristics of the natural membrane. That such a transformation of the bilayer structure is possible is indicated by the modification of permeability properties in model membranes caused by the addition of naturally occurring compounds (see page 63). The proteins of the respiratory chain and phosphorylating mechanisms in the proper spatial arrangement could be bound to the surfaces of the lipid bilayer. Nonlipid links between the two protein surfaces through the lipid bilayer may occur. The existence of a layer of "structural protein" between the lipid and the enzyme proteins that would have specific binding sites for the enzyme proteins and assure their proper spatial arrangement is an attractive idea, but, so far, remains unproven. Thus, we arrive at a membrane model that carries ordered arrays of enzyme systems on its surface, but preserves a continuous lipid bilayer as the structural backbone of the membrane.

As a working hypothesis, such a model has definite advantages over either the simple Danielli model or a subunit model consisting of enzyme protein complexes occupying the whole thickness of the membrane. It is compatible with all functional and morphological observations on mitochondrial membranes, and the properties of mitochondrial lipids and proteins found in model systems. It might be considered as a synthesis between the Danielli and the subunit model. However, it should be pointed out that the Danielli model was never intended to be more than a first approximation to membrane structure. Therefore, I would rather look at the model proposed here as a further development of the Danielli model.

A number of important problems related to the morphology of the mitochondrial membrane are now approached in several laboratories, but so far no conclusive results have been obtained—e.g., the detailed structure of specific sites in the membrane, the distribution of specific enzymes on both sides of the membrane, conformational changes linked to the different functional states, the isolation and structural characterization of isolated enzyme complexes that still retain the spatial arrangement of components as it existed in the intact membrane, and the reconstitution of a fully functional membrane from the isolated components. A solution of these problems should eventually enable us to establish what determines the order of the components in the membrane. This, of course, is of great significance for general biological problems beyond the immediate interest in the structure and function of mitochondria. The inner mitochondrial membrane appears as a highly specialized structure, adapted to carry out only a few comparatively well-characterized functions. It will probably prove to be a very favorable object for the elucidation of some general principles of membrane organization.

*   *   *

Some of the unpublished work described here has been carried out in collaboration with, and using material prepared by, Drs. Y. Kagawa and E. Racker, and Drs. S. and B. Fleischer. I am grateful to Dr. T. E. Thompson and Dr. J. B. Finean for allowing me to include references to their work before publication. The major part of my own work reported here was supported by U.S.P.H.S. Grants RG–6977, RG–8812, and GM–11825.

REFERENCES

1. Andreoli, T. E., Tieffenberg, M., and Tosteson, D. C. (1967), *J. Gen. Physiol.* **50**, 2527.
2. Bangham, A. D., and Horne, R. W. (1964), *J. Mol. Biol.* **8**, 660.
3. Bangham, A. D., Standish, M. M., and Watkins, J. C. (1965), *J. Mol. Biol.* **13**, 238.
4. Bangham, A. D., De Gier, J., and Greville, G. D. (1967), *Chem. Phys. Lipids* **1**, 225.
5. Blasie, J. K., Dewey, M. M., Blaurock, A. E., and Worthington, C. R. (1965), *J. Mol. Biol.* **14**, 143.
6. Branton, D. (1966), *Proc. Natl. Acad. Sci.* **55**, 1048.
7. Brosemer, R. W., Vogell, W., and Buecher, T. (1963), *Biochem. Z.* **338**, 854.
8. Bullivant, S., and Ames, A. (1966), *J. Cell Biol.* **29**, 435.
9. Caspar, D. L. D., and Klug, A. (1963), "Viruses, Nucleic Acids, and Cancer," p. 27, Williams & Wilkins Co., Baltimore.

10. Cass, A., and Finkelstein, A. (1967), *J. Gen. Physiol.* **50**, 1765.
11. Chance, B., and Williams, G. R. (1956), *Advan. Enzymol.* **17**, 65.
12. Chappell, J. B., and Crofts, A. R. (1966), "Regulation of Metabolic Processes in Mitochondria," Vol. 7, p. 293, Elsevier Publ. Co., Amsterdam.
13. Das, M. L., Hiratsuka, H., Machinist, J. M., and Crane, F. L. (1965), *Biochim. Biophys. Acta* **60**, 443.
14. Deamer, D. W., and Branton, D. (1967), *Science* **158**, 655.
15. Doty, P., and Schulman, J. H. (1949), *Discussions Faraday Soc.* **6**, 21.
16. Eley, D. D., and Hedge, D. G. (1956), *J. Colloid Sci.* **11**, 445.
17. Eley, D. D., and Hedge, D. G. (1957), *J. Colloid Sci.* **12**, 419.
18. Engelman, D. M., Terry, T. M., and Morowitz, H. J. (1967), *Biochim. Biophys. Acta* **135**, 381.
19. Fernandez-Moran, H. (1962), *Circulation* **26**, 1039.
20. Fernandez-Moran, H., Oda, T., Blair, P. V., and Green, D. E. (1964), *J. Cell Biol.* **22**, 63.
21. Fleischer, S., Fleischer, B., and Stoeckenius, W. (1967), *J. Cell Biol.* **32**, 193.
22. Green, D. E., Allmann, D. W., Bachmann, E., Baum, H., Kopaczyk, K., Korman, E. F., Lipton, S., MacLennan, D. H., McConnell D. G., Perdue, J. F., Rieske, J. S., and Tzagoloff, A. (1967), *Arch. Biochem. Biophys.* **119**, 312.
23. Gulik-Krzywicki, T., Rivas, E., and Luzzati, V. (1967), *J. Mol. Biol.* **27**, 303.
24. Haldar, D., Freeman, K., and Work, T. S. (1966), *Nature* **211**, 9.
25. Henn, F. A., Decker, G. L., Greenawalt, J. W., and Thompson, T. E. (1967), *J. Mol. Biol.* **24**, 51.
26. Henn, F. A., and Thompson, T. E. (1968), *J. Mol. Biol.* **31**, 227.
27. Hopper, U., Lehninger, A. L., and Thompson, T. E. (1968), *Proc. Natl. Acad. Sci.* **59**, 484.
28. Jacobs, E. E., Andrews, E. C., Cunningham, W., and Crane, F. L. (1966), *Biochem. Biophys. Res. Commun.* **25**, 87.
29. Jacobs, E. E., Andrews, E. C., Wohlrab, H., and Cunningham, W. (1968), "Structure and Function of Cytochromes," p. 114. University Park Press, Baltimore.
30. Kagawa, Y., and Racker, E. (1966), *J. Biol. Chem.* **241**, 2461.
31. Kagawa, Y., and Racker, E. (1966), *J. Biol. Chem.* **241**, 2467.
32. Kagawa, Y., and Racker, E. (1966), *J. Biol. Chem.* **241**, 2475.
33. King, T. E. (1966), *Advan. Enzymol.* **28**, 155.
34. Klingenberg, M., and Pfaff, E. (1966), "Regulation of Metabolic Processes in Mitochondria," Vol. 7, p. 180, Elsevier Publ. Co., Amsterdam.
35. Kreutz, W. (1964), *Z. Naturforsch.* **19b**, 441.
36. Kreutz, W. (1966), "Biochemistry of Chloroplasts," Vol. 1, p. 83, Academic Press, New York.
37. Kreutz, W., and Weber, P. (1966), *Naturwissenschaften* **53**, 11.
38. Lee, C-P., and Ernster, L. (1966), "Regulation of Metabolic Processes in Mitochondria," Vol. 7, p. 218, Elsevier Publ. Co., Amsterdam.
39. Lehninger, A. L. (1964), "The Mitochondrion," W. A. Benjamin, Inc., New York.

40. Lehninger, A. L. (1967), "Molecular Organization and Biological Function," Vol. 5, p. 107, Harper & Row, New York.
41. Levy, M., Toury, R., and André, J. (1967), *Biochim. Biophys. Acta* **135,** 599.
42. Loyter, A., Christiansen, R. O., and Racker, E. (1967), *Biochem. Biophys. Res. Commun.* **29,** 450.
43. Loyter, A., Saltzgaber, J., Steensland, H., and Racker, E. (1969), *Ann. N.Y. Acad. Sci.* (in press).
44. Luzzati, V., and Husson, F. (1962), *J. Cell Biol.* **12,** 207.
45. Malaton, R. U., and Schulman, J. H. (1949), *Discussions Faraday Soc.* **6,** 27.
46. Malhotra, S. K., and Eakin, R. T. (1967), *J. Cell Sci.* **2,** 205.
47. McConnell, D. G., Tzagoloff, A., MacLennan, D. H., and Green, D. E. (1966), *J. Biol. Chem.* **241,** 2373.
48. Menke, W., "Biochemistry of Chloroplasts" (1966), Vol. 1, p. 3, Academic Press, Inc., New York.
49. Mitchell, P., and Moyle, J. (1965), *Nature* **208,** 147.
50. Mitchell, P. (1967), *Fed. Proc.* **26,** 1370.
51. Moor, H., Ruska, C., and Ruska, H. (1964), *Z. Zellforsch.* **62,** 581.
52. Mueller, P., Rudin, D. O., Tien, H. T., and Wescott, W. C. (1964), *Recent Progr. Surface Sci.* **1,** 379.
53. Mueller, P., and Rudin, D. O. (1967), *Nature* **213,** 603.
54. Mueller, P., and Rudin, D. O. (1967), *Biochem. Biophys. Res. Commun.* **26,** 398.
55. Mueller, P., and Rudin, D. O. (1968), *Nature* **217,** 713.
56. Oda, T., Seki, S., and Hayashi, H. (1965), *J. Electronmicroscopy* **14,** 354.
57. Oda, T., and Seki, S. (1966), "Electron Microscopy 1966," Vol. 11, p. 369, Maruzen Co., Tokyo.
58. Oda, T., Hayashi, H., and Seki, S. (1967), *J. Cell Biol.* **35,** 179A.
59. Palade, G. E. (1953), *J. Histochem. Cytochem.* **1,** 188.
60. Papahadjopoulos, D., and Miller, N. (1967), *Biochim. Biophys. Acta* **135,** 624.
61. Papahadjopoulos, D., and Watkins, J. C. (1967), *Biochim. Biophys. Acta* **135,** 639.
62. Park, R. B., and Branton, D. (1966), *Brookhaven Symposia in Biol. No.* **19,** 341.
63. Parsons, D. F. (1965), *Intern. Rev. Exp. Pathol.* **4,** 1.
64. Parsons, D. F., Williams, G. R., and Chance, B. (1966), *Ann. N.Y. Acad. Sci.* **137,** 643.
65. Parsons, D. F., Williams, G. R., Thompson, W., Wilson, D., and Chance, B. (1967), "Mitochondrial Structure and Compartmentation," p. 29, Adriatica Editrice, Bari, Italy.
66. Parsons, D. F., and Yano, Y. (1967), *Biochim. Biophys. Acta* **135,** 362.
67. Pauly, H., Packer, L., and Schwan, H. P. (1960), *J. Biophys. Biochem. Cytol.* **7,** 589.
68. Pauly, H., and Packer, L. (1960), *J. Biophys. Biochem. Cytol.* **7,** 603.
68a. Penefsky, H. S., and Warner, R. C. (1965), *J. Biol. Chem.* **240,** 4694.
69. Pette, D., and Buecher, T. (1963), *Z. Physiol. Chem.* **331,** 180.

70. Pressman, B. C., Harris, E. J., Jagger, W. S., and Johnson, J. H. (1967), *Proc. Natl. Acad. Sci.* **58**, 1949.
70a. Pullman, M. E., Penefsky, H. S., Datta, A., and Racker, E. (1960), *J. Biol. Chem.* **235**, 3322.
71. Quagliariello, E., Papa, S., Slater, E. C., and Tager, J. M., eds. (1967), "Mitochondrial Structure and Compartmentation," p. 126 (Discussion), Adriatica Editrice, Bari, Italy.
72. Racker, E., Tyler, D. D., Estabrook, R. W., Conover, T. E., Parsons, D. F. and Chance, B. (1964), "Oxidases and Related Redox Systems," Vol. 11, p. 1077, John Wiley & Sons, Inc., New York.
73. Reed, L. J., and Cox, D. J. (1966), *Ann. Rev. Biochem.* **35**, 57.
74. Richardson, S. H., Hultin, H. O., and Fleischer, S. (1964), *Arch. Biochem. Biophys.* **105**, 254.
75. Robertson, J. D. (1960), *Progr. Biophys. Biophys. Chem.* **10**, 343.
76. Rogers, J., and Winsor, P. A. (1967), *Nature* **216**, 477.
77. Seki, S., Hayashi, H., and Oda, T. (1967), I. U. B. Symposium Series, 7th Intern. Cong. Biochem., Abstracts, Vol. V, p. 886.
78. Sjöstrand, F. S. (1953), *Nature* **171**, 30.
79. Sjöstrand, F. S. (1963), *Nature* **199**, 1262.
80. Sjöstrand, F. S. (1963), *J. Ultrastruct. Res.* **9**, 561.
81. Sjöstrand, F. S., Andersson-Cedergren, E., and Karlsson, U. (1964), *Nature* **202**, 1075.
82. Skulachev, V. P., Sharaf, A. A., and Liberman, E. A. (1967), *Nature* **216**, 718.
83. Sottocasa, G. L., Ernster, L., Kuylenstierna, B., and Bergstrand, A. (1967), "Mitochondrial Structure and Compartmentation," p. 74, Adriatica Editrice, Bari, Italy.
84. Sottocasa, G. L., Kuylenstierna, B., Ernster, L., and Bergstrand, A. (1967), *J. Cell Biol.* **32**, 415.
85. Steim, J. M., and Fleischer, S. (1967), *Proc. Natl. Acad. Sci.* **58**, 1292.
86. Stoeckenius, W. (1959), *J. Biophys. Biochem. Cytol.* **5**, 491.
87. Stoeckenius, W. (1963), *J. Cell Biol.* **17**, 443.
88. Stoeckenius, W. (1966), "Principles of Biomolecular Organization," p. 418, J. & A. Churchill Ltd., London.
89. Stoeckenius, W. (1966), *Ann. N.Y. Acad. Sci.* **137**, 641.
90. Stoeckenius, W., and Rowen, R. (1967), *J. Cell Biol.* **34**, 365.
91. Stoeckenius, W. (1967), *Protoplasma* **63**, 214.
92. Thompson, J. E., Coleman, R., and Finean, J. B. (1967), *Biochim. Biophys. Acta* **135**, 1074.
93. Thompson, J. E., Coleman, R., and Finean, J. B. (1968), *Biochim. Biophys. Acta* **150**, 405.
94. Tzagoloff, A., MacLennan, D. H., McConnell, D. G., and Green, D. E. (1967), *J. Biol. Chem.* **242**, 2051.
95. Tzagoloff, A., Byington, K. H., and MacLennan, D. H. (1968), *J. Biol. Chem.* **243**, 2405.
96. Ulmer, D. D., Vallee, B. L., Gorchein, A., and Neuberger, A. (1965), *Nature* **206**, 825.

97. Urry, D. W., Mednieks, M., and Bejnarowicz, E. (1967), *Proc. Natl. Acad. Sci.* **57,** 1043.
98. Winkler, H. H., Bygrave, F. L., and Lehninger, A. L. (1968), *J. Biol. Chem.* **243,** 20.
99. Zalkin, H., and Racker, E. (1965), *J. Biol. Chem.* **240,** 4017.

3

Structure and Function of Phospholipids in Membranes

D. Chapman and R. B. Leslie
Unilever Research Laboratory
The Frythe, Welwyn, Herts

I. INTRODUCTION

Recent major advances in molecular biology have revealed important information about the replication processes of the cells, and attention is turning increasingly to problems associated with cell organization. The structure and function of biological membranes are, therefore, topics of great interest.

The idea of a cell membrane has existed for a great number of years but, for a long time, it was almost entirely at the conceptual level. The membrane was considered to act as a barrier controlling diffusion between the interior of the cell and its surroundings. Even before any direct analyses of individual or isolated membranes had been obtained, suggestions had already been put forward as to their structure and organization, and lipid was considered to be an important element in their construction.

The light microscope was unable to resolve such structures as cell membranes because of their thinness, although useful information was obtained by the use of polarization microscopy. The introduction of the electron microscope with its enormous resolving power changed this situation, and showed that there are layered structures on the outsides of cells, and that there are other membranes as well as the plasma membrane. Thus, mitochondria and lysosomes were seen in the electron microscope to be bounded by thin membranes. Membranes were also observed in chloroplasts and in the rods and cones of the eye.

The early idea of the structure of the cell membrane has been dominated by the concept of a bilayer of lipid being the important structural entity. This was first put forward as a result of experiments by Gorter and Grendel [39] using monolayer techniques, and then by Danielli and Davson[18] who added to this original concept the idea that the lipid was sandwiched by protein material interacting electrostatically with the polar head groups of the lipids (Figure 3-1).

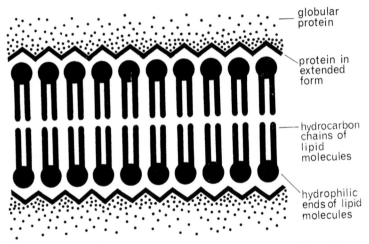

globular protein

protein in extended form

hydrocarbon chains of lipid molecules

hydrophilic ends of lipid molecules

FIGURE 3-1. The classical Danielli-Davson model of cell membrane structure.

Robertson,[87] in particular, partly on the basis of electron microscope studies, suggested that all cell membranes were made from the same basic type of unit. He put forward his concept of a "unit membrane" based essentially on the bilayer of lipid. Recently, doubts have been expressed by many scientists about the universality of this model. A feature of some relevance to a unifying concept of membrane construction is the fact that the protein to lipid ratio appears to vary considerably from one membrane type to another.[56] Furthermore, Korn[56] has attacked the interpretation of electron microscope data. He suggests that there is no way to interpret, in molecular terms, the dense lines seen in electron micrographs of membranes fixed with potassium permanganate and, on the basis of biochemical studies, also doubts the interpretation of membranes "fixed" with osmium tetroxide. Green and co-workers[40] also expressed doubts about the bilayer model. They have put forward a model for mitochondrial membranes in which there is hydrophobic interaction between the lipid fatty acid chains and the hydrophobic amino acids of the protein.

The introduction of analytical techniques, including gas liquid chromatography and thin layer chromatography, has now enabled very careful analyses of the lipid content of cell membranes to be made. These studies have shown that there is a whole variety of classes of phospholipid associated with the structure of cell membranes. Associated with each class of phospholipid is a distribution of fatty acids of varying chain length and unsaturation.[104] Other components of membranes include protein, water, metal ions, and, sometimes, cholesterol, depending upon the membranes which are studied.

In this chapter, we shall discuss what is known at present about the physical chemical properties of phospholipids and their role in biological membranes with particular reference to mitochondrial membranes. First we shall examine the analysis of these membranes.

II. THE COMPOSITION OF MITOCHONDRIAL MEMBRANES

The essential structural features of the mitochondrion are the double membrane and the characteristic cristae. Palade[74] stressed that the number of cristae and their conformation varies considerably from one type of mitochondrion to another. However, the elements that appear to be common to all mitochondria are an outer membrane, an inner membrane, cristae, mitochondrial "sap" or "matrix," dense granules, and DNA. The mitochondrial "sap" or "matrix" is generally regarded as being the space between the cristae and the inner membrane, and is sometimes called the "inner compartment." The space between the inner and outer membranes is best called the "outer compartment." The cristae are double-membrane structures, and the space between these membranes is in connection with the outer compartment. During mitochondrial swelling, this space and the outer compartment are greatly expanded. The nomenclature used is based on that of Whittaker,[109] and a diagram showing these relationships is given in Figure 3-2.

The variations in the morphology of cristae are discussed in detail by Lehninger.[59] They may have the appearance of simple stacked membranes, as in liver and kidney mitochondria, or may be tube-like, branched, or have a variety of other shapes. The inner layer of the mitochondrial membrane is generally assumed to be in continuity with the membranes of the cristae, which may be regarded as invaginations of this inner membrane. However, these invaginations may only occur at certain regions of the inner membrane.

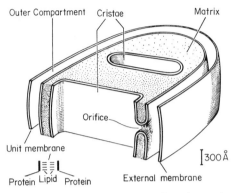

FIGURE 3-2. Diagram illustrating the main structural features of the mitochondrion.[109] (Reproduced with permission)

Recent studies have led to the isolation of both inner and outer mitochondrial membranes from mammalian liver.[36,79] The lipid:protein ratios for these membranes are shown in Table 3-1, and are compared with microsomal membranes. Detailed analysis for whole mitochondria have also been given.

TABLE 3-1
COMPARISON OF MITOCHONDRIAL AND MICROSOMAL MEMBRANES

	Inner Membrane	Outer Membrane	Smooth E.R.
x Density	1.21	1.13	c. 1.13
x Protein: lipid	1:0.275	1:0.829	1:0.385
x Cardiolipin (% of total PpL)	21.5	3.2	0.5
x Phosphatidylinositol (% of total PpL)	4.2	13.5	13.4
x Phosphatidylserine (% of total PpL)	N.D.*	N.D.	4.5
xx Cyt. (a + a₃) μ moles g^{-1}	0.24	<0.02	0.0
xx Cyt. b₅ μ moles g^{-1}	0.17	0.51	0.79

x Thompson, W., and Parsons, D. F., unpublished.
xx Wilson, D., Chance, B., and Parsons, D. F., unpublished.
* Not detected.

A. Lipid Class and Chain Length Distribution

The phospholipid composition of mitochondria is now well established both as to head group type and chain length distribution.[36] Conflict between earlier analyses from different laboratories has been ascribed largely to analytical and methodological difficulties associated with the handling of the labile lipid molecules, rather than to real differences. The very high degree of chain unsaturation found in mitochondrial lipids, compared, for example, to the moderately unsaturated myelin lipids, requires correspondingly greater care in handling procedures. Methods previously applied to mitochondrial lipid analysis, such as silicic acid chromatography and the paper chromatographic technique of Marinetti *et al.*[70] have been shown to lead to high values for phosphatidylinositol and changes in the phosphatidylethanolamine.[36] DEAE cellulose chromatography[89] has been applied to highly purified mitochondria by Fleischer *et al.*[37]

The total phospholipid and neutral lipid (mainly cholesterol, coenzyme Q, α-tocopherol, and carotenoids) content of mitochondria has been determined by several groups[36,37] with good agreement. All mitochondria appear to have a very similar total lipid content (27%) and, of this, greater than 90% is phospholipid.

The lipid composition of mitochondria from different organs differs in detail. All mitochondria, irrespective of source, appear to contain three major phospholipid classes. The phospholipids present are lecithin (PC),

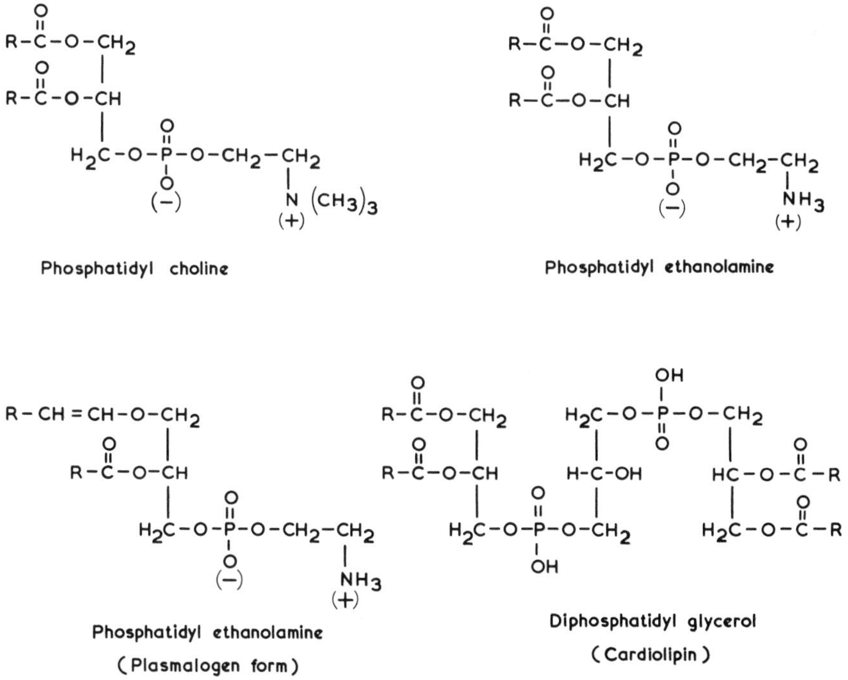

FIGURE 3-3. Molecular structures for the phospholipids found in mitochondria.

phosphatidylethanolamine (PE), and cardiolipin (CL). The structures of the major lipids of mitochondria are shown in Figure 3-3.

Analysis of highly purified mitochondria from beef heart indicate that the PC and PE lipids are present in roughly equal amounts and comprise about 76–78% of the total phospholipid present. The remaining 24% phospholipid consists of cardiolipin 20%, and minor components (mainly phosphatidylinositol ~3–5%). When adjusted for molecular weight, the molar ratios of cardiolipin:lecithin:phosphatidylethanolamine is 1:4:4, and the cholesterol–phospholipid ratio is very low, probably less than 0.05.

A feature of the beef heart lipids is the relatively high plasmalogen content which is found almost exclusively in the ethanolamine and choline fractions to the extent of approximately 0.5 μmole/μmole of phosphorus. This appears to be peculiar to heart mitochondria, since liver mitochondria have little or no plasmalogen.[37]

The most significant feature of the phospholipid composition of mitochondria is the very high degree of unsaturation found in the fatty acyl chains.[36] Though the three lipid classes mentioned above are invariably present, great variability is found in the fatty acid distribution. Chain length and chain unsaturation range between quite wide limits, showing considerable species and organ specificity, and seem susceptible both to dietary and environmental factors. The fatty acid distribution of mitochondria from various organs and species has recently been reviewed in detail,[36] and we shall highlight only a few of the more general features. In Table 3-2 we have collected data relevant to the acyl chain distribution of the cardiolipin, lecithin, and phosphatidylethanolamine fractions of rat liver and beef heart mitochondria.[36] Only the general trends, and not the complexity of the fatty acid chain distribution, are illustrated in Table 3-2. It appears that the cardiolipin is almost invariably associated with unsaturated lipid chains, the major constituent being linoleic. Saturated chains, or chains with more than two double bonds, occur to the extent of one or two per cent at most.

In contrast with the cardiolipin, the PC and PE fractions show considerably more variation in chain characteristics. Saturated, unsaturated, and polyunsaturated acids occur side by side in comparable amounts. Palmitic and stearic are the most abundant saturated lipids; oleic and linoleic are among the unsaturated lipids; polyunsaturated chains are represented by docosahexaenoic acid and arachidonic chains, the latter being present in amounts comparable to linoleic.

Though the cardiolipin chains appear superficially similar in both rat liver and beef heart mitochondria, the chains of both the PE and PC fractions from these two mitochondrial sources are certainly different. The most significant changes are seen in the PC fraction where saturated, mildly unsaturated, and polyunsaturated acids are present in comparable amounts in the liver mitochondria. In the heart mitochondria, on the other hand,

TABLE 3-2*
ACYL CHAIN DISTRIBUTION OF RAT LIVER AND BEEF HEART MITOCHONDRIA

Fatty acid	Cardiolipin fraction		Phosphatidyl-ethanolamine fraction		Phosphatidyl-choline fraction	
	Rat liver	Beef heart	Rat liver	Beef heart	Rat liver	Beef heart
Palmitic 16:0	— 4		17.5 20	38	13 20	24
Stearic 18:0			30.4 19.5		22 19	5
Oleic 18:1	10 13	5	5 10	4	14 12	19
Linoleic 18:2	84 74	84	4 16		20 20	37
Linolenic 18:3		6		15		
Arachidonic 20:4			21 21	33		4
Docosahexaenoic 22:6			14 9		14 19.5 / 8 3.4	4

* Table compiled from data given in Fleischer and Rouser.[38] All figures expressed as area percentages, and only those components present to the extent of 4% included. All data rounded off to nearest whole number.

the polyunsaturated chains are present in relatively small amounts, but this is compensated by a very significant increase in the mildly unsaturated oleic and linoleic chains.

B. Mitochondrial Proteins

There are a very large number of different species of protein components associated with the mitochondrial membrane. It is not yet clear whether some of them have a purely catalytic role, and some a purely structural role, or whether some may encompass both these roles.

C. Cholesterol Content

Mitochondrial membranes differ in the molar ratio of cholesterol to phospholipid. This is 1:53 for the inner membrane and 1:9 for the external membranes.[61]

III. PHYSICO-CHEMICAL BEHAVIOR OF PHOSPHOLIPIDS

A. Phase Transitions and Liquid Crystalline Behavior

It is only in recent years that detailed studies of the physical properties of phospholipid molecules have been carried out. Phospholipids of the type found in biological membranes are usually insoluble in water. This points to the fact that information about the "solid-state properties" of these molecules may be relevant to an understanding of their physical behavior in membrane systems. While x-ray studies of phospholipid are presently in progress,[10] no single crystal x-ray structure of a phospholipid has, as yet, been reported.

Various studies using a range of physical techniques have, however, been carried out with phospholipids extracted from natural sources and with pure synthetic materials. These studies show a number of interesting facts which we shall summarize.

1. Phospholipids can exhibit solid-state polymorphism, i.e., the existence of different crystal phases. Lipids commonly pack in a layered or lamellar (or bilayer type) structure. However, other types of organization have been shown to exist. Thus, lipids can occur in a sheet network system of the type shown in Figure 3-4. This is known as the rhombohedral phase.[63] This arrangement consists of a two-dimensional lattice made up of rod-like structures. It is uncertain whether this rhombohedral phase is relevant to membrane systems. Nevertheless, it is of interest that lipids can organize themselves into arrangements other than the normal bilayer type structure.

2. Phospholipids, either in the anhydrous condition or in the presence of water, exhibit the property of thermotropic mesomorphism, i.e. the property of forming a liquid-crystalline phase. A transition temperature exists for each phospholipid, or group of phospholipids, below which it

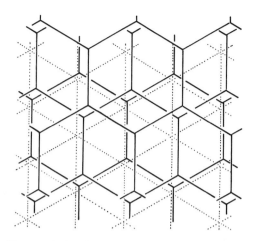

FIGURE 3-4. Rhombohedral phase shown by the lecithin water system. A two-dimensional hexagonal lattice is made up of linked rod-like structures.[63] (Reproduced with permission)

exists in a crystalline form, and above which it exists in a liquid-crystalline type of organization. At the transition point, a marked absorption of heat occurs (Figure 3-5). In the liquid-crystalline phase, the hydrocarbon chains of the phospholipid appear to be in a highly mobile condition, as indicated by results obtained by infrared spectroscopy and nuclear magnetic resonance (n.m.r.) spectroscopy. The x-ray diffraction patterns of the materials also show a marked change. The long spacings, which are to some extent a function related to the length of the molecule, fall dramatically at this temperature, while the short spacings related to the packing of the hydrocarbon chains also change dramatically. The new short spacing is a diffuse spacing at 4.6 Å similar to that observed with liquid paraffin.

The presence of water causes the transition temperature to fall, but it does not fall indefinitely; instead, it reaches a limiting value which is characteristic for each particular phospholipid. A phase diagram for dipalmitoyl lecithin–water is shown in Figure 3-6. With soaps or detergents at low lipid to water concentration, the transition is directly from a crystal or gel to a micellar structure. This does not occur with lecithins; here the transition is directly to liquid-crystalline aggregates. The transition temperature from crystalline (or gel) to liquid-crystalline phase depends upon:

(a) the length of hydrocarbon chain associated with the lipid,
(b) the degree of unsaturation or branching in the chain,
(c) the class of phospholipid,
(d) the presence of cations associated with the phospholipid,
(e) the presence or absence of water.

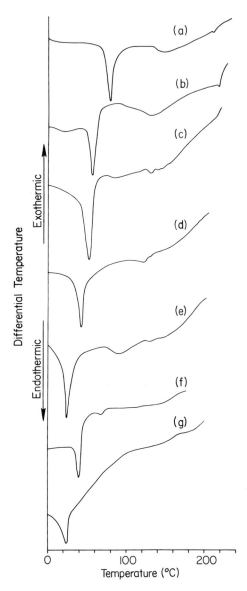

FIGURE 3-5. Differential thermal analysis for various 1,2-diacyl-L-phosphatidyl-choline monohydrates (α_1 form). (a) distearoyl, (b) dipalmitoyl, (c) dimyristoyl, (d) dilauroyl, (e) dicapryl, (f) 1-stearoyl-2-oleoyl, (g) egg yolk.[12]

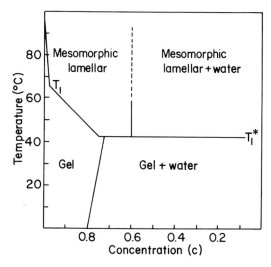

FIGURE 3-6. Phase diagram of the 1,2-dipalmitoyl-L-phosphatidylcholine/water system.[12]

The transition temperature for fully saturated phospholipids, even in the presence of water, can be quite high. Thus, distearoyl lecithin in water has a transition temperature of about 60°C. On the other hand, dipalmitoyl lecithin in water has a transition temperature of 41°C.[12] With highly unsaturated phospholipids, the transition temperature from crystalline (or gel) to liquid-crystalline phase occurs at very low temperatures (below 0°C). We can appreciate that, at body temperature, highly unsaturated phospholipids will exist in a highly mobile condition with the chains waggling and twisting fairly freely. Inhibition of chain movement by interaction with other molecules may, of course, occur in a natural membrane.

Above the transition temperature when water is present, phospholipids can also form different types of phase organization, e.g., some phospholipids give lamellar and hexagonal type structures. Recent studies have shown that the lecithins appear to exhibit only a lamellar type structure over a large range of water concentration. Other phospholipids, such as phosphatidylethanolamines and samples of brain lipid, appear to be able to exist in both the hexagonal and lamellar types of organization, depending upon the concentration in water. Recent studies by Gulik-Krzywicki, Rivas, and Luzzati[43] have shown that four phases can be distinguished with a mitochondrial lipid extract. The composition of the extract from beef heart mitochondria was 34% lecithin, 29% phosphatidylethanolamine, 10% phosphatidylinositol, 20% cardiolipin, 2% cholesterol, and 5% neutral lipids. The phases observed were hexagonal as well as various lamellar

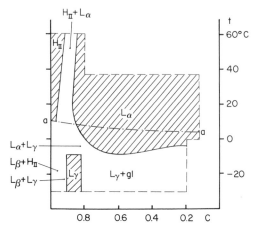

FIGURE 3-7. Phase diagram of mitochondrial lipid-water system. Five phases are observed: one hexagonal (H_{11}), three lamellar (L_α, L_β, $L\gamma$) and ice (gl).[43]

phases. The phase diagram is shown in Figure 3-7 and the structures of some of the lamellar phases in Figure 3-8. As yet no relationship between these phases and their biological function has been made. Nevertheless, these phase properties have to be considered and held in mind in discussions of lipid behavior.

B. The Effect of Cholesterol

The presence of cholesterol, which occurs in some membranes, has been discussed many times in terms of its condensing effect on the monolayer properties of phospholipids. That there can indeed be an interaction between cholesterol and the hydrocarbon chains of phospholipids has also been recently demonstrated by n.m.r. spectroscopic techniques.[14] Calorimetric studies[57] have shown that cholesterol has a marked effect upon the transition temperature of different phospholipids. When cholesterol is present in equal molar ratios with phospholipid, the transition between gel and liquid-crystalline phase is removed. This has led to the suggestion that one of the purposes for the existence of cholesterol in cell membranes may be to prevent the lipid chains from crystallizing either into a crystalline or gel condition.[57] This appears to be important with myelin lipids where the lipids would be in a gel or in a crystalline condition at body temperature, except for the presence of cholesterol. Since the mitochondrial lipids are highly unsaturated and have very low transition temperatures, the presence of cholesterol should not be required for this purpose in mitochondrial systems operating at 37°C. Consistent with this idea is the

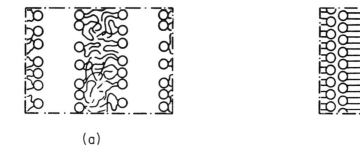

(a)   (b)

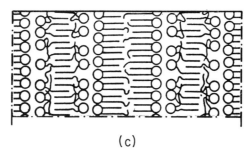

(c)

FIGURE 3-8. Structures of the lamellar phases found in the mitochondrial lipid:water system. The hydrocarbon chains are represented by lines, and the hydrophilic end groups by an open circle respectively. (a) $L\alpha$, high temperature form (found above the line a—a in Figure 3-8). The chains are completely disordered. (b) $L\beta$. The hydrocarbon chains are rigid and packed in a two dimensional hexagonal lattice. Because the chains differ in length, a thin disordered layer is present in the centre of the hydrocarbon layer. (c) $L\gamma$. This is formed by the alternate sequence of L_α, L_β layers.[43] (Reproduced with permission, © Academic Press, Inc.)

fact that the cholesterol composition of mitochondrial membranes as we have seen from the analysis is quite small.

Phospholipid dispersions in water have been used to study permeability behavior towards glycerol and erythritol.[23] These studies have shown that an introduction of double bonds into the lipid chains increases permeability. Decreases of chain length also increase permeability. Furthermore, at lower temperatures, phospholipids with asymmetrical chains are more permeable than those of lipids with chains of equal length but with the same number of paraffin carbon atoms. When cholesterol is present with the phospholipid dispersion there is a decrease in permeability which is proportional to the concentration of cholesterol.[23,24] Cholesterol also lowers the water permeability of thin lipid membranes prepared from egg lecithin.[35]

C. Lipids and Metal Ions

The interaction of metal ions with phospholipids has been studied for many years, from the earliest times of Thudichum[101] in 1901 to the present day.

The monolayer technique has been used to study the interaction of lipids with metal ions. Anderson and Pethica[2] studied the association of synthetic lecithin with metal ions and showed, indirectly, that there was no binding of sodium, potassium, or lithium. Hauser and Dawson[46] have also made monolayer studies on the binding of calcium to monomolecular films of lipid. They showed that the adsorption of calcium is largely independent of the chemical nature of the phospholipid. They suggested that the affinity is controlled by Coulombic forces and is directly related to the net excess negative charge of the lipid molecule. No measurable adsorption of calcium was found to occur at pH 5.5 with phosphatidyl choline, phosphatidylethanolamine, and sphingomyelins. The phosphatidylinositol, phosphatidyl serine, and phosphatidic acid were found to bind calcium at a concentration corresponding to approximately one calcium ion distributed between 5.2–5.8 negative charges on the phospholipid.

D. Lipid–Protein Interactions

An important question underlying the structure of cell membranes is the way in which proteins and lipids interact. The protein–lipid ratio has been found to vary quite considerably, and we know that the protein content of myelin appears to be sufficient to cover only 43% of the area occupied by the lipid. With erythrocyte membranes there is sufficient protein to give a monolayer film some two to five times the area of the lipid. As we have seen (Table 3-1) with mitochondrial membranes, the lipid–protein ratio appears to depend upon whether one is discussing the inner or outer membrane.

When we consider the various ways in which lipids and proteins might interact, then we have to consider the following types of binding forces:

(a) covalent bonding,
(b) electrostatic bonding,
(c) polarization interaction,
(d) dispersion interaction,
(e) hydrophobic bonding,
(f) other types of bonding.

Covalent bonding. There is little evidence for covalent bonding between phospholipids and proteins, although some lipo-amino acids are known to exist. Of the other forces, electrostatic and hydrophobic bonding are probably the most important.

Electrostatic bonding. Electrostatic bonding has frequently been postulated, and certain complexes are observed which are probably of this type, such as those which are formed between phosphatidyl serine and basic proteins.

The first significant x-ray diffraction experiments attempting to understand the structure of lipoproteins and the mode of interaction between lipids and proteins were made by Palmer *et al.* in 1941.[75]

These workers studied the insoluble complexes formed when cephalin reacts with certain basic proteins such as histone. From our present knowledge of the composition of earlier cephalin fractions, it can be assumed that the product they used probably consisted of ethanolamine-containing phosphoglycerides contaminated with considerable amounts of the more acidic phospholipid, such as phosphatidyl serine and phosphatidyl-inositol. The x-ray diffraction data suggest that, in water, the "cephalin" forms bimolecular leaflets separated from one another by extensive aqueous layers of about 80 Å thickness. This is presumably due to the repulsion between the negatively charged phospholipid head groups separating the sheets of phospholipid bilayers. When the basic protein is added to the cephalin, a collapse of the hydrated cephalin sol occurs, presumably because of the positively charged groups on the protein reacting with the negatively charged phospholipid head groups. This results in a dehydration equivalent to that produced in similar sols by metallic cations. The diffraction pattern of the final structure is best interpreted by assuming the complex consists of bimolecular leaflets of cephalin separated by a mono-layer of slightly hydrated protein. The protein was presumed to be partially unfolded and electrostatically interacting with both lipid surfaces on either side of the hydrated layer.

Recent studies[95] in our laboratory have been made of the complexes formed between cytochrome c and phospholipids such as phosphatidyl-choline and phosphatidyl serine. When the protein and dispersed phospholipids are mixed within a certain pH range, precipitation occurs, and the precipitate can be extracted into isooctane.

The stoichiometry of the complexes, and the extractability into isooctane appear to be controlled by several factors; the ratio of neutral to acidic lipid appears quite critical for maximum extractability. They have a lipid to protein stoichiometry of about 20 or 30 lipid molecules per protein molecule. Preliminary low angle x-ray diffraction on the spun-down water insoluble complexes indicate a lamellar structure, with one or two layers of cytochrome c molecules going between the lipid bilayers, as has also been suggested by the recent work of Papahadjopoulos and Miller,[76] and Stoeckenius.[98]

Studies of the complexes in isooctane by the spin labelling technique of Griffith and McConnell [42] directly confirmed that the major bonding force was electrostatic in nature.[5] Fluorescence polarization techniques using DNS labelled lysozyme (which appears similar in behavior to cytochrome *c*) and analytical ultracentrifugation with isooctane as the solvent, have indicated that complexes between phospholipids and lysozyme have a very high molecular weight, probably of the order of a million or so. The application of low-angle x-ray scattering methods to the complexes in isooctane have indicated [95] that there are two types of complexes with radii of gyration of about 65 and 85 Å. These dimensions are consistent with the large molecular weights indicated by ultracentrifugation studies.

Hydrophobic bonding. It is now usual to consider that, when molecules are in an aqueous environment, hydrophobic bonding can be a significant interaction. The hydrophobic bond was considered in some detail by Kauzmann[52] with reference to the structure of proteins. It has been suggested by Green and co-workers[40] to play a role in mitochondrial membranes.

Kauzmann[52] points out that nearly all proteins contain a relatively high proportion of amino acids with nonpolar side chains, such as the isopropyl of valine, the secondary and isobutyl groups of the leucines, and the benzyl of phenylalanine. When proline, alanine, and tryptophan are included, the nonpolar amino acids amount to some 35–45% of the amino acids. The nonpolar side chains have a low affinity for water, and so those polypeptide configurations, which bring large numbers of these groups into contact with each other and are removed from the aqueous phase, will be more stable than other configurations (other things being equal). This tendency of nonpolar groups to adhere to one another in aqueous environments is referred to as hydrophobic bonding. It is considered to be one of the more important factors stabilizing the folded configuration in many native proteins. To some extent this is confirmed by the recent x-ray investigations of hemoglobin,[81] myoglobin,[54] lysozyme,[7] and cytochrome *c*.[25] In each case, the charged groups of the amino acids are on the surface of the protein. The nonpolar groups are usually found in the interior of the molecule and are closely packed together. Kendrew and Watson[55] point out that in myoglobin there is a great variability in the contact-forming capacity of different amino acids and groups. A highly specific conformation is apparently maintained in which specific interactions, such as covalent bonds, hydrogen bonds, and salt linkages appear to play no part. In the case of cytochrome *c*, its structure[25] is said to be a good example of the "hydrophobic drop" structure. All the contacts inside the molecule are nonpolar and nonspecific. Here the molecule contains a center of loosely packed hydrophobic side chains with a dense framework of polypeptide chains and, outside this, a less dense packing of hydrophilic side chains.

Kauzmann[52] suggests that these hydrophobic bonds are stabilized largely by entropy effects. He suggests that hydrocarbon groups in an aqueous environment increase the order or quasi-crystalline order of water molecules in their immediate vicinity. When the nonpolar groups leave the aqueous environment, a large gain in entropy therefore occurs. The free energy change in the transfer of the nonpolar group into the aqueous environment is estimated to be about 3,000–5,000 cal per nonpolar amino acid side chain, or 1,000 cal/CH$_2$ group.

Data substantiating the importance of hydrophobic bonds have come from the theoretical work of Scheraga[91] and Scheraga *et al.*[92] As well as experimental approaches, the latter emphasizes the effects of nonaqueous solvents,[96] hydrocarbons,[110,111] and fatty acids on proteins and the interactions of nonpolar polypeptides.[62] The latter studies on model polypeptides with nonpolar side chains indicate that, in aqueous solvents, the α-helix is a favored conformation for hydrophobic polypeptides.

Now the configurations adopted in water by phospholipids and polar lipids forming micelles and liquid crystalline aggregates, are also such that the molecules have their polar groups facing the aqueous environment and the hydrocarbon groups clustered together. Hydrophobic bonding is considered to take place in this situation as well. The idea that the hydrocarbon chains are squeezed out of water, rather than attracted to each other, is a well-known concept in colloid chemistry.[45]

When lipid and protein are brought together it seems reasonable that both lipid and protein will try to maximize the hydrophobic-bonding possibilities, and some new configuration may be adopted. The lipid–protein interactions in mitochondrial membranes may involve hydrophobic bonding.

Other types of bonding. In addition to the processes which we have just discussed, there are other ways in which lipids and proteins may interact. These include hydrogen bonding, entropy effects, the presence of metal ions, etc. The binding of lipid to protein could take place by a bridging divalent metal ion.

It is difficult to assess the importance of hydrogen bonding between the groups involved in lipid–protein complexes. It is known that water competes for the hydrogen bonding sites.[52] The carboxylic and phosphoric acid groups in phospholipids are completely ionized in the physiological pH range.

The molecular details of the interaction. Having considered the various forces and types of interaction which *a priori,* can occur between lipids and proteins, let us consider the interaction between phospholipids, e.g., lecithins, phosphatidylethanolamines, phosphatidyl serines, and proteins.

First we have to examine whether the reaction between the phospholipid and protein is to be considered one involving, say, a single lipid molecule and the protein. When phospholipids are biosynthesized, it is not known whether a single phospholipid molecule is produced which can react with protein, or whether a number of phospholipid molecules are produced before the interaction occurs. What we do know is that, at very low concentrations in water, lecithin molecules are already in a bimolecular form, i.e., there is a very low critical micelle concentration for this type of phospholipid.[88] At a higher lipid concentration, lecithins in water adopt a lamellar-type structure consisting of large liquid-crystalline aggregates of bilayers of lipid. Each bilayer is separated by water. The lecithins have a polar head group which is in a zwitterion arrangement, and the charges are neutralized. The lecithins are considered to be isoelectric over a wide range of pH.[2]

The electrostatic field at the interface between the phospholipid and the bulk aqueous phase is an important property which has to be considered. Phospholipid molecules orientate themselves so that their ionic portion (head group) is directed towards the aqueous phase. The electrostatic charge on the head groups attract oppositely charged counter ions in the bulk aqueous phase. The distribution of these counter ions determines the potential gradient from the plane of the head groups into the aqueous phase. The total potential difference is known as the ψ-potential. Some of the counter ions will travel with the particle when it moves in an applied electrical field. The potential (relative to that of the bulk aqueous phase) at the plane of shear between the particle with its associated counter ions and the bulk phase is known as the ζ-potential (Figure 3-9). This can be calculated from the rate of movement of the particle in the applied electrical field, and is a measure of the surface-charge density.

The potential gradient at the phospholipid–water interface can play a major part in determining (a) the packing and alignment of adjacent phos-

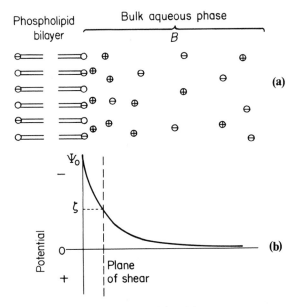

Figure 3-9. (a) Phospholipids with some negatively charged head groups in a bilayer arrangement in water. (b) The ψ- and ζ-potentials which can affect the rate of phospholipase attack on lecithin.

pholipid molecules in the interface and whether dispersion of the laminated form into spherical micelles or transitional states takes place, and (b) the approach and orientation of protein molecules. The effective charge at the interface can be varied by mixing with the phospholipid amphipathic anions or cations, or by adding suitable water-soluble counter ions to the aqueous medium to mask the effect of the surface charge.

All membrane lipids are either isoelectric or negatively charged at physiological pH values. The protein will, therefore, approach an organized lipid structure with a surplus of negative sites on its surface. The diffuse double layer will have a negative potential, unless it is reversed by a high concentration of divalent or multivalent cations in the fixed ionic environment immediately surrounding the lipid structure.

The relative signs and magnitudes of the electrostatic fields surrounding the protein and lipid interfaces will influence the probability of a protein molecule coming into close proximity to bind with the surface and form a stable complex. This can happen even when the net charge on the protein is opposite to that in the diffuse double layer. Dawson[20] has pointed out that phospholipase D below its isoelectric point (positively charged) can approach and be adsorbed on a surface composed of lecithin plus an anionic amphipathic substance where the ζ potential has suffered charge reversal by the addition of a high concentration of calcium or magnesium. In such cases, the electrostatic field in the diffuse double layer can attract and orientate negative sites and areas on the protein molecule, even though the total net charge of the whole is positive. This could be affected by a divalent ion shrinking the effective thickness of the diffuse electrical double layer around the lipid to below the diameter of the globular protein molecule.

A lamellar structure is the feature of the lecithins. They retain this structure over a wide range of concentration in water and do not appear to exhibit any other liquid crystalline phase (except in the monohydrate form).[12] Once the water content is more than ~5%, other phospholipids, e.g., the phosphatidylethanolamines or mixtures of phospholipids and other polar lipids, exhibit a hexagonal as well as this lamellar phase.[64,83] The existence of a particular liquid-crystalline phase depends upon the concentration of lipid in water and the temperature. It is not yet known whether the particular phase of the lipid can determine the interaction with protein. However, the structure of the lipid phase can be affected by such an interaction, in principle.

In many studies of lipid–protein complexes, the phospholipid is dispersed by sonication, e.g., many phospholipases which hardly attack coarse lecithin dispersions nevertheless hydrolyse sonicated lecithins at an appreciable rate.[19,22] The apoprotein of the mitochondrial enzyme β-hydroxybutyrate dehydrogenase does not react with coarse lecithin dispersions, but does so when the lecithin has been sonicated.[50] Sonication of the lipid breaks up the aggregates, and can produce small particles which are, however, considered to retain a lamellar-type structure.[90]

Summarizing this, we can say that, if we bring lipids and protein together in water, various consequences may occur dependent upon the resultant free energy of the product and involving some, or all, of the types of binding forces which we have considered. These are:

(a) No interaction and no complex formation. Both lipid and protein retain their original configuration;

(b) Interaction with the lipid losing all, or part of, its configuration, i.e., a lipid phase change but with the protein retaining its configuration;

(c) Interaction with the protein losing all, or part of, its configuration while the lipid retains its configuration;

(d) Interaction with both lipid and protein losing all, or part of, their original configuration;

(e) Interaction with both lipid and protein (largely) retaining their original configuration.

As an example of (b), we can imagine charged groups on the protein being neutralized by a few lipid molecules with the protein (largely) retaining its original configuration. As an example of (c), we can envisage a protein unfolding its amino acids at the lipid surface while the lipid configuration remains largely unchanged. As an example of (d), we can envisage a gross rearrangement with the hydrocarbon chains of the lipid being rearranged so that they are now within the hydrophobic central region of the protein associated with the nonpolar amino acids, and now both polar groups of lipid and protein are on the outside of the complex. An example of (e) would be a small interaction involving a sheet of lipid and protein in which both largely retain their original configuration.

E. Membrane Lipid–Protein Associations

Schmitt *et al.,*[93] were the first to study a natural lipoprotein system, the nerve myelin sheath, and this work was considerably extended by Finean and Robertson.[34] A combination of electron microscopy and x-ray diffraction methods shows a radially oriented lamellar structure, the dimensions of which (~180 Å) are consistent with an alternation of double layers of lipid 50–55 Å thick, and thin (~30 Å) layers of protein, alternate layers probably differing in terms of some specific groups identified as the "difference factor." The importance of the water present in myelin was considered and, at a certain hydration level, a breakdown of the structure of myelin into a mixed lipoprotein–lipid system was demonstrated.[28]

The changes in the x-ray intensities which accompany the swelling of the myelin sheath of the peripheral nerve by hypotonic solutions can be used to deduce the shape of the Fourier transform of the myelin layer which, in turn, allows the calculation of an electron density distribution curve through the layer.[32] Despite the limitation imposed by the limited intensity data and its correction, the myelin structure is reasonably consistent with the electron density curve derived.

Other membrane systems, e.g., erythrocyte ghosts,[31,48] plasma membranes,[33] and mitochondrial outer membrane[100] have been studied by x-ray diffraction methods, but in these cases it has been necessary to order the membrane preparation by ultracentrifugation and then to allow some drying of the membrane to occur before discrete x-ray reflections could be observed. Only lamellar spacings, ∼120 Å, are observed at this hydration level, but further changes similar to those observed in myelin occur on further dehydration.

In all the diffraction experiments described so far, no indication is present of an arrangement of subunits of macromolecular dimension often shown by electron microscopy. The study of outer segment membranes from the frog retina[6] can be interpreted in terms of arrays of spherical particles of diameter 40 Å, either in the surface of the membrane or within it, but one awaits a demonstration of the generality of lipoprotein subunits in membrane systems. The x-ray diffraction method has given information which has hinted at but, as yet, has not provided precise information about the nature of lipid–protein associations.

X-ray studies of the model lipoprotein complexes formed between phospholipids and basic proteins and serum lipoproteins, either by conventional x-ray diffraction techniques and/or by small angle x-ray scattering may provide useful information. In particular, small-angle x-ray scattering, a technique used successfully to derive size, shape, molecular weight, hydration properties, etc. of proteins in solution, may prove fruitful. In our laboratory,[95] we have obtained scattering curves for isooctane soluble cytochrome c/phospholipid complexes (see Figure 3-10).

From the Guinier plots, radii of gyration have been derived of the scattering particle in isooctane indicating that the complex is aggregated and not monomeric. These studies have also shown that at least two complexes exist within the concentration range studied. Further physical studies on this and related model lipoprotein complexes will provide information on the mode of interaction between lipids and proteins in these model systems. It may be possible to relate their size and shape to membrane parameters derived by more conventional x-ray diffraction methods.

Infrared spectroscopy has been applied to the study of lipoprotein complexes present in erythrocyte and other cell membranes. The technique depends upon the absorption of infrared radiation because of the vibrations of the atoms of the molecule. In the fundamental region 2–15 μ, the absorption bands have been associated with specific chemical groups. The difficulty of applying this technique is that the presence of water gives broad absorption bands which dominate the spectrum. This can only be partly overcome by the use of D_2O.

The main use of this technique to date has been to provide information about the protein conformation. This makes use of the fact that the positions of certain bands, referred to as amide I and amide II bands in the infrared spectra of peptide chains, differ, depending upon the conforma-

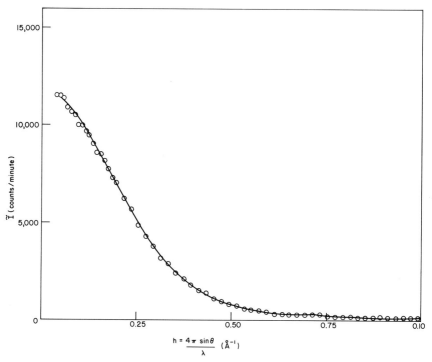

$$h = \frac{4\pi \sin\theta}{\lambda} \quad (\text{Å}^{-1})$$

FIGURE 3-10. X-ray small angle scattering curve of an isooctane soluble phospholipid-cytochrome c complex.

tion of these chains. Thus, the amide I band is located at 1652 cm^{-1} and is associated with an α-helical and/or random coil conformation of peptide chains. The amide II band at about 1535 cm^{-1} does not allow distinction between the α- or β-conformation. A band at 1630 cm^{-1} is correlated with a β-conformation. The infrared spectra of erythrocyte[66] and Ehrlich ascites carcinoma membranes[107] do not show a strong band at 1630 cm^{-1}. This suggests that there is no extensive β-conformation of the protein **structure** in these membranes.

The n.m.r. technique appears to have considerable potential for the detailed study of lipid–protein interactions. Chapman *et al.*[11] have studied membrane fragments of the erythrocyte membrane (Figure 3-11). A particular feature of the spectra is that, while signals are sharp from the

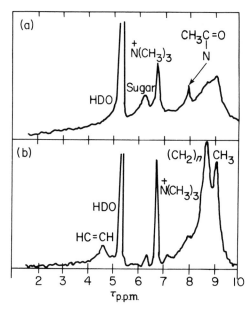

FIGURE 3-11. N.m.r. spectrum of (a) erythrocyte membrane dispersed in D_2O; (b) total lipid from membrane dispersed in D_2O.

protons in the choline group — O — CH_2 — $N^+(CH_3)_3$ the expected signal from the protons in the $[CH_2]_n$ lipid chain is not apparent unless the membrane fragment is heated to high temperatures (120°C) or treated with bile salts, detergents, or trifluoroacetic acid. It is suggested that this is due to interaction between the hydrocarbon chain of the lipid and the nonpolar amino acids of the membrane protein. Cytochrome *c*-phospholipid and lysozyme–phospholipid complexes have also been studied by this method, and studies have also been made with the apoprotein–lipid complex from the high density serum lipoproteins.[13]

Another magnetic resonance technique, the electron-spin resonance (e.s.r.) spectroscopic method, also has potential for the study of lipoprotein complexes. In this case, an unpaired electron is required in the molecule before a signal can be observed. Since this does not usually occur naturally with lipids and proteins, spin labels containing an unpaired electron are prepared. These can be attached either to the protein, or, alternatively, to the lipid, and the e.s.r. spectrum of the spin label studied before and after complex formation. Recent studies[5] have been published of the cytochrome *c*-lipid complex where the protein was labelled with N-(1-oxyl-2,2,5,5-tetramethyl-pyrrolidinyl)-maleimide, and preliminary studies in our laboratory have been made with erythrocyte membranes.

Polar lipids form and orient spontaneously on a water surface giving a monomolecular film whose area can be accurately determined, and which can be submitted to varying pressures giving rise to pressure-area curves. The opportunity to have such a well controlled and oriented layer has led to many studies, particularly at air–water interfaces, but also at oil–water interface as well, both of lipids themselves and also of the interaction between lipids and proteins. Thus, Schulman and Rideal,[94] Doty and Schulman,[26] and Eley and Hedge[27] have studied the interaction between lipids and proteins by measuring the changes in surface pressure produced on injection of various quantities of protein solution under the film.

Most early papers regarded the process of spreading a protein at an interface to involve a complete denaturation of the molecule, i.e., the protein would form a two-dimensional disordered chain with all secondary structure broken. This is not now generally regarded as true,[68,69] although there is still a lack of clear-cut experimental evidence. Proteins which contain covalent linkages in their secondary structure retain a high degree of their original identity when spreading at an interface, and are now envisaged as having a central core which is unchanged, surrounded by a more easily disarranged outer shell. This applies particularly to molecules such as bovine serum albumin (S-S links) and some enzymes which have been shown to regain their activity slowly after an interfacial experience. Our knowledge of the interaction between lipids and proteins at interfaces is in a fairly unsatisfactory state. Elkes *et al.*[29] clearly demonstrated that there is a strong interaction between lipid and proteins of opposite charge, and that, under these conditions, protein will coagulate at an interface. Despite the results of Eley and Hedge,[27] there is no strong evidence that neutral lipids and proteins interact. The results of these authors can be explained in terms of the protein filling gaps left by the lipid film.[105] Interaction does occur when a charged matrix is presented to the absorbing protein, such as that created by a mixed monolayer of fatty acid and monoglyceride or negatively charged phospholipid, even when the protein and interface are of like charge.

The changes which a protein undergoes at an interface can be summarized as follows:

(a) No interaction, protein adsorbs reversibly and in equilibrium with bulk, no denaturation;

(b) Interaction with lipid, protein adsorbs, denatures, and coagulates as lipid–protein complex;

(c) At low-lipid coverage, protein adsorbs, denatures, and coagulates between the lipid molecules.

Further studies with monolayer techniques and supporting evidence from other techniques such as infra red and fluorescence spectroscopy may be useful.

There is little evidence that electrostatic interactions alone are responsible for the lipid–protein associations in cell membranes, except for some of the phosphoinositides of myelin which can be extracted only by acidified organic solvents.[21,38,58] Indeed, neither the acidic phosphatides of *H. halobium*[8] nor the mixed phospholipids of ascites carcinoma membranes[107] are dissociated noticeably from their membrane proteins by manipulations of ionic strength and pH which disrupt the artificial ionic complexes discussed above. In contrast to the synthetic model lipid–protein complexes, the phospholipids of these and other membranes are readily separated from the proteins by extraction with 2:1 chloroform:methanol, a solvent which should stabilize ionic interactions. The model ionic phosphatide–protein complexes dissociate under unfavorable ionic conditions, but this does not occur with cellular membranes. This does not necessarily indicate lack of ionic bonding; it suggests that nonpolar interactions, and in particular hydrophobic bonding, may be important.

Green and Fleischer[40] have concluded from their studies of the interaction of mitochondrial "structural protein," the electron transport enzymes, and phospholipids with each other and with detergents, bile salts, and extraneous phospholipids, that hydrophobic bonding has an important role in mitochondrial membrane structure and function. "Structural protein" is insoluble in aqueous media except at extremes of pH, but can be solubilized by anionic detergents.[40] It forms salt-insensitive complexes with all mitochondrial phosphatides, in contrast to cytochrome c which, as discussed earlier, binds ionically to negatively charged phosphatides (mainly diphosphatidylglycerol). Studies on the binding of synthetic alkyl phosphates to "structural protein" indicate that the binding of these compounds depends only on the length and nature of the hydrocarbon chain. The ionic groups of the bound alkyl phosphates remain free to react with basic proteins (e.g., cytochrome c) to form salt-sensitive ternary complexes.

It seems that there is increasing evidence for hydrophobic interactions between proteins and lipids in membranes. The traditional models of membrane structure, in which the protein is spread (in β-conformation) over the polar surfaces of a lipid bilayer, do not permit such interactions, since extension of apolar amino acid side chains into the apolar core of the bilayer[47] has been suggested to be energetically improbable and sterically impossible. However, the available infrared spectroscopic evidence and the ORD and CD studies on the conformation of proteins in membranes[53,60,66,67,103,106,108] and the n.m.r. studies of erythrocyte membranes[11] indicate that the spatial relationships of proteins and lipids may be other than those proposed in the classical model.

A new membrane model has been put forward by Wallach and co-workers[106] in which, (a) membrane peptide is envisaged to be located on both of the membrane surfaces and also within the apolar core of the membrane; (b) surface-located peptide is considered to be coiled irregularly; (c) penetrating protein segments are considered to be predominantly helical rods, each, like hemoglobin helices, with a hydrophobic face packed to form subunit assemblies whose exterior circumferential aspects are organophilic. The axes of the subunit assemblies are thought to lie normal, or nearly normal, to the membrane surfaces; (d) the apolar amino acid residues which make up the external surfaces of the subunit assemblies are considered to comprise specific binding sites for the hydrocarbon residues of tightly retained membrane lipids, whose head groups can also participate in polar associations with surface-located protein side chains. The association of tightly bound lipids with membrane proteins is considered to be analogous to the haem–protein interactions in hemoglobin. Additional lipid, lying more distant from the protein, is considered to be bound less tightly and less specifically, and would provide a bridge between adjacent lipoprotein units; the loosely bound lipid may be in a bilayer arrangement. It is suggested that the distribution of polar amino acids is such that their side chains lie at the membrane surfaces and/or cluster around the central axis of each subunit, producing aqueous channels which penetrate the membrane. This scheme is analogous to that actually found in hemoglobin[80] and provides a conceptual basis for membrane "pores."

The complex structure of the proteins within membranes is considered to depend upon their association with appropriate lipids, and to represent the state of lowest free energy and maximal entropy of the system protein—lipid–water. Membrane proteins may be released biosynthetically in a conformation other than that existing in membranes (i.e., polar exterior, apolar core). Only after combination with lipids will the membrane conformation be attained.

IV. SOME PHYSICO-CHEMICAL STUDIES OF MITOCHONDRIAL MEMBRANES

1. Some efforts have been made to investigate the protein surface layer component of the inner membrane of mitochondria.[78] The inner membrane is separate from the outer membrane, and its surface is thrown into folds or tubes. On applying the negative staining method to partly fragmented mitochondria, a striking morphological difference is seen between the inner and outer membranes. The surface of the outer membrane is relatively smooth, while the cristae, or tubes, and peripheral part of the inner membrane are covered with regular knob-like structures. These structures consist of a polyhedral head some 90 Å in diameter mounted on a stem 35 Å

wide and 45 Å tall. These spherical particles, lining the inner mitochondrial membrane, have been shown by Racker and co-workers (see Chapter 4) to represent the morphological expression of the mitochondrial ATPase (F_1). If the soluble ATPase (F_1) is recombined with inner membrane preparations stripped of knob-like projections, the reconstituted preparations appear to have regained the projecting head pieces when examined in the electron microscope.[51]

Similar subunits have been visualized in thin sections where the preparative treatment of the membrane is quite different from that used in negative staining methods. The subunits are strictly attached to membranes having electron transport and oxidative phosphorylation functions, mitochondrial inner membranes, chloroplasts, and the plasma membranes of some bacteria. The particles are not seen on the outer membranes of mitochondria of endoplasmic reticulum or plasma membranes.[77]

2. Measurement of the circular dichroism and optical rotary dispersion of the peptide chromophores has become extremely valuable in elucidating the conformations of soluble polypeptides and proteins, because in such polymers, the optical activity of the peptide transitions is not simply the summed contribution of single peptide linkages, but depends critically upon their spatial relationships. Thus, polypeptides in known α-helical, β- or unordered conformations exhibit distinctly different ORD and CD spectra.

Detailed ORD spectra have been obtained from the following membrane types: plasma membranes and endoplasmic reticulum of Ehrlich ascites carcinoma,[106,108] erythrocyte ghosts and *Bacillus subtilis* and mycoplasma membranes,[60] and mitochondrial fragments.[103] Comparison of these spectra with those obtained from polypeptides of known conformation reveals that they all have the following anomalous features: (a) a shape closely approximating that of pure, right-handed α-helix, (b) low amplitude, (c) displacement of the entire spectrum to longer wavelengths than that observed for α-helix. It should be noted, however, that the ORD spectra of soluble lipoproteins and of ionic complexes between phosphatides and cytochrome c[102] do not show the peculiarities of the membrane spectra. There is considerable ambiguity in the interpretation of ORD spectra. More decisive information concerning protein conformations can be obtained from measurements of CD.

When polypeptides are in unordered conformation, their CD spectra are dominated by an intense negative band at 198 mμ (π^0–π^- transition). The small positive band at 223 mμ is due to the n–π^- transition. Polypeptides in right-handed α-helical conformation have two large minima at 222.5 mμ and 208 mμ, and a maximum at 192 mμ. The two extremes at shorter wavelengths arise from the π^0–π^- transition, which is split into components (at 206 mμ and 192 mμ) in the α-helical conformation. The 222.5 mμ minimum is due primarily to the strong n–π^- transition at 224 mμ. The CD spectra of polypeptides in β-conformation show a single negative band at 218 mμ, attributable to the n–π^- transition and a maximum at 195 mμ. Thus, conformational analysis of proteins should, in principle, be possible from comparisons of their CD spectra with those of synthetic polypeptides of known conformation.

Membrane CD spectra have been reported by Lenard and Singer,[60] Urry et al.,[103] and Wallach and Gordon[106] and have the following characteristics: (a) "α-helical" shape, with a broad, negative band at 223–225 mμ, (b) low amplitude and (c) a red shift of the spectrum, at least at shorter wavelengths, with a shoulder at 210–212 mμ, and the crossover to positive CD at 205 mμ. More recent data also reveal a maximum near 195 mμ.

As far as the π^0–π^- transition is concerned, all of the data show red displacement. This could be due to high polarizability of the medium surrounding the helical portions of membrane proteins.[106,108] Wallach and Gordon[106] suggest that the bandwidths of the π^0–π^- transitions are also increased, a condition necessary to account completely for the observed amplitudes and shapes of the CD and ORD spectra. Urry et al.[103] do not discuss polarizability, but point out that, in the case of helical polypeptides, media of low polarity would also cause a red shift of the π^0–π^- transition. Both groups come to the same tentative conclusion; namely, that helical segments of membrane protein lie in a hydrophobic environment.

Steim and Fleischer[97] have studied mitochondrial membranes. These authors suggest that the red shift observed with these membranes can be accounted for in terms of mitochondrial "structural protein" alone, without invoking lipid–protein interactions. They suggest the presence of considerable α-helix structure in the "structural protein," and point out that the existence of helical proteins in membranes is not consistent with the simple bilayer model.

3. Recent work on the physical properties of phospholipids has led to the suggestion that a possible function of fatty-acid chain distribution is to match membrane fluidity, environmental temperature and the rate of membrane mediated processes.[9] There is now some support for this suggestion with respect to the gross lipid composition of poikilothermic and homeothermic organisms. For example, Marr and Ingraham[72] showed that the proportion of unsaturated fatty acid of *Escherichia coli* decreased as the temperature of the growth medium was increased. Similarly, the brain lipids of goldfish acclimatized to decreasing temperatures showed increased amounts of chain unsaturation.[49] Richardson et al.,[86] in an interesting paper concerned with the influence of diet, compared mitochondrial lipid patterns found in fish, fish-eating birds, rat livers, and beef hearts.

Table 3-3 is a compilation from the data of Richardson et al.[85,86] of the major fatty acids found in mitochondria from one organ (liver), from a variety of animal species. The significant feature emerging from Table 3-3 is the higher content of unsaturation in the fish and fish-eating animals. Un-

TABLE 3-3

SPECIES VARIATION OF THE MAJOR LIVER MITOCHONDRIAL FATTY ACIDS

	16:0	18:0	18:1	18:2	20:4	20:5	22:5	22:6	
Rat	20.4	22.3	11.6	16.3	24.7	—	—	—	land based
Chicken	26	7.5	38	16	6	—	—	—	animals
Seal	14.5	21.2	14.6	2.0	14.3	10.2	3.4	10.7	fish
Pelican	17.0	19.6	19.6	2.2	8.3	14.2	0	11.9	eating
Murre	15.6	22.2	15.5	2.5	11.7	9.3	2.4	12.6	animals
Cormorant	20.2	19.5	16.7	2.2	12.2	6.8	0	12.7	
Bass	19.2	2.7	22.6	1.9	2.1	10	1.9	15.2	
Sturgeon	18.8	2.7	23.8	0	3.5	7.2	2.3	15.3	
Flounder	19.2	2.0	15.4	0	4.6	9.9	3.8	20.0	fish
Salmon	20.9	11.2	19.9	0.9	4.0	16.5	6.5	15.6	
Catfish	19.1	6.1	41.2	0	4.5	8.4	1.8	11.8	
Carp	25.5	6.9	18.3	11.4	9.2	1.0	3.3	11.8	

All data are mole percentages. Table compiled from data given in Richardson, T., Tappel, A. L., Smith, C. M., and Houle, C. R. (1962). *J. Lipid Research* **3**, 344. Richardson, T., Tappel, A. L., and Gruger, E. R. (1961). *Arch. Biochem. Biophys.* **94**, 1.

fortunately, one is not able to separate the possible dietary and temperature effects responsible for this higher content of unsaturation from the available data. The virtual absence of the polyenoic very long-chain acyl residues (20:5, 22:5, 22:6) in the rat and chicken compared with the other species is noteworthy. Fish mitochondria were also found to have very little or no linoleic, linolenic, and very little arachidonic acid, in contrast to the rat. The appearance of significant amounts of the rather unusual eicosapentaenoic acid in the fish-eating animals is probably an indication of dietary influences.

From the data in Table 3-3, it is clear that there are considerable differences in the fatty acid patterns of the fish and mammals. Fish mitochondria function at lower temperatures than the land based animals, and there is more unsaturation in the lipids of the former. Richardson and Tappel [84] also showed a correlation between the highly unsaturated polyenoic fatty acyl chains of the lipids found in fish, and the faster rate of swelling of the fish liver mitochondria compared to those from rat liver. These authors offer the speculation that the different physical properties of the more highly unsaturated fish lipids may enable them to perform their various functions at lower temperatures.

Over all, the mitochondrial membranes of beef hearts show about 3.2 double bonds per atom of lipid phosphorus, making them one of the most unsaturated membrane types. There is some evidence that the lipids of mitochondria follow the general tendency for many membranes to maintain their over-all fluidity by increasing the degree of phospholipid chain unsaturation, as the environmental temperature is lowered. The high degree of phospholipid unsaturation, and, consequently, lipid fluidity, may be related to the fact that mitochondrial membranes are the site of such intense metabolic and mechano-chemical activity. In contrast, the less active myelin contains more saturated lipid.

4. Mitochondria also participate in energy transduction in active ion transfer as discussed in greater detail in Chapter VI. The amount of Ca^{++} taken up at saturation levels in the presence of inorganic phosphate is far in excess of that which may be accounted for by assuming combination with the phospholipid or proteins of the mitochondria. It is deposited as an insoluble salt in the matrix. The question arises then, how are the phospholipids of the mitochondrial membrane involved in this ion translocation? Several ion transporting systems in other membranes associated with a Mg^{++} stimulated ATPase have a specific phospholipid requirement.[99] Martonosi[71] reported recently that ATPase activity and Ca^{++} transport in skeletal muscle microsomes were inhibited by treatment with phospholipase C, parallel with the hydrolysis of lecithin; Fenster and Copenhaver[30] reported that the best reactivation of a lipid-depleted ATPase from brain microsomes was achieved with an acidic lipid, phosphatidylserine.

Perhaps the fact that model lipoprotein complexes apparently "invert" when going from water to isooctane[40] may be pertinent to their role in ion translocation. ATP may cause a water-in-oil to oil-in-water micellar inversion according to Maas and Coburn,[65] and some recent magnetic resonance experiments[1] on the binding of Mn^{++}, phosphatidylserine and adenosine triphosphate appear to be explicable in terms of a structural inversion.

Mitochondrial membranes have rather specific ion permeability characteristics with ion size having little, if any, importance.[17] Thus, of the anions, chloride penetrates slowly, whereas phosphate penetrates quite rapidly. Cations, with specific exceptions such as Ca^{++} and Mn^{++}, penetrate very slowly. Chappell and collaborators,[15,16,17] and Harris *et al.,*[44] Pressman,[87] and Mueller and Rudin[73] have studied the influence of the mitochondrial inhibitors, gramicidin and valinomycin, on ion transport in isolated mitochondria and appropriate model systems. These investigators[17,44,82] have demonstrated that normal mitochondrial membranes have a very low permeability to sodium and potassium ions. The antibiotics cause an increased permeability to both sodium and potassium (gramicidin) and increased permeability to K^+, but not Na^+ (valinomycin). It was shown[16] that gramicidin A caused a K^+/H^+ exchange in red cells, and both gramicidin and valinomycin caused K^+ efflux from phospholipid micelles prepared according to Bangham *et al.*[3,4] The effect of gramicidin A on the phospholipid dispersions was not enhanced by uncoupling agents, whereas the K^+ efflux in the presence of valinomycin was markedly accelerated by addition of dinitrophenol. The difference was ascribed to the ability of valinomycin to cause a selective increase in the permeability of the lecithin membrane to K^+, but not to protons. Gramicidin appears to enhance proton and K^+ permeability in both the lecithin dispersion and the mitochondrial membrane.

A detailed understanding of how these compounds cause these dramatic permeability changes in mitochondrial membranes and phospholipid bilayers is clearly desirable.

V. CONCLUSIONS

Interest is turning increasingly to bridging the gap between studies related to pure physiological activities, such as the measurement of substances passing through membranes, with other studies concerned with the physics and chemistry of membranes and their constituent molecules.

Analytical techniques have revealed that membranes usually contain various phospholipid classes and, associated with each class, a distribution of fatty acids of varying chain length and unsaturation. The lipid–protein ratio has been found to vary from membrane to membrane, and information about the protein material of membranes is still rather fragmentary. Cholesterol has been found to occur in some, but not all, membranes.

Because of the knowledge available about membrane constituents, it is becoming possible to investigate the physical and chemical properties of these molecules, and to study their mutual interaction. Summarizing the conclusions obtained from the recent physical studies we know that:

(a) Pure phospholipids have characteristic transition temperatures corresponding to endothermic processes at which the hydrocarbon chains of the lipid become highly mobile. When the phospholipids contain shorter chain lengths or unsaturated bonds, these phase transitions occur at lower temperatures. In the presence of increasing amounts of water, this transition temperature for a given phospholipid decreases progressively, but then reaches a limiting value. Some of the water associated with the polar groups appears to be "bound" and does not freeze at 0°C.

(b) The lipid is more readily dispersed when heated above this limiting transition temperature, and spontaneously forms myelin structures. Phases other than the lamellar arrangement are observed with some types of phospholipid.

(c) The effect of cholesterol present at equimolar quantities with a phospholipid is to remove the transition region so that an abrupt change from crystalline gel to liquid crystal does not occur. Cholesterol also affects the chain mobility. A rough correlation between chain length, unsaturation, and cholesterol content in membranes occurs. The mitochondrial membranes contain little, if any, cholesterol, and have much more unsaturation in their hydrocarbon chains; in myelin, cholesterol is present to a much higher degree, and the hydrocarbon chains are more saturated.

We also know that proteins and polypeptides can affect the lipid chain mobility.

These variations in lipid transition temperature lead to the idea that the particular distribution of fatty acid chains observed in cell membranes may be regulated so as to provide the correct fluidity and permeability at a particular environmental temperature to match the required rate of diffusional and metabolic processes for the tissue. Thus, in mitochondrial membranes where these processes are likely to be of a rapid nature, the average transition temperature of the phospholipid present is low compared with the environmental temperature. In other membranes, for example, the myelin of the central nervous system where the over-all metabolic processes are slower, the average transition temperature for the phospholipids is higher and is close to that of the biological environmental temperature. There is also some evidence that some organisms and bacteria may have a feedback mechanism locked to the environmental temperature which enables the fatty acid residues of the phospholipids to be altered, to maintain a suitable membrane fluidity.

Despite the possibility, in principle, of controlling the fluidity by varying chain length, unsaturation, or even lipid class, it seems that cholesterol and proteins, can also control the fluidity and, therefore, perhaps the permeability of membranes.

When we ask how do the phospholipids and proteins of the membranes interact, we see that *in principle,* a whole range of binding forces can contribute and that these may occur separately or collectively. The three main interactions which are likely to be involved in membranes are hydrophobic, electrostatic and, in some membranes, metal–ion mediation. There is evidence from some of the new physical techniques applied to membranes that hydrophobic interactions may be particularly important, i.e., involving nonpolar interactions between lipid chains and apolar amino acids.

Many more studies using the new physical techniques, e.g., i.r. and n.m.r. spectroscopy and calorimetric techniques, etc. need to be carried out on well-characterized mitochondrial membrane systems. Combined with model membrane studies, the nature of the excitable characteristics of cell membranes may be revealed by future physical studies of lipid–polypeptide interactions.

REFERENCES

1. Allen, B. T., Chapman, D., and Salsbury, N. J. (1966), *Nature* **212**, 282.
2. Anderson, P. J., and Pethica, B. A. (1955), Biochemical Problems of Lipids, Interscience Publishers, Inc., New York.
3. Bangham, A. D., Standish, M. M., and Watkins, J. C. (1965), *J. Mol. Biol.* **13**, 238.

4. Bangham, A. D., Standish, M. M., and Weissman, G. (1965), *J. Mol. Biol.* **13,** 253.

5. Barratt, M. D., Green, D. K., and Chapman, D. (1968), *Biochim. Biophys. Acta* **152,** 20.

6. Blaisie, J. K., Dewey, M. M., Blaurock, A. E., and Worthington, C. R. (1965), *J. Mol. Biol.* **14,** 143.

7. Blake, C. C. F., Koenig, D. F., Mair, G. A., North, A. C. T., Phillips, D. C., and Sarma, V. R. (1965), *Nature* **206,** 757.

8. Brown, A. D. (1965), *J. Mol. Biol.* **12,** 491.

9. Chapman, D. (1966), *Annals N.Y. Acad. Sci.* **137,** 745.

10. Chapman, D., Byrne, P., and Shipley, G. G. (1966), *Proc. Roy. Soc.* **290A,** 115.

11. Chapman, D., Kamat, V. B., de Gier, J., and Penkett, S.A.P. (1968), *J. Mol. Biol.* **31,** 101.

12. Chapman, D., Ladbrooke, B. D., and Williams, R. M. (1967), *Chem. Phys. Lipids* **1,** 445.

13. Chapman, D., Leslie, R. B., Hirz, R., and Scanu, A., (1969) *Biochim. Biophys. Acta.* **176,** 524.

14. Chapman, D., and Penkett, S. A. P. (1966), *Nature* **211,** 1304.

15. Chappell, J. B., and Crofts, A. R. (1965), *Biochem. J.* **95,** 393.

16. Chappell, J. B., and Crofts, A. R. (1966), "Regulation of Metabolic Processes in Mitochondria," B.B.A. Library, p. 292, (J. M. Tager, S. Papa and E. Quagliariello and E. C. Slater, eds., Elsevier Publ. Co., Amsterdam.

17. Chappell, J. B., and Haarhoff, K. N. (1967), "Biochemistry of Mitochondria," p. 75, Academic Press, New York.

18. Danielli, J. F., and Davson, H. A. (1935), *J. Cellular Comp. Physiol.* **5,** 495.

19. Dawson, R. M. C. (1963), *Biochim. Biophys. Acta* **70,** 697.

20. Dawson, R. M. C. (1968), "Biological Membranes," D. Chapman ed., Academic Press, London.

21. Dawson, R. M. C., and Eichberg, J. (1965), *Biochem. J.* **96,** 634.

22. Dawson, R. M. C., and Hemington, N. (1967), *Biochem. J.* **102,** 76.

23. De Gier, J., Mardersloot, J. G., and van Deenen, L. L. M. (1968), *Biochim. Biophys. Acta* **150,** 666.

24. Demel, R. A., Kinsky, S. C., Kinsky, C. B., and van Deenen, L.L.M. (1968), *Biochim. Biophys. Acta* **150,** 655.

25. Dickerson, R. E., Kopka, M. L., Borders, C. L. Varnum, J., Weinzierl, J. E., and Margoliash, E. (1967), *J. Mol. Biol.* **29,** 77.

26. Doty, P., and Schulman, J. H. (1949), *Disc. Farad. Soc.* **6,** 21.

27. Eley, D. D., and Hedge, D. G. (1956), *Disc. Farad. Soc.* **21,** 221.

28. Elkes, J., and Finean, J. B. (1949), *Disc. Farad. Soc.* **6,** 134.

29. Elkes, J., Fraser, A. C., Schulman, J. H., and Stewart, H. C. (1945), *Proc. Roy. Soc.* **184A,** 102.

30. Fenster, L. J., and Copenhaver, J. H. (1967), *Biochim. Biophys. Acta* **137,** 406.

31. Finean, J. B., Coleman, R., Green, W. G., and Limbrick, A. R. (1967), *Biochim. Biophys. Acta* **135,** 1074.

32. Finean, J. B., and Burge, J. B. (1963), *J. Mol. Biol.* **7,** 672.

33. Finean, J. B., Coleman, R., and Green, W. A. (1966), *Ann. N.Y. Acad. Sci.* **137,** 414.
34. Finean, J. B. and Robertson, J. D. (1958), *Brit. Med. Bull.* **14,** 267.
35. Finkelstein, A., and Cass, C. A. (1967), *Nature* **216,** 717.
36. Fleischer, S., and Rouser, G. (1965), *J. Amer. Oil Chem. Soc.* **42,** 588.
37. Fleischer, S., Rouser, G., Fleischer, B., Casu, A., and Kritchevsky, G. (1967), *J. Lipid Res.* **8,** 170.
38. Folch, J. (1942), *J. Biol. Chem.* **146,** 35.
39. Gorter, H. F., and Grendel, F. (1925), *J. Exp. Med.* **41,** 439.
40. Green, D. E., and Fleischer, S. (1963), *Biochim. Biophys. Acta* **70,** 554.
41. Green, D. E., and Perdue, J. F. (1966), *Proc. Natl. Acad. Sci.* **55,** 1295.
42. Griffith, O. H., and McConnell, H. M. (1966), *Proc. Natl. Acad. Sci.* **55,** 8.
43. Gulik-Krzywicki, T., Rivas, E., and Luzzati, V. (1967), *J. Mol. Biol.* **27,** 303.
44. Harris, E. J., Cockrell, R., and Pressman, B. C. (1966), *Biochem. J.* **99,** 200.
45. Hartley, B. S. (1955), *Progr. Chem. Fats Lipids* **3,** 20.
46. Hauser, H., and Dawson, R. M. C. (1967), *European J. Biochem.* **1,** 61.
47. Haydon, D. A., and Taylor, J. (1963), *J. Theoret. Biol.* **4,** 281.
48. Husson, F., and Luzzati, V. (1963), *Nature* **197,** 822.
49. Johnston, P. V., and Roots, B. I. (1964), *Comp. Biochem. Physiol.* **11,** 303.
50. Jurtshuk, P., Sekuzu, I., and Green, D. E. (1963), *J. Biol. Chem.* **238,** 3595.
51. Kagawa, Y., and Racker, E. (1966), *J. Biol. Chem.* **241,** 2475.
52. Kauzmann, W. (1959), *Advan. Protein Chem.* **14,** 1.
53. Ke, B. (1965), *Arch. Biochem. Biophys.* **111,** 544.
54. Kendrew, J. C., Dickerson, R. E., Strandberg, B. E., Hart, R. G., Davies, D. R., Phillips, D. C., and Shore, V. C. (1960), *Nature* **185,** 422.
55. Kendrew, J. C., and Watson, H. C. (1967), "Principles of Biomolecular Organisation," Ciba Foundation, J. A. Churchill, London.
56. Korn, E. D. (1966), *Science* **153,** 1491.
57. Ladbrooke, B. D., Williams, R. M., and Chapman, D. (1968), *Biochim. Biophys. Acta* **150,** 333.
58. Le Baron, F. N. (1963), *Biochim. Biophys. Acta* **70,** 658.
59. Lehninger, A. L. (1964), in "The Mitochondrion," W. A. Benjamin, New York.
60. Lenard, J., and Singer, S. J. (1966), *Proc. Natl. Acad. Sci.* **56,** 1552.
61. Levy, Marianne, and Sauner, M.-T. (1968), *Chem. Phys. Lipids* **2,** 291.
62. Lotan, N., Yaron, A., and Berger, A. (1966), *Biopolymers* **4,** 365.
63. Luzzati, V., Gulik-Krzywicki, T., and Tardieu, A. (private communication) 1968.
64. Luzzati, V., and Husson, F. (1962), *J. Cell. Biol.* **12,** 207.
65. Maas, J. W., and Coburn, R. S. (1965), *Nature* **208,** 41.
66. Maddy, A. H., and Malcolm, B. R. (1965), *Science* **150,** 1616.
67. Maddy, A. H., and Malcolm, B. R. (1966), *Science* **153,** 213.
68. Malcolm, B. R. (1962), *Nature,* **195,** 901.

69. Malcolm, B. R. (1964), "Surface Activity and the Microbial Cell," p. 102, Symposium Soc. Chem. Ind. London.
70. Marinetti, G. V., Erbland, J., and Kochen, J. (1957), *Fed. Proc.* **16**, 837.
71. Martonosi, A. (1967), *Biochem. Biophys. Res. Commun.* **29**, 753.
72. Marr, A. G., and Ingraham, J. L. (1962), *J. Bact.* **74**, 1260.
73. Mueller, P., and Rudin, D. O. (1967), *Biochem. Biophys. Res. Commun.* **26**, 398.
74. Palade, G. E. (1953), *J. Cell Biol.* **1**, 188.
75. Palmer, K. J., Schmitt, F. O., and Chargaff, E. (1941), *J. Cell. Comp. Physiol.* **18**, 43.
76. Papahadjopoulos, D., and Miller, N. (1967), *Biochim. Biophys. Acta* **135**, 624.
77. Parsons, D. F., (1966), Proc. 7th Canad. Cancer Res. Conf. Ontario.
78. Parsons, D. F., Bonner, W. D., and Verboon, J. G. (1965), *Can. J. Bot.* **43**, 647.
79. Parsons, D. F., Williams, G. R., and Chance, B. (1966), *Ann. N.Y. Acad. Sci.* **137**, 643.
80. Perutz, M. F. (1965), *J. Mol. Biol.* **13**, 646.
81. Perutz, M. F., Rossmann, M. G., Cullis, A. F., Muirhead, H., Will, G., and North, A.C.T. (1960), *Nature* **185**, 416.
82. Pressman, B. C. (1965), *Proc. Natl. Acad. Sci.* **53**, 1076.
83. Reiss-Husson, F. (1963), *J. Mol. Biol.* **25**, 363.
84. Richardson, T., and Tappel, A. L. (1962), *J. Cell Biol.* **13**, 43.
85. Richardson, T., Tappel, A. L., and Gruger, E. R. (1961), *Arch. Biochem. Biophys.* **94**, 1.
86. Richardson, T., Tappel, A. L., Smith, L. M., and Houle, C. R. (1962), *J. Lipid Research* **3**, 344.
87. Robertson, J. D. (1966), "Principles of Biomolecular Organization," Ciba Foundation Symposium, J. & A. Churchill, London.
88. Robinson, N. (1960), *Trans. Farad. Soc.* **56**, 1260.
89. Rouser, G., and Fleischer, S. (1967), "Methods in Enzymology," Vol. 10, p. 385, Academic Press, New York.
90. Saunders, L. (1966), *Biochim. Biophys. Acta* **125**, 70.
91. Scheraga, H. A. (1961), *J. Phys. Chem.* **65**, 1071.
92. Scheraga, H. A., Nemethy, G., and Steinberg, J. Z. (1962), *J. Biol. Chem.* **237**, 2560.
93. Schmitt, F. O., Bear, R. S., and Palmer, K. J. (1941), *J. Cell Comp. Physiol.* **18**, 31.
94. Schulman, J. H., and Rideal, E. K. (1937), *Proc. Roy. Soc.* **122B**, 29.
95. Shipley, G. G., Leslie, R. B., and Chapman, D. (1969), *Biochim. Biophys. Acta* **173**, 1.
96. S. J. (1962), *Advan. Protein Chem.* **17**, 1.
97. Steim, J. S., and Fleischer, S. (1967), *Proc. Natl. Acad. Sci.* **58**, 1292.
98. Stoeckenius, W. (1966), in "Principles of Biomolecular Organization," p. 418, G. E. W. Wolstenholm and Maeve O'Connor, eds. J. & A. Churchill, London.
99. Tanaka, R., and Strickland, K. P. (1965), *Arch. Biochem. Biophys.* **111**, 583.

100. Thompson, J. E., Coleman, R., and Finean, J. B. (1966), *J. Cell Sci.* **1,** 287.
101. Thudichum, J. L. W. (1901), in "Die Chemische Konstitution des Gehrins des Menschen und der Tiere," pp. 80, 130, 270, F. Pietzcker, Tübingen.
102. Ullmer, D. D., Vallee, B. L., Gorchein, A., and Neuberger, A. (1965), *Nature* **206,** 825.
103. Urry, D. W., Mednieks, E., and Bejnarowiecz, E. (1967), *Proc. Natl. Acad. Sci.* **57,** 1043.
104. van Deenen, L. L. M. (1965), *Prog. Chem. Fats Lipids* **8,** 1.
105. Villalonga, F., Altschul, R., and Fernandez, M. S. (1967), *Biochim. Biophys. Acta* **135,** 406.
106. Wallach, D. F. H., and Gordon, A. (1967), 15th International Symp. on the Protides of Biol. Fluids, Brugge, Belgium.
107. Wallach, D. F. H., Kamat, V. B. (1964), *Proc. Natl. Acad. Sci.* **52,** 721.
108. Wallach, D. F. H., and Zahler, H. P. (1966), *Proc. Natl. Acad. Sci.* **56,** 1552.
109. Whittaker, V. P. (1966), "Regulation of Metabolic Processes in Mitochondria," Vol. 7, p. 1., J. M. Tager, S. Papa, E. Quagliariello and E. C. Slater, eds., Elsevier Publ. Co. (BBA Library) Amsterdam.
110. Wishnia, A. (1962), *Proc. Natl. Acad. Sci.* **48,** 2201.
111. Wishnia, A., and Pinder, J. W. (1966), *Biochemistry* **5,** 1534.

4

Function and Structure
of the Inner Membrane
# of Mitochondria
and Chloroplasts

E. Racker

Section of Biochemistry and Molecular Biology
Division of Biological Sciences
Cornell University
Ithaca, New York

I. INTRODUCTION

The major pathway of biological energy production takes place in a multi-enzyme system which is associated with the inner membrane of mitochondria and chloroplasts. These structures contain an assembly of enzymes which catalyze the transfer of electrons and protons from hydrogen donor to an acceptor. The energy generated during this oxidation process is captured and transformed into ATP, the most versatile energy donor in living cells.

In both mitochondria and chloroplasts the major hydrogen donor is water, but the final acceptor is different. In mitochondria it is oxygen, in chloroplasts it is TPN. The importance of water as a hydrogen donor in mitochondrial oxidations is not frequently emphasized. During the metabolic acrobatics of pyruvate oxidation in the Krebs cycle, water is incorporated into intermediates at three steps: in the hydration of fumarate to malate, and in the course of the utilization of acetyl CoA and succinyl CoA. Thus, for each molecule of pyruvate which donates four hydrogens, water contributes six additional hydrogens which are channeled into the mitochondrial oxidation chain. Mitochondria, therefore, like chloroplasts, cleave water, albeit by a different mechanism. In chloroplasts the oxygen of water is liberated as molecular oxygen; in mitochondria it is incorporated into the CO_2 liberated during the oxidations of the Krebs cycle intermediates. The differences in the fate of oxygen and hydrogen in mitochondria and chloroplasts facilitate the operation of the carbon cycle as illustrated in Figure 4-1.

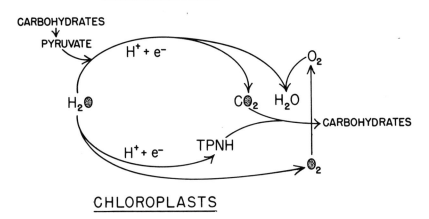

FIGURE 4-1. The role of water in the carbon cycle.

In both mitochondria and chloroplasts, the oxidation chains contain flavoproteins and cytochromes which transfer the electrons down the steps of a thermodynamic ladder to a final electron sink. At three steps of this process in mitochondria, the energy of oxidation is captured by a coupling device consisting of several proteins called coupling factors which are required for the conversion of phosphate and ADP into ATP. A similar device operates in chloroplasts.

Attempts to resolve the multienzyme system of oxidative phosphorylation into individual components have yielded considerable information on the properties of the participating catalysts as well as of the inner membrane proper. With the aid of biochemical and morphological assays, a mitochondrial membrane has been isolated from which most of the oxidation catalysts have been removed. This membrane has been further resolved and shown to consist of proteins and phospholipids that can be reassembled to yield structures which are functionally defective but, morphologically, remarkably similar to submitochondrial particles which catalyze oxidative phosphorylation.[40]

II. THE MECHANISM OF OXIDATIVE PHOSPHORYLATION

A. The Fate of the Oxygen of Water during Oxidative Phosphorylation

The generation of ATP from ADP and Pi takes place with the removal of water, resulting in the formation of the anhydride bond of the terminal pyrophosphate group of ATP. A critical question in the mechanism of this process is linked to the fate of the oxygen atom of the departing water molecule. In substrate-level oxidative phosphorylation, as represented by glyceraldehyde-3-P dehydrogenase, the key enzyme of glycolysis, this question is readily answered. During the oxidation of the aldehyde to the acid, the oxygen of phosphate is incorporated into the carboxyl group of the acid. Although in oxidative phosphorylation the answer is still unknown, it is remarkable that in most formulations of the mechanism the problem is ignored. First comes the question whether the oxygen is derived from orthophosphate or from ADP. Until recently the consensus has been[77,91] that the oxygen is removed from phosphate presumably during the formation of a phosphorylated high-energy intermediate $X \sim P$. Virtually all chemical formulations of the energy-yielding process during electron transfer have been based on this assumption. The evidence for this concept has rested on Boyer's experiments with ^{18}O demonstrating that the bridge oxygen in the terminal phosphate of ATP generated during oxidative phosphorylation in mitochondria is derived from ADP.[9] Although the data are unambiguous, the interpretation is not. It was pointed out by Löw et al.[52] that the experiments yield information only on the origin of the bridge oxygen of the extra-mitochondrial ATP which may be one or two steps removed from the primary P transfer reaction. For example, the ^{18}O data cannot be used to differentiate between the following mechanisms:

$$(a) \quad X \sim Y + NDP \rightleftharpoons NDP \sim X + Y$$
$$NDP \sim X + Pi \rightleftharpoons NTP + X$$
$$NTP + ADP \rightleftharpoons ATP + NDP$$
$$(b) \quad X \sim Y + Pi \rightleftharpoons X \sim P + Y$$
$$X \sim P + ADP \rightleftharpoons X + ATP$$

In mechanism (a), NDP may be a nucleoside diphosphate other than ADP, or it may be compartmentalized ADP or AMP (e.g., protein bound). It is apparent from these considerations that with the uncertainty of the fate of oxygen the nature of the high-energy intermediates is in doubt as well. It can be stated, however, that a nonphosphorylated, high-energy intermediate or state does indeed exist, and that it can be formed in the absence of orthophosphate and without a functional coupling device.[24,77] This conclusion is based on the experimental observations that energy formed during substrate oxidation can be utilized to drive energy requiring reactions, such as ion transport and the reduction of DPN by succinate.[91, 24,73,48] Since oxidative energy can be utilized in the presence of oligomycin, which inhibits ATP formation, and in the absence of Pi, all recent formulations[73,59,14,76] of oxidative phosphorylation have included a nonphosphorylated high-energy intermediate $(X \sim Y)$. In the following discussion of the formation and utilization of $X \sim Y$, the fate of oxygen will be taken under consideration.

B. The Chemical and Chemiosmotic Hypotheses

There are two major formulations that account for the formation and utilization of $X \sim Y$. Since the properties of the inner mitochondrial membrane bear on the validity of these two hypotheses, a discussion of the salient points of differences in the formulations seem appropriate.

The first formulation is usually referred to as the chemical hypothesis. It was first proposed by Slater[93] mainly in analogy with the mechanism of action of glyceraldehyde-3-P dehydrogenase,[80] and was discussed in detail elsewhere.[77,91] An abbreviated version is shown in Figure 4-2A. The second formulation, Figure 4-2B or chemiosmotic hypothesis, was elaborated by Mitchell.[58,60]

There are two important differences between these two formulations. The chemical hypothesis includes a chemical intermediate of the respiratory

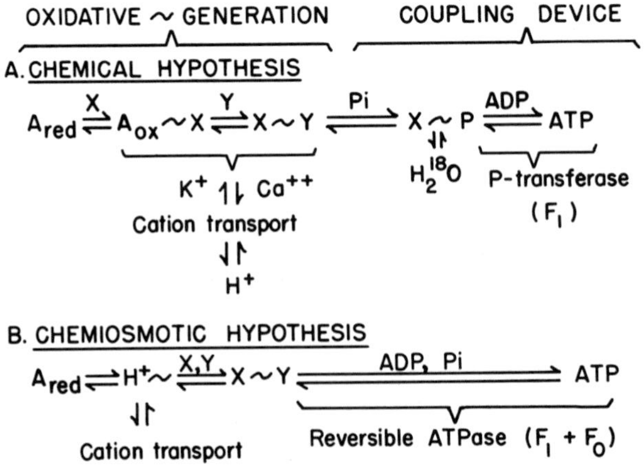

FIGURE 4-2. The chemical and the chemiosmotic hypothesis.

chain $A \sim X$.* Mitchell rejects such an intermediate and substitutes for it a pH gradient and a membrane potential which are created by the translocation of protons during the respiratory process. According to both formulations, there is a high-energy intermediate $X \sim Y$, but the mechanisms for its production and utilization are quite distinct and invoke different functions for the membrane. According to the chemical hypothesis, the membrane serves as an organizer of the catalysts which participate in the generation and utilization of $X \sim Y$. In the chemiosmotic hypothesis, the role of the membrane is much more elaborate.

During respiration, the oxidation chain is visualized to develop a "proton-motive" force across the membrane, the protons being removed from the substrate at one side of the membrane and translocated to the other side. A cyclic "proton current" is established by driving the protons back through the reversible ATPase, thereby accomplishing the dehydration of ADP and Pi and the formation of ATP. A more detailed scheme of this process is shown in Figure 4-3. Mitchell proposed that the anisotropic coupling membrane of mitochondria which contains a relatively ion-impermeable layer (M-phase), separates the inside and the outside. Two proton-translocating systems are embedded in the membrane: one consists of three "loops" of the oxidation chain, the other is the oligo-

COUPLING
MEMBRANE

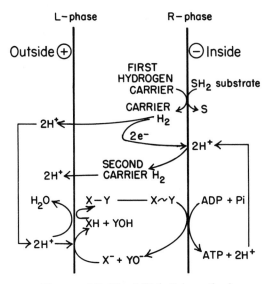

FIGURE 4-3. The Mitchell hypothesis.

* Recent formulations of a "conformational" hypothesis are variants of the chemical hypothesis substituting a change in protein conformation for $A \sim X$.

mycin-sensitive ATPase ($F_1 + F_0$). Each loop contains a hydrogen carrier and an electron carrier which accomplish the proton-translocation as outlined in Figure 4-3.

The substrate donates its hydrogen at the R-phase of the membrane to the hydrogen carrier of the first loop. This carrier releases 2 protons to the medium at the L-phase of the membrane, while transferring the electrons to an electron carrier that returns them to the R-phase of the membrane, thus completing the first loop. The protons at the L-phase are taken up by the coupling device (reversible ATPase), and incorporated into the water that is released during the formation of X–Y. Protons are released again, now at the R-phase, when the formation of ATP from ADP and Pi is taking place. These protons together with the electrons of the first loop, reduce the hydrogen carrier of the second loop. This cyclic process is repeated until six protons are translocated for each DPNH oxidized, and three molecules of ATP are formed. The polarity of the membrane is such that in mitochondria the protons transverse from the inside to the outside during oxidation; in chloroplasts from the outside to the inside.

During ATP formation (shown in Figure 4-3), X–Y is formed by dehydration of XH + YOH, and is transformed to a high-energy level $X \sim Y$ by moving from the L-phase (high potential) of the membrane to the R-phase (low potential). The driving force for the ATP synthesis is visualized to be due partly to the low H^+ concentration in the R-phase, and partly to the low concentration of X^- and YO^-, the negative potential in the R-phase being primarily responsible for the movement of these negatively charged groups toward the L-phase of the membrane.

The ion impermeability of the membrane and the requirements of pH control, osmotic stability and accessibility of substrate at the appropriate side of the membrane, made it necessary to postulate a very rapid diffusion–exchange system for certain ions (H^+/cation antiport and H^+/anion symport).

The assumptions made in this ingenious scheme are formidable and controversial.[14,19,92] Some of the controversies center around minor features such as the identity and locations of the hydrogen and electron carriers in the three loops. Even though some of these carrier assignments may be incorrect, hydrogen and electron carriers are present in mitochondria and their interplay in the oxidation chain has not been previously taken sufficiently into consideration. More serious objections are based on experiments on ion transport and on the exchange reactions which are discussed in greater detail below. On the other hand, the chemiosmotic hypothesis has already proved its value by stimulating important experimental approaches.[39]

A brief evaluation of some of the major issues pertaining to the two hypotheses follows.

1. The role of the membrane.   The chemiosmotic hypothesis requires a vesicular structure of the membrane which allows for a clear separation of the two sides of the membrane; the chemical hypothesis does not. A demonstration of oxidative phosphorylation in a vesicle-free system would eliminate the major feature of the Mitchell hypothesis. In spite of numerous attempts, such a system has not been experimentally established. Moreover, certain submitochondrial particles (A-particles) that have a large amount of a catalytically active coupling factor (F_1) do not generate ATP unless more F_1 is added.[76] The additional F_1 need not be catalytically active, and can be replaced by low concentrations of oligomycin or DCCD which interact with the membrane.

These experiments seem to indicate that a degree of membrane integrity is essential for the phosphorylation process, but it remains to be established whether this requirement is primary or secondary to the mechanism. For example, an ion pump, associated with the intact vesicle, may become uncontrolled when the membrane is damaged and drain the energy produced during oxidation. Nevertheless, it appears from the above considerations that on a score board (Table 4-1) the information available on

TABLE 4-1

SCOREBOARD OF EXPERIMENTAL EVIDENCE FOR THE CHEMICAL AND CHEMIOSMOTIC HYPOTHESIS OF ENERGY GENERATION DURING OXIDATIVE PHOSPHORYLATION AND PHOTOPHOSPHORYLATION

	Chemical	Chemiosmotic
Role of the membrane	−	+
Ion transport	+	×
Action of uncoupling agents	−	+
Isolation of high-energy intermediates	−	±
^{32}Pi–ATP exchange ⎫ ADP–ATP exchange ⎬ $H_2{}^{18}O$ exchanges ⎭	+	−

the role of the membrane should be placed in favor of the Mitchell hypothesis.

2. Ion transport. Serious discrepancies in the stoichiometry between proton translocation and cation transport have been reported.[14,19,92] Mitchell [59] has discussed some of these findings, but some of the questions need to be answered by decisive experiments. For example, is there a process of proton translocation, mediated by ATPase or by the respiratory chain, which does not require the presence of either divalent or monovalent cations? If there is indeed a primary hydrogen movement, the problem of the compatibility of the kinetics and thermodynamic parameters of the translocation process with electron transport and ATP generation[14,19] will still have to be further analyzed, although it may be extremely difficult to resolve.[58]

Submitochondrial particles obtained by sonication of mitochondria have been stated to lack the facility of Ca^{++} accumulation.[16] Mitchell has pointed out[58] that in submitochondrial particles which are "inside out," i.e., the side of the membrane facing the mitochondrial matrix becomes exposed to the outside medium,[47] the polarity of the membrane should be reversed, thus explaining the lack of cation transport. Since proton translocation now takes place from outside to inside, a secondary accumulation of Ca^{++} should not be expected. On the other hand, submitochondrial particles prepared with digitonin which are capable of Ca^{++} accumulation, should have the same polarity as intact mitochondria and should not show morphological inversion.[58] Unfortunately, convincing morphological evidence is not available on this point. In fact, recent experiments in our laboratory[53] have established that submitochondrial particles from beef heart which are morphologically inverted, yet are capable of accumulating Ca^{++}, can be prepared by either sonication or with digitonin. These observations are very difficult to explain in terms of the Mitchell hypothesis. They are somewhat more readily accommodated within the chemical hypothesis which is more flexible with regard to structural organization.

The experiments of Jagendorf and his collaborators[39] on ATP formation in chloroplasts during an acid–base transition have been frequently cited in favor of the Mitchell hypothesis. It is apparent from an inspection of the formulation of the chemical hypothesis (Figure 4-2) that ions are in a reversible equilibrium with the high-energy intermediate of the oxidation chain. By establishing an ion gradient, energy production could be achieved whether the ion is a proton, or a mono or divalent cation. Indeed, experiments by Cockrell and his collaborators[20] have demonstrated ATP production due to a potassium gradient in mitochondria. Thus, it cannot be decided on the basis of such experiments whether a proton or an ion gradient is on the direct pathway of energy production. Moreover, it was shown recently by McCarty[55] that the light induced pH change due to H^+ translocation can be completely eliminated in subchloroplast particles by addition of 2 mM NH_4Cl, without any effect on photophosphorylation. These findings show that photophosphorylation can take place without a pronounced pH gradient, but they do not eliminate the contribution of a membrane potential which, as repeatedly emphasized by Mitchell, is a major component in his hypothesis.

3. Uncoupling agents.   There are a vast number of agents of divergent chemical properties which uncouple oxidative phosphorylation. Oxidation proceeds without ATP formation, and the energy is being dissipated as heat. There is no unified hypothesis available to explain the mode of action of these agents in terms of the chemical hypothesis. According to Mitchell, these agents affect the permeability of the membrane, thus abolishing the proton gradient and membrane potential. Indeed, uncouplers such as valinomycin and dinitrophenol have been shown to have a pronounced effect on the permeability of artificial membrane models (see Chapters 1 and 6). Although it is possible to invoke again, secondary energy drainage due to alterations of the membrane, at the present time, the mode of action of uncouplers appears to be more readily explained in terms of the Mitchell hypothesis as indicated on the scoreboard.

4. The exchange reactions.   (a) The ^{32}Pi–ATP exchange in mitochondria is readily explained by either of the two hypotheses. The lack of a ^{32}Pi–ATP exchange in chloroplasts is more difficult to understand in terms of the reversible ATPase in the Mitchell hypothesis. If the conversion of $X \sim Y$ to ATP is a concerted reaction, the ^{32}Pi–ATP exchange should take place in either mitochondria or chloroplasts. The chemical hypothesis is more versatile in this respect since it includes several discrete steps and intermediates. As will be elaborated later, a simple assumption with regard to the availability of one of the postulated intermediates in photophosphorylation explains the lack of several partial reactions in chloroplasts and their activation by dithiothreitol.

The finding[69] that oxidative phosphorylation as well as the ^{32}Pi–ATP exchange takes place in particles in the presence of an inhibitor of mitochondrial ATPase is also difficult to reconcile with Mitchell's formulation of water participation in the reversible formation of X–Y.

(b) The dinitrophenol and oligomycin-sensitive ADP–ATP exchange in mitochondria is more readily understood in terms of a phosphorylated intermediate postulated in the chemical hypothesis.[13,98]

(c) The H_2O^{18} exchange reactions also differ in chloroplasts and mitochondria. Whereas in mitochondria there is a rapid energy dependent exchange between water and either Pi or ATP in chloroplasts only the H_2O–ATP exchange takes place at a rapid rate under conditions of photophosphorylation.[3,90]

According to the chemiosmotic hypothesis, one can readily visualize an exchange between the oxygen of water and that of either phosphate or ADP, depending on the source of oxygen in the formation of YO^- (Figure 4-3). If it is derived from ADP, an $H_2^{18}O$–ATP exchange might be expected, if it is derived from phosphate an $H_2^{18}O$–Pi exchange should take place. The experimental observation of a rapid exchange with both phosphate and ATP in mitochondria is more difficult to explain by a concerted mechanism, although such attempts have been made.[8] Two sites of entry of water during oxidative phosphorylation have been proposed,[9,21] but during photophosphorylation in chloroplasts only the $H_2^{18}O$–ATP exchange takes place at a rapid rate.[3,90] Current formulations of the chemical hypothesis (Figure 4-4, scheme I) readily explain the $H_2^{18}O$–Pi exchange, but it should be independent of adenine nucleotides which is contrary to experimental findings.[37,61] We, therefore, propose a modification of the formulation of the chemical hypothesis which eliminates this discrepancy. As shown in scheme II of Figure 4-4, the removal of oxygen takes place

SCHEME I

1. $A_{red} + B_{ox} + XH + YOH \rightleftharpoons A_{ox} + B_{red} + X{\sim}Y + H_2O$

2. $X{\sim}Y + H_3PO_4 \rightleftharpoons X{\sim}P + YOH$

3. $X{\sim}P + ADP \rightleftharpoons XH + ATP$

SCHEME II

1. $A_{red} + B_{ox} + X\Theta H + YH \rightleftharpoons A_{ox} + B_{red} + X{\sim}Y + H_2O$

2. $X{\sim}Y + H_3PO_4 \rightleftharpoons X\Theta{\sim}PO_3H_2 + YH$

3. $X\Theta{\sim}P + ADP \rightleftharpoons ATP + X\Theta H$

FIGURE 4-4. Variants of the chemical hypothesis.

from XOH rather than from YOH. This minor alteration changes the fate of oxygen and includes it during step 3 which requires ADP. In order to explain the oxygen exchange between water and ATP, we proposed [37] a phosphorylated high-energy intermediate which can exchange its oxygen with that of water. Such an oxygen exchange would be analogous to the chemical exchange that takes place between $H_2{}^{18}O$ and arsenate,[42] or between $H_2{}^{18}O$ and phosphate esters during conditions of hydrolysis.[33]

Scheme II also explains the absence of the energy-dependent $H_2{}^{18}O$–Pi and ^{32}Pi–ATP exchange in chloroplasts, if it is assumed that under conditions of photophosphorylation step 2 is kinetically irreversible. This may be due to a lack of YH, either because of its instability or due to its removal by the reaction catalyzed in step 1. Irreversibility of step 2 is in line with the absence of ATPase activity under conditions of photophosphorylation. The induction of ATPase activity and of a ^{32}Pi–ATP exchange in chloroplasts by addition of dithiothreitol may be due to either maintenance or substitution of YH by the sulfhydryl compound.[57] These findings are difficult to explain in terms of the chemiosmotic hypothesis, particularly since it is necessary to assume that the formation of the low-energy ester X–Y from XH and YOH can take place in chloroplasts under conditions when the thermodynamically more favored hydrolysis of $X \sim Y$ is absent. And if the hydrolysis does not take place, how can one account for the exchange between $H_2{}^{18}O$ and ATP? On the scoreboard, the exchange reactions weigh heavily in favor of a chemical hypothesis which includes discrete steps.

5. Intermediates. The failure to isolate high-energy intermediates of the respiratory chain cannot be used fairly as an argument[58] against the chemical hypothesis, while the chemiosmotic hypothesis includes a high-energy intermediate ($X \sim Y$) that has equally resisted isolation. Most of the experimental efforts have been directed toward the isolation of a phosphorylated high-energy intermediate because of the availability of highly radioactive orthophosphate. In view of the lack of success of these attempts, the task of isolating a high-energy intermediate of a member of the oxidation chain without a label seems rather formidable considering the tenacity of union between the membrane and oxidation catalysts on the one hand and the low concentration of the latter on the other hand. In spite of these dim prospects, the outlook for an investigator who is concentrating his effort in the direction of the resolution of the system of oxidative phosphorylation seems somewhat brighter with the proposed intermediates of a chemical hypothesis, than with a concerted reaction calling for a direct interaction between $X \sim Y$, ADP, and phosphate. In any case, on the scoreboard, the failure to isolate a high-energy intermediate, particularly a phosphorylated one, should be somewhat more distressing to the proponents of the chemical hypothesis.

It is doubtful that the discrepancies and controversies outlined above will be settled by arguments or calculations. In reading the discussions of the proponents of the two hypotheses[14,59] one gains the impression that the evidence against the formulations of the opponent is overwhelming. As pointed out previously,[76] the presence of a proton pump linked in parallel to the phosphorylation mechanism can explain many experiments that now favor the Mitchell hypothesis. On the other hand, modifications of the current chemiosmotic hypothesis could be introduced without changes in the basic concept to explain findings on exchange reactions which do not seem to fit the present scheme.

Having been raised by the music of substrate-level phosphorylation my own prejudices induce me to lean toward some aspects of the chemical hypothesis, e.g., the phosphorylated high-energy intermediate. In its favor is the tendency of nature to repeat itself with respect to mechanisms. It also permits a brighter experimental outlook than the concerted reaction formulated by Mitchell. On the other hand, the experimental evidence in favor of an intimate relation between oxidative phosphorylation and the formation of a membrane potential and proton movements is mounting.

C. A New Proposal

In the past, controversies between outstanding scientists have usually been resolved with each of the opponents being partly right. It is tempting, therefore, in the current dilemma of oxidative phosphorylation to make a compromise proposal which includes features of both hypotheses. Such a formulation is given in Figure 4-5 (scheme III), and will undoubtedly be vigorously rejected by both opposing parties. The scheme has certain features that should be discussed.

STEP

I a. $AH_2 + 2Fe^{+++} \rightleftharpoons A + 2H^+ + 2Fe^{++}$

b. $2Fe^{++} + XOH + YH + Q \rightleftharpoons 2Fe^{+++} + XO^- + Y^- + QH_2$

c. $XO^- + Y^- + 2H^+ + Pi \rightleftharpoons XO{\sim}P + YH + H_2O$

d. $XO{\sim}P + ADP \rightleftharpoons XOH + ATP$

2 a. $QH_2 + 2cyt_b^{+++} \rightleftharpoons Q + 2H^+ + 2cyt_b^{++}$

b. $2cyt_b^{++} + XOH + YH + Q \rightleftharpoons 2cyt_b^{+++} + XO^- + Y^- + QH_2$

3 a. $QH_2 + 2cyt_c^{+++} \rightleftharpoons Q + 2H^+ + 2cyt_c^{++}$

b. $2cyt_c^{++} + XOH + YH + O \rightleftharpoons 2cyt_c^{+++} + XO^- + Y^- + H_2O$

FIGURE 4-5. A compromise (compromising) proposal.

The first step (1a) consists of the separation of electrons and protons which are donated by the substrate (AH_2). In the second step (1b), the reduced electron carrier is oxidized by quinone which, at the same time, accepts protons from XOH and YH. This formulation resembles, in part, an ingenious scheme proposed by Davis many years ago[23b] involving participation of Q as proton acceptor and a separation of charges within the structured system of oxidative phosphorylation. The resulting negatively charged XO^- and Y^- are shielded in the membrane and do not interact with protons in tightly coupled mitochondria, unless phosphate participates (step 1c) giving rise to a phosphorylated intermediate $XO \sim P$. In step 1c, the phosphorylated intermediate donates its phosphate to ADP. At the second phosphorylation site, the oxidation process is repeated with the reduced quinone as hydrogen donor and cytochrome b as electron acceptor (step 2a). At the third phosphorylation site, reduced quinone is again the hydrogen donor, cytochrome c the electron acceptor, (step 3a) but oxygen is now the final proton and electron acceptor (step 3b). Steps c and d are the same at all three sites. This scheme accounts for virtually all partial reactions including the ^{18}O exchanges which cannot be satisfactorily explained by the chemiosmotic hypothesis. It includes the negatively charged compounds (XO^- and Y^-) proposed by Mitchell, but without a transformation into low-energy (X–Y) and high-energy (X $\sim$ Y) forms. XO^- and Y^- are visualized to interact either with phosphate and protons to yield a high-energy phosphorylated intermediate $(XO \sim P)$, or by reversal of steps 1a and b to yield reduced DPN. It was pointed out by Boyer[8] that the nucleophilicity required for a direct loss of the oxygen of phosphate to form water may arise during electron transport by proton loss from an appropriate group. Butow and Racker[12] as well as Boyer[8] suggested the possible participation in oxidative phosphorylation of a non-heme iron which binds the phosphate, thus favoring departure of the hydroxyl group. The chemical nature of XO^- is not known, but it could be the enolate ion of an oxidized hydroxyl group of a serine or threonine residue in a protein or lipoprotein.[30] Equally unknown is the nature of Y^-, but it may be an S^- group of the same, or of an adjacent protein.

Uncoupling of oxidative phosphorylation according to this scheme can occur by several mechanisms as in the case of substrate-level oxidative phosphorylation.[77] It can take place by entrance of water at Steps b or c, or by an alteration of the membrane to make it permeable to H^+ so that XO^- and Y^- are no longer shielded and have little chance of participating in step c. Thus, XOH and YH required for oxidation are continuously regenerated. According to this formulation, the lack of site I phosphorylation in baker's yeast can be explained by the entrance of water as a substitute for XOH at step 1b. The presence of respiratory control in beef heart particles with DPNH, but not with succinate, may be due to tight coupling at step 1b but water entrance at steps 2b and 3b.

Oligomycin at high concentrations is visualized to alter the membrane in such a manner that phosphate has no longer access to XO^- in step c, whereas at low concentrations the entrance of water at this step is hindered.[45,81]

One of the less attractive, but not unreasonable features, of Scheme III is the participation of Q at the first as well as the second phosphorylation site. The basic assumption is made that Q interacts with different proteins of the oxidation chain in such a manner as to affect its redox potential. An analogous example is the well-known variation in redox potential of different flavoproteins. Experimental evidence[87] will be presented in the section on the oxidation chain showing that Q_0 (an artificial hydrogen acceptor) can operate in oxidative phosphorylation both at the first and second site.

III. THE OXIDATION CHAIN OF THE INNER MITOCHONDRIAL MEMBRANE

It was pointed out by Slater[94] that the term "electron transport chain of mitochondria" prejudges the importance of the role of the electrons. In view of the present state of uncertainty and the central role of proton participation in the chemiosmotic hypothesis and in scheme III, the use of a noncommittal term seems advisable. Slater recommended the old term "respiratory chain," but in the current article, the term "oxidation chain" will be used preferentially, since it applies also to the multienzyme system in chloroplasts.

The oxidation chains of mitochondria and chloroplasts are intimately associated with the inner membrane of these structures. In fact, until recently it was assumed that the oxidation enzymes are an integral part of the inner membrane. This view was challenged by the experimental findings[40] that structures retaining some of its properties, i.e., catalysis of an oligomycin-sensitive ATP hydrolysis, can be isolated. These virtually colorless membranes contained neither cytochrome *c* nor cytochrome oxidase and only very small amounts of cytochrome *b* and flavoproteins. Yet in electron micrographs they resembled submitochondrial particles capable of oxidative phosphorylation. It was, therefore, proposed that the membrane proper consists of colorless proteins and phospholipids, and that the enzymes of the oxidation chain represent secondary attachments which take place during the morphogenesis of the inner membrane. This suggestion[82] was at first violently opposed,[31] but confirmations of the experimental findings are now forthcoming.[96]

Our knowledge of the assembly and sequence of the catalysts of the oxidation chain is still incomplete. In Figure 4-6, our currently most-favored

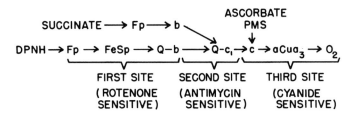

FIGURE 4-6. The oxidation chain in mitochondria.

version is illustrated with inclusion of the sites of phosphorylation and site of action of some widely used inhibitors.

A. The Segment between DPNH and Cytochrome *b* and the First Site of Phosphorylation

Although definite advances in this area have been made in recent years, the role of Q and nonheme iron is still controversial. Extraction of lyophylized beef heart mitochondria with pentane yielded particles that required addition of Q and cytochrome *c* for DPNH oxidation.[95] Studies of the oxidation–reduction of intramitochondrial Q[43] suggest the participation of this carrier between the flavoproteins and the cytochromes. Chance[14] still challenges the role of Q as a member of the oxidation chain.

The exact role of the nonheme iron is equally uncertain. The EPR signal at $g = 1.94$ in submitochondrial particles is related to the presence of non-heme iron, but only a small portion of it seems to contribute to the signal.[4] A communication[35] on the partial resolution of the DPNH-cytochrome b segment deserves attention, since this approach seems to be the most promising at the present time. The proposed function of a nonheme iron protein between DPNH dehydrogenase and Q is of particular interest.

Schatz and Racker have shown[87] that the first phosphorylation site can be accurately measured with DPNH as hydrogen donor and Q_1 as acceptor. DPNH oxidation was measured in the presence of antimycin or of cyanide and antimycin to block the oxidation via the cytochromes. The oxidation of DPNH was highly sensitive to rotenone at the low concentrations of Q_1 (about 0.1 mM) used in the assay, but at increasingly higher levels of Q_1 both the rotenone sensitivity and the P:O steadily declined. There seems to be general agreement now[14] that the site of rotenone action is on the oxygen side of DPNH dehydrogenase rather than on the substrate side. The earlier assignment of the first phosphorylation site between DPNH and DPNH dehydrogenase, which was based on studies of crossover points,[17] should therefore be abandoned in favor of a localization of the site between the flavoprotein of DPNH dehydrogenase and cytochrome b.[87] It has been recently suggested that a second flavoprotein operates in the oxidation chain on the oxygen side of the rotenone inhibition[16a] thus allowing a reinterpretation of the earlier data on the crossover point. The localization of the energy conservation site between DPNH dehydrogenase and cytochrome b was firmly established by experiments[36] on the energy-dependent reduction of DPN by succinate, a reaction believed to represent a reversal of the first site of phosphorylation. In submitochondrial particles, in the presence of sufficient amounts of succinate, malonate, antimycin and KCN to keep cytochrome b reduced, the addition of ATP resulted in an oxidation of cytochrome b. This oxidation of cytochrome b was sensitive to rotenone and oligomycin, but did not result in the reduction of the flavin of DPNH dehydrogenase unless DPN as well as ATP were added. Since in the presence of excess ATP but absence of DPN, the oxidation of the reduced cytochrome b continued for a considerable time, it became evident that an electron sink must be present. In view of an observed energy-dependent accumulation of hydrogen peroxide, it was proposed that an unidentified carrier was reduced by cytochrome b and reoxidized by air. When ferricyanide was added, it was used as the preferential electron sink as indicated by its rapid and ATP-dependent reduction which took place in the absence of DPN. Since ferrocyanide was not oxidized by the submitochondrial particles, the energy-dependent reduction of ferricyanide represents a most convenient assay for the first coupling site. Moreover, the rotenone sensitivity of the ferricyanide reduction suggests that this inhibitor acts either on the unknown carrier itself, or at a site between it and cytochrome b as shown in Figure 4-7. The unknown carrier may be Q_{10}

or a nonheme iron protein. The latter possibility is particularly attractive in view of the recent work of Hatefi and Stempel [35] mentioned earlier. The experiments of Butow and Racker[12] and Schatz *et al.*,[88] moreover, suggest an intimate relationship between nonheme iron and the first phosphoryla-

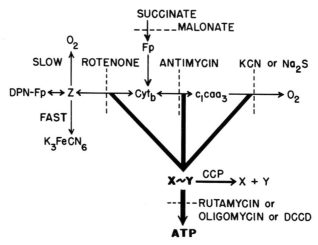

FIGURE 4-7. Localization of the action of some inhibitors of the oxidation chain.

tion site. The lack of a phosphorylating site I in mitochondria and sub-mitochondrial particles from baker's yeast[63,86] was shown to be associated with a lack of an EPR signal g = 1.94,[88] which is characteristic for the DPNH-cytochrome *b* region of mammalian mitochondria.

B. The Segment between Succinate and Cytochrome *b*

Recent studies were carried out in our laboratory[10] on the resolution and reconstitution of the respiratory segment between succinate and Q. As guideline in this work, we used the phenomenon of allotopy,[76] a term which refers to the pronounced differences in the properties of enzymes and of the catalytic processes when they become associated with a membrane. Thienyltrifluorobutanedione (TTB) was chosen as allotopic indicator since it inhibits succinate oxidation in the particles, but does not inhibit soluble succinate dehydrogenase. A reconstitution of a TTB-sensitive segment was achieved by the combination of three purified proteins with phospholipids and Q_2 or Q_{10}. The proteins were: succinate dehydrogenase,[41] cytochrome b,[29] and F_4.[22] Zalkin and Racker[100] have shown that F_4 resembles a "structural protein" from mitochondria[23,83] in its ability to combine with phospholipids and cytochrome b. Addition of phospholipids to a solution of F_4

results in the formation of a precipitate which consists of membranous vesicles.[40] This F_4–phospholipid complex was used by Bruni and Racker[10] as a biological scaffold for the organization of the oxidation chain. The reconstituted segment catalyzed the reduction of dichlorophenolindophenol by succinate provided succinate dehydrogenase, Q_2 and cytochrome *b* were added to the complex. The requirement for phospholipids was not specific, but complexes that were made with phospholipids from mammalian mitochondria were considerably more stable than those with soybean phospholipids. The reduction of dichlorophenolindophenol was as sensitive to TTB in these reconstituted particles as in submitochondrial particles. Of particular interest was the requirement for cytochrome *b* which appeared to function as a structural rather than as a catalytic component, since a reduction of cytochrome *b* by succinate could not be detected. The dependency of the membrane-linked oxidation chain on a component that has lost its catalytic activity, is analogous to the structural role played by catalytically inactive F_1 which will be discussed in detail later.

C. The Segment between Cytochrome *b* and Cytochrome *c* and the Second Site of Phosphorylation

Resolution and functional reconstitution of the oxidation chain in this segment has been recently accomplished.[99] After cleavage of a crude b–c_1 complex by 1.3 M guanidine, two fractions were isolated. One fraction was insoluble and contained cytochrome *b*, the other fraction contained cytochrome c_1 as well as a nonheme iron protein. On further purification, the nonheme iron protein was removed from the c_1 preparation. Reconstitution of the complex was achieved by allowing cytochrome *b*, cytochrome c_1, phospholipids, and Q_{10} to interact. After addition of purified succinate dehydrogenase, the complex catalyzed the reduction of cytochrome *c* by succinate. The reaction was completely inhibited by antimycin A. If the fraction containing cytochrome *b* was exposed prior to reconstitution to a very low concentration of antimycin, which had little or no effect when added to the reconstituted complex, a considerable inhibition of cytochrome *c* reductase activity was observed. Exposure of cytochrome c_1 to these low antimycin concentrations had no effect. It seems likely, therefore, that antimycin interacts either with cytochrome *b* directly or with a component closely associated with it. This is in line with observations on absorption changes of cytochrome *b* on addition of antimycin (see Ref. 15.). Since the reconstituted b–c_1 complex was very active in spite of the lack of a nonheme iron protein, the question of the role of nonheme iron in this segment must be raised. It may participate as suggested earlier in the phosphorylation mechanism rather than in electron transport. The role of Q_{10} in this segment also requires comment. It was shown[49] that succinate oxidation in mitochondria that were extracted with acetone requires addition of Q_{10}. However, these mitochondria were damaged, and did not catalyze oxidative phosphorylation. The exact function of Q_{10} in oxidative phosphorylation remains unknown. In scheme III (Figure 4-5), it was assigned the role as a proton acceptor at the first and second phosphorylation site. Although this may not seem to be an appealing solution, experimental evidence is available which supports this formulation. It was shown[87] that Q_0, in contrast to Q_1 and Q_2, can serve as a hydrogen acceptor at the second site of phosphorylation, as well as at the first site. With Q_0 as hydrogen acceptor, the oxidation of DPNH as well as the P:O ratio was partially inhibited by antimycin which had no effect on oxidative phosphorylation with Q_1 as acceptor. The rate of Q_0 reduction at the second site was about one-half that at the first site, indicating efficient mediation at both sites. Although it is recognized that Q_0 is not a physiological constituent, it is tempting to propose that Q_{10}, or a related compound, may serve as hydrogen acceptor at the second as well as at the first phosphorylation site.

A satisfactory assay for Site II phosphorylation has not been developed as yet. However, phosphorylation associated with the reduction of Q_0 by DPNH in the presence and absence of antimycin[87] on the one hand [sites (I + II) − I], and P:O measurements of succinate and ascorbate oxidation on the other hand [sites (II + III) − III] serve to evaluate phosphorylation associated with this segment.

D. The Segment between Cytochrome c and Oxygen

At the third phosphorylation site, copper may fulfill the role played by nonheme iron at the first two sites.[62] The question of the mechanism of the interaction of electrons of the respiratory chain with molecular oxygen has puzzled investigators for many years. Kinetic evidence suggests[38] that the cytochromes act in pairs. In tightly coupled mitochondria the reactions were found to be second order and in the presence of dinitrophenol, first order.

It should be pointed out that the various preparations of cytochrome oxidase described in the literature are probably either severely damaged representatives of the terminal oxidation catalyst, or are lacking an essential ingredient. The major reason for this statement is the fact that unphysiologically large concentrations of cytochrome c are required to obtain rates of oxidation with cytochrome oxidase preparations comparable to rates observed with mitochondria. Recently Dr. W. Arion in our laboratory has obtained crude preparations of cytochrome oxidase which responded to physiological concentrations of cytochrome c to yield oxidation rates comparable to those in mitochondria.

Site III phosphorylation in submitochondrial particles is satisfactorily assayed with reduced phenazinemethosulfate as substrate. The dye can be reduced either by ascorbate or by DPNH in the presence of antimycin. It was shown[51] and confirmed [87] that the latter assay indeed measures phosphorylation associated with the oxidation step between cytochrome c and oxygen.

IV. THE OXIDATION CHAIN OF THE INNER CHLOROPLAST MEMBRANE

There are striking similarities between the oxidation chain of mitochondria and chloroplasts. Pyridine nucleotides, quinones, flavoprotein, and cytochromes participate in both systems and catalyze electron and proton transfers that are coupled to a device which generates ATP.[2,84] Some uncoupling agents such as carbonylcyanide p-trifluoromethoxyphenylhydrazone (CCP) and inhibitors such as N,N¹-dicyclohexylcarbodiimide (DCCD) are effective in both systems, and electron flow is controlled by the availability of ADP.

The cleavage of water by System II yields oxygen and a reduced compound (reduced plastoquinone?) which is oxidized by the oxidation chain, accompanied by ATP generation. The components of the oxidation chain include plastoquinone, cytochrome f, plastocyanine, P_{700}, ferredoxin, a flavoprotein, and TPN, the final hydrogen acceptor. Activation of System I provides the electron acceptor for System II (a weak oxidant) as well as a strong reductant (ferredoxin) which is capable of reducing TPN via a flavoprotein. ATP and TPN, which are required for starch formation, are produced by the combined action of System I and II (Figure 4-8).

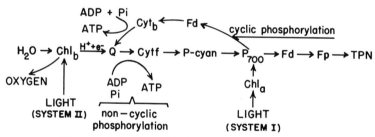

FIGURE 4-8. The oxidation chain in chloroplasts.

A. Noncyclic and Cyclic Photophosphorylation

Artificial electron acceptors can be substituted for TPN giving rise to oxygen evolution in a smaller segment of the oxidation chain (Hill reaction). With TPN and with some acceptors (e.g., ferricyanide), ATP generation can be demonstrated to take place during the Hill reaction (noncyclic phosphorylation).

Activation of System I by light in the presence of PMS or ferredoxin catalyzes cyclic phosphorylation. There are differences in opinion whether cyclic phosphorylation utilizes the same phosphorylation site operative in noncyclic phosphorylation associated with the Hill reaction,[2] or whether it is catalyzed by a separate phosphorylation system as proposed by Arnon and collaborators.[1] Evidence for a separate system is supported by data showing that cyclic phosphorylation is sensitive to dinitrophenol and antimycin, whereas noncyclic phosphorylation is resistant. Although assay conditions for cyclic and noncyclic phosphorylation are different, the susceptibility of the system to various inhibitors, makes it highly probable that different phosphorylation sites do, indeed, exist. Over ten years ago,[89] we made the observation that subchloroplast particles, obtained by disruption of chloroplasts in a Nossal shaker, maintained a considerable proprotion of the capacity for cyclic phosphorylation, while noncyclic phosphorylation with TPN rapidly declined to zero. These findings strongly supported the view that there is a more stable phosphorylation site which functions only in cyclic photophorylation, but not in phosphorylation associated with TPN reduction.

One of the possible formulations consistent with these, as well as other findings,[2,84] is given in Figure 4-8. In this scheme, a formulation which is not generally accepted, a cytochrome *b* is placed in a sidepath between P_{700} and plastoquinone. Cyclic phosphorylation is shown to operate with two phosphorylation sites, one between cytochrome *b* and Q, the second between Q and cytochrome *f*. The latter also participates during phosphorylation associated with the Hill reaction. The dual function of this site could help to explain many of the puzzling discrepancies in the field of photophosphorylation.

An alternative possibility is that cyclic phosphorylation represents a sidepathway within System I with a phosphorylation site between P_{700} and the electron acceptor (e.g. PMS).

B. Resolution of the Oxidation Chain

Important advances are being made in the resolution of the oxidation chain of chloroplasts and of photosynthetic microorganisms by fractionations with digitonin and triton.[84] Heavy and light subchloroplast particles have been obtained by differential centrifugation. The heavy particles lacked P_{700}, but were enriched in chlorophyll *b* and Mn^{++} and catalyzed the Hill reaction. The reduction of TPN by ascorbate-DCIP, catalyzed by light-activated System I, was very slow. The light particles lacked Mn^{++}, but were enriched in chlorophyll *a* and β carotene. They contained P_{700} and catalyzed the light-activated TPN reduction by ascorbate-DCIP, but not the Hill reaction.

The most successful resolution of the chloroplast oxidation chain has been obtained in the region of TPN reduction. Two participating proteins, ferredoxin and the flavoprotein, have been isolated in pure form.[1,2]

Plastocyanine has been highly purified, and was found to be a copper-containing protein. It appears to act as a mediator between System II and I.[84] Cytochrome f has been partially purified and described to have properties similar to cytochrome c. There are two species of cytochrome b and their location in the oxidation chain is controversial. P_{700}, a most interesting component, is related to chlorophyll a, but only represents a minute portion of the total chlorophyll. It appears to serve as the reaction center of System I, and undergoes absorption changes during oxido-reduction. Similar reaction centers with different absorption spectra have been observed in photosynthetic microorganisms.

V. RESOLUTION OF THE COUPLING DEVICE

A. Apo-particles: Submitochondrial and Subchloroplast Particles Deficient in Coupling Factors

Various procedures have been employed to obtain resolution of the coupling device from the oxidation chain. Usually, most members of the oxidation chain have remained attached to the inner membrane, while proteins of the coupling device (coupling factors) have been dissociated. It is well known that the resolution of holo-enzymes (enzyme–coenzyme complexes) into apo-enzymes and coenzymes alters the properties of the partners of the complex. Changes in redox potential and absorption changes are among common alterations of the coenzymes; increased lability and decreased interaction with substrate are among the alterations encountered with apo-enzymes. Similarly, resolved submitochondrial particles (apo-particles) are often less stable, and particularly more sensitive to trypsin than functionally active holo-particles. The coupling factors are also less stable after resolution, they react differently with inhibtors, and exhibit frequently decreased solubility following purification and storage.

Thus far, the two most successful means of resolution of mitochondria, particularly if used sequentially, are sonic oscillation in a salt-free medium at an alkaline pH and treatment with urea at $0°$ or below. Various apo-particles that have been useful in our laboratory are listed in Table 4-2.

N-particles were the first apo-particles obtained [67,70,71] that exhibited a requirement for F_1. It was shown that pure F_1 catalyzed the hydrolysis of ATP, and that removal of ATPase resulted in a loss of coupling activity of the apo-particles. It was curious to note, therefore, that N-particles had considerable amounts of residual ATPase activity, and that from these particles more ATPase could be extracted with properties identical to those of F_1. Moreover, stimulation of phosphorylation in N-particles or A-particles was also obtained with catalytically inactive F_1,[65] or even by addition of unrelated compounds such as oligomycin[46] and DCCD.[78] These and other observations led to the discovery of two functions of F_1, a catalytic and a structural, which will be discussed later.

TABLE 4-2

APO-PARTICLES FROM MITOCHONDRIA AND CHLOROPLASTS

Apo-particles	Procedure of Preparation	Factor Dependency for Phosphorylation
N-particles	Exposure of mitochondria to Nossal shaker	F_1
SMP	Exposure of mitochondria to sonic oscillation at neutral pH	F_1, F_2
A-particles	Exposure of mitochondria to sonication at pH 9.2	F_1, F_3, F_2
ASU-particles	Exposure of A-particles to urea after passage through Sephadex	F_1, F_2, F_3, F_5
TU-particles	Exposure of SMP to trypsin and urea	particles confer sensitivity to added ATPase (F_1)
TUA-particles	TU-particles exposed to sonic oscillation at pH 10.0	particles $+F_c$ confer oligomycin sensitivity to added ATPase (F_1)
PC-particles	Exposure of chloroplasts to sonication in the presence of P-lipids	CF_1
EDTA-particles	Exposure of chloroplasts to salt-free media containing 0.5 mM EDTA	CF_1

Apo-particles that were completely resolved with respect to F_1 were first obtained [74] by exposing SMP to trypsin and urea. Treatment with trypsin activated a masked ATPase activity presumably by the digestion of a small molecular mitochondrial protein which specifically inhibits the ATPase activity of F_1, and is highly sensitive to trypsin.[69] This inhibitory protein combined with F_1 and protected it against cold-inactivation. Urea, which increased the cold-lability of the ATPase[77] was also less effective against the F_1-inhibitor complex.[69] Therefore, activation of ATPase activity prior to urea treatment proved to be a very essential step in the complete removal of F_1 from particles. Unfortunately, particles that were exposed to trypsin frequently lost their ability to catalyze oxidative phosphorylation even after addition of coupling factors, probably because of damage to those segments of the inner mitochondrial membrane that were depleted of F_1.[74] A much superior method to activate the masked ATPase was filtration of A-particles through a column of Sephadex G-50.[78] These AS-particles contained ATPase activity approaching 12 μ moles/min/mg protein as compared to 1.5–3 of the starting A-particles. From this high specific activity, it can be calculated that about 20% of the protein of submitochondrial particles is F_1.

Following urea treatment of AS-particles, the resulting ASU-particles lost 99% of the ATPase activity and required the addition of several coupling factors[78] for phosphorylation.

The submitochondrial particles listed in Table 4-2 are "loosely coupled," even after reconstitution with coupling factors. The difference between tight coupling, loose coupling, and uncoupling is illustrated in Figure 4-9. The tightly coupled systems catalyze electron transport only in the

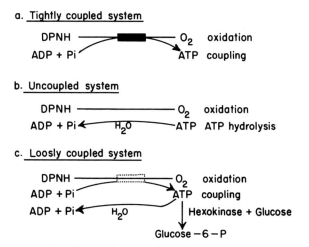

FIGURE 4-9. Coupling and uncoupling of oxidative phosphorylation.

presence of Pi and ATP. This ingenious control mechanism gears the production of energy to its utilization, thereby avoiding heat production and waste of energy. Carefully prepared mitochondria are tightly coupled: They do not exhibit much ATPase activity, and they do not respire without addition of ADP and Pi (Figure 4-9a). They can be uncoupled by numerous agents ranging from nonspecific detergents to specific uncouplers such as carbonylcyanide p-trifluoromethoxyphenyl hydrazone, which is effective at concentrations below 10^{-7} M. In the presence of uncouplers, respiration takes place without addition of either Pi or ADP, and added ATP is hydrolyzed (Figure 4-9b). In loosely coupled preparations, the oxidation chain and ATPase are also independently active (Figure 4-9c), but in contrast to uncoupled particles, the loosely coupled particles are capable of forming ATP, which is rapidly hydrolyzed unless it is utilized by an appropriate P acceptor system, such as hexokinase and glucose. Since the degree of coupling varies with different submitochondrial particles, it is necessary to adjust the energy trapping system to a level adequate to compete with the ATPase activity. In terms of the hypothesis of Scheme III, the degree of uncoupling is a function of the exposure of XO^- and Y^- to hydrogen ions. Thus, the transitions from tight to loose coupling and to uncoupling is a quantitative rather than a qualitative change.

Two types of apo-particles derived from spinach chloroplasts are listed in Table 4-2. PC-particles are obtained by sonication of chloroplasts; EDTA-particles are chloroplasts that have been extracted with dilute EDTA (below 1 mM) in the absence of any other salt. These apo-particles are stimulated by CF_1.[56,57,97] PC-particles usually show somewhat less resolution than EDTA particles, but in contrast to the latter, they can be stored for several months at $-70°$ with little loss of activity.

B. Coupling Factors

The coupling device (Figure 4-2), which transforms the oxidative energy into ATP energy, appears to reside in the inner membrane itself. Several proteins have been separated from mitochondria and were called coupling factors simply because they were required for the operation of the coupling device. The noncommittal term "factor" was chosen because these components may be structural components of the membrane participating in the proper organization of the catalysts, rather than being catalysts themselves.[22,76,100] It is possible, moreover, that the function of some of the coupling factors may be neither catalytic nor structural, but regulatory. For example, they may act by counteracting natural uncouplers which control energy metabolism. Some natural uncouplers found in mitochondria were counteracted by addition of serum albumin.[72] Since this observation was made, serum albumin has been always added during assay of oxidative phosphorylation.[67] Recently, brown fat mitochondria from cold-adapted rats and newborn rabbits were isolated [32] which required extraordinarily large amounts of serum albumin (5–10 mg per mg of mitochondrial protein) to counteract the "natural uncouplers," presumably long-chain fatty acids.

Another complication was introduced into the concept of "coupling factors" with the important discovery of Lee and Ernster[46] that low concentrations of oligomycin (0.4 μg per mg protein) stimulated phosphorylation in submitochondrial particles that were depleted in coupling factors. The authors, therefore, suggested that coupling factors may act like oligomycin by inhibiting a side reaction (hydrolysis) of the high-energy intermediate. This possibility was ruled out, however, by demonstrating with the aid of an antibody against coupling factor 1 (F_1) that the stimulatory effect of oligomycin was dependent on the presence of residual F_1 in the partially depleted submitochondrial particles.[28] Further, decisive evidence was obtained [78] when submitochondrial particles were completely resolved with respect to F_1. Oligomycin (or rutamycin) had little or no effect on such particles unless a certain amount of *catalytically active* F_1 was added. Great caution in the interpretation of the mode of action of coupling factors is required, since, in some cases, an indirect effect such as suggested by Lee and Ernster,[46] or a structural effect,[22,76,100] or a regulatory effect, as suggested above, may be operative. It is precisely because of these uncertainties that the term "coupling factors" should be retained until more precise information with regard to the mode of action of each coupling factor will be available.

1. Coupling factor 1 (F_1) from mitochondria.

PROPERTIES OF F_1. Coupling factor 1 (F_1) was first isolated from extracts obtained by mechanical disintegration of beef heart mitochondria in a Nossal shaker.[67,70] Recently, a simpler procedure for the large-scale preparation of F_1 from beef heart mitochondria after disruption by sonic oscillation has been developed in our laboratory. The protein appears homogeneous in the ultracentrifuge, and sediments with a sedimentation coefficient $S_{20,w}^{\circ}$ of 13. The molecular weight of the native enzyme is 284,000. The protein dissociates into inactive subunits when exposed to temperatures below 10° to give an equilibrium mixture of 3.5 S, 9.1 S, and 12 S components.[68] If the exposure to low temperatures was short, the reaction was reversible at 30° to yield active enzyme which sedimented like the native enzyme. The amino acid analysis of the enzyme revealed the absence of tryptophan, and the presence of eleven half-cystines per mole. The composition of other amino acids are shown in Table 4-3. Subunits are clearly seen in electron micrographs of F_1, and the size of the molecule (85 Å in diameter) is consistent with the calculated molecular weight. It should be emphasized that the cold-labile enzyme is not a lipoprotein, since no fatty acids are present.

The mechanism of the cold lability is unknown, but it is likely that hydrophobic bonds are labilized at the lower temperatures. The marked effect of salt and H^+ concentration on the rate of cold inactivation[68] suggest, however, that electrostatic forces make an important contribution in holding the subunits together.

The ATPase activity of F_1 was inhibited by a number of inhibitors of oxidative phosphorylation, such as azide,[70] and also by ADP. Although several other trinucleotides (GTP, ITP etc.) were rapidly cleaved by the enzyme, other dinucleotides were not inhibitory. This is in line with the specificity of ADP in oxidative phosphorylation, which in our experience also extends to a considerable degree to submitochondrial particles.

THE CATALYTIC ROLE OF F_1. The role of F_1 in oxidative phosphorylation is visualized to be at the last step of transphosphorylation (Figure 4-10) resulting in ATP formation. In reverse, XOH is phosphorylated to yield $X \sim P$ which is transformed to $X \sim Y$. In loosely coupled or uncoupled

TABLE 4-3

AMINO ACID COMPOSITION OF F_1 AND CF_1

	Residues per half cystine	
Amino Acid	F_1	CF_1
Lysine	15	11
Histidine	4	2
Arginine	14	15
Aspartic Acid	19	18
Threonine	14	19
Serine	15	20
Glutamic Acid	28	36
Proline	10	10
Glycine	22	22
Alanine	25	25
Half Cystine	1	1
Valine	18	20
Methionine	5	6
Isoleucine	15	19
Leucine	21	27
Tyrosine	7	7
Phenylalanine	7	7

mitochondria, or in submitochondrial particles, this process results in ATP hydrolysis. With the soluble enzyme, H_2O is visualized to substitute for XOH. Such substitution for substrates by water is well known to occur with many enzymes which catalyze a group transfer reaction.

In intact mitochondria which are tightly coupled and do not exhibit

1. TRANS PHOSPHORYLATION

$$XO \sim P + ADP \rightleftharpoons XOH + ATP$$

2. ATP HYDROLYSIS

a. In phosphorylating particles

$$ATP + XOH \longrightarrow ADP + XO \sim P$$

$$XO \sim P + YH \longrightarrow XO^- + Y^- + 2H^+ + P_i$$

b. With soluble enzyme

$$ATP + HOH \longrightarrow ADP + Pi$$

FIGURE 4-10. Catalytic functions of F_1.

ATPase activity, the entrance of water is blocked, presumably by the combination of F_1 with a small molecular protein which specifically inhibits mitochondrial ATPase.[69] Recent studies with this inhibitor* have revealed that it requires both Mg^{++} and ATP for the inhibition of ATPase activity. The enzyme-inhibitor complex, which cannot hydrolyze ATP, is still capable of catalyzing oxidative phosphorylation.[69] It is likely, therefore, that the inhibitor which prevents the interaction of F_1 with water plays an important role in the phenomenon of respiratory control.

In line with the proposed site of action of F_1, it was shown that F_1-depleted particles require addition of F_1 for all reactions that are associated with oxidative phosphorylation and are dependent on ADP or ATP. For example, $H_2^{18}O$–ATP, $^{32}P_i$–ATP exchange reactions or oxidative phosphorylation at each of the three sites do not take place in the absence of F_1.[28,37]

THE STRUCTURAL ROLE OF F_1. As mentioned earlier, in addition to F_1, a large variety of chemically unrelated compounds stimulated oxidative phosphorylation in partially resolved A-particles. Oligomycin,[46,78] dicyclohexylcarbodiimide,[78] yeast F_1,[85] and chemically modified F_1[65,66,79] were effective. Since the chemically modified preparations of F_1 had no ATPase activity, and since chemicals like oligomycin or DCCD could substitute for catalytically active F_1, it became apparent that there must be a structural function for F_1 which is independent of its catalytic role. This phenomenon of a structural role of a membrane component has now been observed with other membrane components as well, and will be discussed again later.

2. Coupling factor 1 (CF_1) from spinach chloroplasts. In many respects, CF_1 is remarkably similar to F_1, e.g., in electron micrographs, sedimentation in the ultracentrifuge, amino acid composition,** and cold-lability in the presence of salt.[56] However, the protein exhibits specificity with regard to immunological and functional reactivity. CF_1 reacted only with an antibody prepared against CF_1, but not with an antibody against F_1.[56] Stimulation of photophosphorylation by CF_1 could not be duplicated by addition of F_1, nor did CF_1 stimulate oxidative phosphorylation. Furthermore, rather interesting differences were found in the susceptibility of the factors to heat with resulting changes in the interaction with the membranes. There was no effect of heat (65°) on the ATPase activity of F_1 or on its ability to stimulate oxidative phosphorylation. In contrast, CF_1 exhibited little or no manifest ATPase activity, but, following exposure to trypsin or heat, ATPase activity appeared whereas coupling activity disappeared.[97] A loss of ability to adsorb to the membrane was associated with the absence of a coupling activity.[5] Mirror-image properties were exhibited by the membranes. Whereas the chloroplast membrane, even after

* Horstman, L. and Racker, E., unpublished experiments.
** Farron, F., McCarty, R. E., and Racker, E., unpublished experiments (Table 4-3).

exposure to trypsin or heat, combined with CF_1, the mitochondrial membrane was found to be very sensitive to exposure to temperatures above 55° or to trypsin.[11]

The activation of ATPase of CF_1 by heat or trypsin was paralled by the appearance of SH groups that became susceptible to alkylation.* The ATPase activity of CF_1 could be also unmasked by prolonged treatment with dithiothreitol. In contrast to the heat-treated or trypsinized CF_1 which no longer adsorbed to the membrane, the ATPase activated by dithiothreitol and released with EDTA was readily adsorbed by the chloroplast membrane, but, thereby, lost its ATPase activity.[57] The components in the membrane that are responsible for interaction with CF_1 have been recently investigated.[50] It was observed that a crude extract of chloroplast interacted with CF_1, conferring upon it greater stability towards heat and cold. Purification of the extract revealed that polynucleotides from spinach chloroplasts, as well as from other sources, greatly increased the heat stability of CF_1 at low concentrations of ATP. Of several enzymatically synthesized polynucleotides, by far the most active one was polycytidylate. Various phospholipids, sulfolipids, and galactolipids protected soluble CF_1 against inactivation at both high and low temperature. The protection against cold inactivation is particularly interesting since CF_1, associated with the chloroplast membrane, was quite stable. It was of significance, therefore, to find that the heat-activated enzyme which was equally cold labile but did not react with membrane, was also considerably less stabilized by lipids than was CF_1.

CF_1 like F_1 appeared to be required for all reactions involving phosphorylation, but had no effect on electron transport. Accordingly, the specific antibody against CF_1 interfered with all light catalyzed reactions that involved the production or utilization of ATP.[56] Even the ATPase activity of heat or trypsin-activated ATPase was fully susceptible to inhibition by the specific antibody.

Finally, it should be mentioned that a coupling factor from Euglena[18] has been isolated with properties between those of F_1 and CF_1. Although the ATPase activity of this protein was not masked, it functionally substituted for CF_1 in stimulating photophosphorylation of spinach chloroplasts.

3. Coupling factor 4 (F_4) and "structural protein." It was mentioned previously that crude preparations of F_4 and of "structural protein" combined with members of the respiratory chain as well as with phospholipids. Although F_4 had the distinct advantage that it was soluble, neither of these two preparations should be regarded as single entities. Exposure of mitochondria to either dodecylsulfate or organic solvents which are used for the preparation of "structural protein" denature several mitochondrial proteins (including F_1). The preparations of "structural protein" described in the literature, therefore, contain, without doubt, mixtures of denatured

* Farron F., McCarty, R. E., and Racker E., unpublished experiments.

proteins. Although preparations of F_4 are considerably less heterogenous,[34] they were shown to contain multiple coupling factors which can be separated.[26] Indeed, it was recently possible to eliminate the need for F_4 preparations altogether in the reconstitution of oxidative phosphorylation by addition of F_1, F_2, F_3, and F_5. Although some of these coupling factors have not been purified to homogenity, it is quite likely that the original suggestion[22] that F_4 preparations contain a protein of low solubility to which coupling factors were strongly adsorbed, was correct. It also appears that several members of the inner membrane are capable of interaction with phospholipids as well as with each other. We are, therefore, not only left with no evidence for the existence of the "structural protein," but are even lacking an adequate assay to search for it. Moreover, it was pointed out before that components of the membrane which can be solubilized have a structural as well as a catalytic role. It would be better, therefore, to avoid the term "structural protein" until evidence for such an entity is available.

4. Other coupling factors (F_2, F_3, F_5). Numerous coupling factors have been reported from various laboratories. A few years ago the situation was very confusing since findings by different investigators appeared incompatible, particularly with respect to the site-specificity of coupling factors. In contrast to the report from other laboratories, all coupling factors we have examined stimulated phosphorylation at each of the three coupling sites.[28] Recently, the fog has cleared when some of the site-specific coupling factors have faded away. Since these and other coupling factors recorded in the literature are being reviewed elsewhere,[64] this section will be restricted to a brief summary of most recent developments in our laboratory.

As mentioned earlier, F_2 and F_3 preparations replaced the need for F_4 in A-particles.[26] On further experimentation, a stimulation of phosphorylation by F_4 was again noted, particularly in assays of oxidative phosphorylation in ASU-particles.[76] Recently, however, a purified coupling factor (F_5) was shown to replace this requirement for F_4.[27] The P:O ratios with DPNH as substrate was raised from 0.03 to over 2.0 on addition of F_1, F_2, F_3, and F_5. It is possible, as will be discussed later, that F_5 is identical with F_c[11] a mitochondrial factor which was required for the conferral of rutamycin sensitivity to TUA-particles (see Table 4-2). At present, attempts are being made to separate the factors completely from each other by starting with a different extraction procedure for each factor. Active extracts of F_2 have been obtained by alkaline extraction of an acetone-dried preparation of beef heart mitochondria. The extraction and purification procedure follows essentially the method described [6] for the preparation of succinate dehydrogenase. Although these preparations are not as active as some other fractions containing F_2, they are being used at present because they are less contaminated with other factors. F_2 was shown to be very sensitive to exposure to iodine.[26] F_3 has been purified from extracts obtained by sonication of light-layer mitochondria, followed by fractionation on a DEAE-cellulose column. Extracts from heavy-layer mitochondria are actually more active, but are more heavily contaminated with other factors. F_5 has been purified by fractionation of an extract obtained by sonication of mitochondria in the presence of 2 M urea. Whereas, in ASU particles, F_1, F_2, F_3, and F_5 were shown to stimulate phosphorylation associated with the oxidation of DPNH, only F_1, F_3, and F_5 appeared to be required for the ATP-driven reduction of DPN by succinate.[27] F_2 may, therefore, be either more rate limiting in the forward reaction, or may not participate in the back reaction. Since F_5 may be identical with F_c, a component of the oligomycin-sensitive ATPase[11] to be discussed later, it is likely that this factor affects the final step of transphosphorylation. It may participate in this step either directly (by supplying or modifying X), or indirectly by serving as a structural component required for the proper orientation of F_1.

VI. PARTIAL REACTIONS CATALYZED BY THE INNER MITOCHONDRIAL MEMBRANE

As in most investigations of multienzyme systems, partial reactions of oxidative phosphorylation represent useful probes in the analysis of the pathway. We can differentiate two categories of partial reactions associated with oxidative phosphorylation that have been studied.

The first category contains reactions that represent sections of the pathway proper, and which are measured either in the forward or backward direction of electron and energy flow. We include in this group the segments of the oxidation chain discussed earlier, and reactions catalyzed by the coupling device. The oligomycin- or rutamycin-sensitive ATPase, represents a partial reaction of the coupling device operating backwards. The ^{32}Pi–ATP exchange is a partial reaction of oxidative phosphorylation which includes steps in both the forward and backward direction. This is an important aspect of this reaction which helps to explain several discrepancies in the literature which have served to cast doubts on the direct relationship between the ^{32}Pi–ATP exchange and oxidative phosphorylation. The oligomycin and dinitrophenol sensitive ^{14}C–ADP–ATP exchange which takes place in intact mitochondria is a partial reaction of oxidative phosphorylation which, unfortunately, cannot be readily measured in submitochondrial particles, and has, thus far, been of no value in reconstitution studies. The $H_2^{18}O$ exchange reactions with Pi and ATP should also be listed here, although it is quite possible that they may involve side reactions.

The second category contains reactions that are associated with oxidative phosphorylation, but include steps that are clearly concerned with side-pathways. We may include here the energy-dependent transhydrogenase, the accumulation of ions, and also, the oligomycin-insensitive ATPase activity of F_1. In view of the consideration discussed earlier ATP formation due to some ion movements (either H^+ or K^+) belong to this category.

There is only one purpose that justifies the division of partial reactions into these two categories. The properties of the reactions of the first category must be related to the process of oxidative phosphorylation, and no modern formulation of a mechanism of oxidative phosphorylation can afford to ignore them.

VII. ISOLATION OF PARTICLES WHICH CATALYZE A $^{32}P_i$–ATP EXCHANGE

Extraction of submitochondrial particles with 0.5% cholate in the presence of 0.15 M ammonium sulfate yielded very small and light particles which had the following properties.* On addition of F_1 they catalyzed a ^{32}Pi–ATP exchange which corresponded to about 40% of the parent submitochondrial particles, and which was sensitive to oligomycin and uncouplers. On the other hand, DPNH and succinate were oxidized at rates less than 1%, and spectral analysis showed the virtual absence of cytochrome a, a_3, and c, whereas cytochrome b and c_1 were only partially depleted. The very low rate of respiration with succinate as substrate was accompanied by phosphorylation with very low efficiency (P:O of 0.1). Addition of a purified preparation of cytochrome oxidase and cytochrome c markedly accelerated the respiration rate, but the additional uptake of oxygen was not associated with phosphorylation resulting in a drastic reduction of the already low P:O.

It is apparent from these preliminary findings that it is possible to remove from the coupling device at least some of the respiratory catalysts. Yet the morphological appearance of the particles is quite similar to that of small submitochondrial particles capable of catalyzing oxidative phosphorylation. Attempts can now be made to reconstitute oxidative phosphorylation by addition of a cytochrome oxidase preparation that can interact with these particles.

VIII. RESOLUTION AND RECONSTITUTION OF THE INNER MITOCHONDRIAL MEMBRANE

It was shown by Lardy and his collaborators[44] that the ATPase activity of mitochondria is very sensitive to oligomycin. In contrast, soluble ATPase (F_1) was insensitive to this antibiotic as well as to several other compounds (rutamycin, DCCD) which inhibit at low concentration ATP hydrolysis in mitochondrial or submitochondrial particles. This altered sensitivity of the .enzyme depending on its association with the membrane has been called an allotopic property.[76] The allotopic property of F_1 was used to develop an assay for the isolation of the membrane components required for the conferral of oligomycin or rutamycin sensitivity. It was first necessary to prepare submitochondrial particles from which all bound oligomycin-sensi-

* Arion, W., and Racker, E., unpublished experiments.

tive ATPase had been removed. This was accomplished by first treating the particles with trypsin, which greatly increased the ATPase activity,[74] presumably because of the removal of a small molecular protein which is an inhibitor of ATPase,[69] and which protects it against inactivation by urea. Exposure of the trypsin-treated particles to 2 M urea at 0° yielded TU-particles (see Table 4-2) which were virtually devoid of ATPase activity. When F_1 was added to TU-particles, the enzyme was bound to the particles and the ATPase activity became sensitive to oligomycin, rutamycin or DCCD. If TU-particles were subjected in the absence of salts to sonic oscillation in 0.25 M sucrose and then to centrifugation at $100,000 \times g$ for one or two hours, an "extract" was obtained which conferred oligomycin sensitivity on F_1. This "extract" contained the entire electron transport chain, and was readily converted to a particulate preparation when salt was added.[75] These observations prompted us to examine these preparations in the electron microscope. It was found that TU-particles or sonic "extracts" were devoid of the "elementary particles" discovered by Fernandez-Moran.[25] Green, Fernandez-Moran, and their collaborators[7] had arrived at the conclusion that these elementary particles contained members of the electron transport chain. However, we noted that TU-particles, which no longer contained the elementary particles, catalyzed the oxidation of DPNH at a very rapid rate and contained all the members of the respiratory chain.[75,82] It was apparent that the contention that the elementary particles represent electron transport enzymes was untenable. Moreover, calculations of the amount of protein removed by urea during the last step of the preparation of TU-particles yielded values that indicated that most of the missing protein could be accounted for by loss of ATPase. The elementary particles, or the inner membrane spheres as we prefer to call them, were, therefore, proposed to be identical with F_1.[40,82] Convincing evidence for this conclusion was obtained by reconstitution experiments with homogenous preparation of F_1 and ASU-particles which were completely resolved with respect to F_1.[78]

These findings led to further investigations of the properties of the membrane components that interact with F_1 and confer to it sensitivity to oligomycin. It was shown[40] that fractionation of the sonic extract of TU-particles with ammonium sulfate in the presence of 2% cholate yielded a fraction (CF_0) capable of binding F_1 and inhibiting its ATPase activity. On addition of phospholipids, the ATPase activity was fully restored and was sensitive to oligomycin or rutamycin. Parallel electron microscopic examinations revealed that CF_0 was an amorphous precipitate (Figure 4-11). On addition of F_1 (Figure 4-12), the structure was covered with inner membrane spheres without changing its amorphous appearance. On addition of phospholipids to the CF_0–F_1 complex, the particles became vesicular (Figure 4-13) and a distinct membrane with attached inner membrane spheres became visible. This was a rather startling discovery

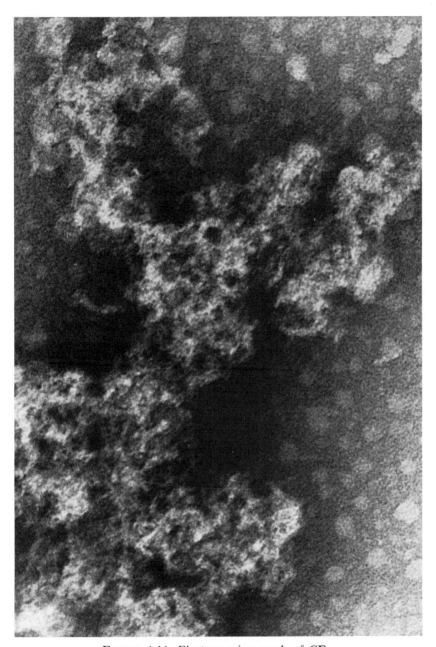

FIGURE 4-11. Electron micrograph of CF_0.

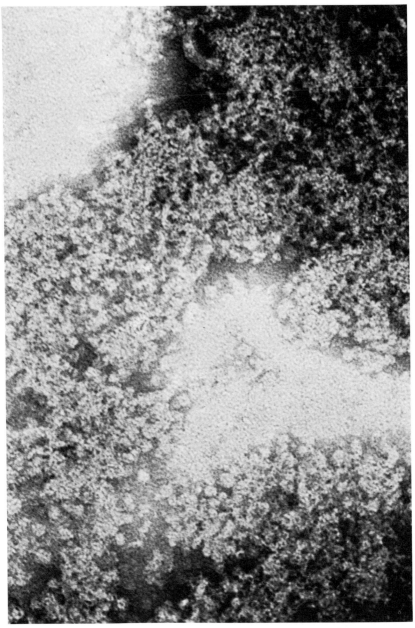

FIGURE 4-12. Electron micrograph of $CF_0 + F_1$.

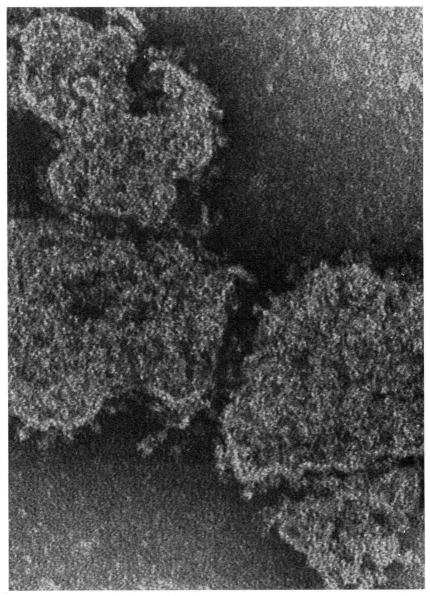

FIGURE 4-13. Electron micrograph of $CF_0 + F_1 +$ P-lipids.

since CF_0 preparations were completely lacking some members of the oxidation chain such as cytochrome a, a_3, and c which were hitherto considered to be integral components of the inner membrane proper.

Attempts to further resolve CF_0 into individual components were unsuccessful because of the instability of the products. However, exposure of TU-particles to sonic oscillation at pH 10.4 followed by centrifugation, yielded new particles (TUA-particles) which did not confer rutamycin sensitivity to added F_1, but which regained this activity on addition of a crude preparation of F_4.[40] Recent studies in our laboratory with TUA-particles and with CF_0 prepared from TUA-particles (TUA-CF_0) revealed the following facts.[11] Binding of F_1 to TUA-particles took place in the presence of low concentrations of Mg^{++} (0.8 mM), or of somewhat higher concentrations of other divalent cations. This requirement had escaped detection in earlier experiments because at high concentrations monovalent cations substitute for Mg^{++}. The buffers which had been used in the assay system contained sufficient monovalent cations to obscure the Mg^{++} requirement. After adsorption of F_1 to TUA-particles in the presence of Mg^{++}, the ATPase activity was insensitive to rutamycin. However, on addition of crude preparations of F_4, the rutamycin-conferral activity of the particles was restored. Purification of a protein factor (F_c) from F_4 based on an assay of conferral activity yielded a preparation which exhibited similarities with F_5. Highly purified preparations of F_5, moreover, contained F_c with a very high specific activity. These preparations of F_5, were, however, only partially resolved with respect to F_3. MacLennan and Tzagoloff[54] have independently purified a mitochondrial factor (OSCP) which conferred oligomycin sensitivity to F_1 added to depleted particles. In recent experiments, we have found that OSCP prepared from F_4 as described [54] contained both F_3 and F_5 activity.

Exposure of either TUA-particles or of F_c to mild heating (55°) or to trypsin resulted in loss of conferral activity. Yet the reconstituted complex of TUA-particles with F_c and F_1 was completely resistant to the same treatments. These additional examples of the allotopic properties of the various membrane components could serve as probes in future investigation of the inner membrane.

The curious observation that addition of F_1 to CF_0 resulted in a CF_0–F_1 complex which was virtually inactive as an ATPase was further examined.[11] As mentioned earlier, further fractionation of CF_0 was unsuccessful because of the considerable lability of the resulting fractions. However, preparations of CF_0 obtained from TUA-particles were found to be quite stable (TUA–CF_0). F_1 added to these particles became attached, but retained its ATPase activity which was insensitive to rutamycin. F_c, which has no effect on soluble F_1, caused a pronounced inhibition of the ATPase activity of the F_1–TUA–CF_0 complex. TUA–CF_0 first reconstituted with F_c reacted with F_1 exactly like CF_0, i.e., it inhibited the ATPase activity.

In both cases, addition of phospholipids reactivated the ATPase activity and induced rutamycin sensitivity. In earlier experiments, crude mixtures of phospholipids were used, but more recent experiments revealed considerable variations in the capacity of different phospholipids to reactivate rutamycin-sensitive ATPase activity. It seems that some phospholipids (e.g., phosphatidylserine) eliminated the masking of ATPase activity by F_c without inducing appreciable rutamycin sensitivity. Other phospholipids (e.g., phosphatidylethanolamine) gave rise to only moderate activation of ATPase activity, which, however, was almost completely rutamycin-sensitive. It is apparent from these studies that phospholipids play an essential part in the assembly of the oligomycin-sensitive ATPase. Indeed, the possibility may now be considered that rutamycin and similar inhibitors actually do not affect ATPase activity directly, but may prevent the reactivation of the latent ATPase of F_1 (owing to interaction with F_c) by the mitochondrial phospholipids.

IX. THE STRUCTURAL ROLE OF CATALYTIC MEMBRANE COMPONENTS

It was proposed [67] that F_1 acts as a phosphotransferase catalyzing the last step in the formation of ATP from ADP and phosphate. The ATPase activity represents a reversal of this process whereby water substitutes for the natural phosphate acceptor XOH. In submitochondrial particles which were only partially depleted of ATPase activity, addition of F_1 markedly raised the P:O at all three phosphorylation sites.[28] Lee and Ernster[46] showed, however, that low concentrations of oligomycin markedly stimulated phosphorylation in A-particles without the addition of coupling factors. Previously, Penefsky[65] observed that F_1 treated with iodine lost all its ATPase activity, yet was still capable of stimulating phosphorylation in N-particles. On the other hand, antibodies against F_1 added to A-particles[28] abolished phosphorylation in the presence of stimulating concentrations of oligomycin without impairing oxidation. These observations suggested that the stimulation of phosphorylation in A-particles by F_1 is not dependent on a catalytic activity of F_1, but may be an expression of a structural role of F_1 in the reconstitution of the membrane.

The first direct evidence for a structural role was obtained in the course of studies on the succinate-driven transhydrogenation of TPN by DPNH. Unexpectedly, this reaction showed a requirement for F_1 in F_1-depleted ASU-particles although no ATP formation was involved (cf 56). The second example was observed during studies of light-induced H^+ translocation in subchloroplast particles which were partially resolved with respect to CF_1.[56] Unexpectedly, the light-induced proton translocation which did not involve formation of ATP required addition of the chloroplast coupling factor. Nevertheless, in this system as well as in the transhydrogenase reaction mentioned above, addition of the specific antibody against the coupling factor had no effect, although under similar conditions the antibody abolished ATP formation. A striking illustration for the selective effect of the antibody was obtained in experiments in which hybrid particles were used which consisted of N-particles which had been fortified with yeast F_1.[85] In these particles, phosphorylation was inhibited by the antibody against beef F_1, but not against yeast F_1. On the other hand, the yeast antibody was fully effective in depressing the P:O in yeast particles which contained yeast F_1 exclusively, and it also inhibited the ATPase activity of the yeast enzyme attached to the beef heart particles. Thus, it appears that the yeast ATPase can only function as a structural component, and that the structural function of F_1 is not inhibited by the specific antibody, but the catalytic function is highly susceptible.

An interesting example for the participation of a catalyst in the organization of the membrane not involving a coupling factor was observed in studies of the reconstitution of the oxidation chain.[10] The respiratory segment between succinate and Q_2 was reconstituted with succinate dehydrogenase, Q_2, F_4, and phospholipids. This system did not exhibit the sensitivity to trifluorothienylbutanediol (TTB) which is characteristic for submitochondrial particles. However, on addition of cytochrome b, the oxidation rate increased and was rendered sensitive to TTB in spite of the fact that the added cytochrome b did not undergo oxidation–reduction. The preparation of cytochrome b used in these studies was inactive as a biological oxido-reduction catalyst as shown by subsequent studies on the reconstitution of the oxidation chain between succinate and oxygen.[99] A preparation of cytochrome b which was comparatively less pure spectroscopically but had not been exposed to dodecylsulfate, was far more active on a heme basis in the reconstitution experiment than the more highly purified and soluble preparation.[10]

These examples and the experiments on the reconstitution of the membrane amply illustrate the complexity of the inner mitochondrial membrane. There are several proteins and phospholipids which contribute to the basic structure of the membrane; there are several catalysts that are associated with the membrane and participate in its organization. Finally, there are membrane appendices like F_1 which attach themselves specifically to the inner mitochondrial membrane. We begin to suspect that the complexity of the biochemical process of energy generation during oxidative phosphorylation is reflected by an equally perplexing complexity of structural organization.

X. STRUCTURAL ORGANIZATION OF THE INNER MEMBRANE OF MITOCHONDRIA

Numerous reviews have been written on the composition of the oxidation catalysts in mitochondrial membranes (cf 48a), and some of the structural aspects of the mitochondrial membrane, particularly the interaction between proteins and phospholipids, are discussed in Chapters 1 and 2. Rather than repeat these data and speculations, it may be worthwhile to emphasize some of the limits of our knowledge in this area.

There can be little doubt that the multi-enzyme system which catalyzes oxidative phosphorylation is in close association with the inner membrane of mitochondria. But if we define the inner membrane in terms of the structure seen by different staining methods in the electron microscope, the relationship of the various components of the oxidation chain and of the coupling device to the membrane is by no means clear. Even the state of molecular assembly of the components themselves is unknown. For example, it would be of interest to learn whether isolated cytochrome oxidase, which is catalytically active as a polymer,[23a] is representative of the oxidase which is associated with the membrane. Since the dimensions of the active polymer exceed by far the thickness of the membrane, the oxidase of the native membrane must be either in a depolymerized form which is activated by the membrane, or a polymer which is not located entirely within the membrane proper. It is of interest to note in this connection that a vesicular submitochondrial membrane, from which cytochrome oxidase was removed, did not exhibit an altered appearance in electron micrographs.[40]

Of particular interest in relation to the chemiosmotic hypothesis is the problem of the topography of the membrane components in relation to each other. The location of the inner membrane spheres (F_1) is clearly established in submitochondrial particles obtained by sonic disruption of beef heart mitochondria. The inner membrane spheres are on the surface, facing the medium. In intact mitochondria, the location of the spheres is, however, not as definite, although the majority of investigators favor the view that the spheres face the matrix. Succinate dehydrogenase can be removed from submitochondrial particles, and it is possible to reconstitute such depleted particles by addition of the soluble enzyme. It, therefore, appears that these dehydrogenases are located at the same side of the membrane as the inner membrane spheres.

Although our knowledge is obviously still rather limited, one could propose a tentative topography of the membrane based partly on the considerations mentioned above, and partly on the susceptibility of submitochondrial particles to externally added trypsin. In the first layer of submitochondrial particles (facing the medium), we find the mitochondrial ATPase inhibitor in a complex with F_1. Digestion with trypsin destroys the inhibitor without affecting F_1, as indicated by marked increase in ATPase activity. In negative stains, F_1 appears to be connected to the membrane by a stalk. Whether this appearance is an artifact induced by the negative stain or is due to the presence, even under physiological conditions, of a connecting structure between F_1 and the membrane still remains to be established. Next on the surface of the membrane is succinate dehydrogenase and DPNH dehydrogenase. Within the membrane is cytochrome b and c_1, and cytochrome oxidase. The location of cytochrome oxidase is of particular interest because, according to Mitchell, it is responsible for the translocation of electrons from the outside surface of the membrane to the inside surface where it interacts with oxygen and because previously mentioned experiments indicate that it is possible to remove cytochrome oxidase from membranes without apparent damage to its appearance. In line with the presence of cytochrome oxidase on the matrix surface of the membrane which corresponds to the medium surface of "inside-out" particles, is the observation that externally added ferrocytochrome c is readily oxidized by submitochondrial particles, albeit without phosphorylation. As pointed out by Mitchell,[58] the location of cytochrome c on the inner membrane surface facing the outer membrane is indicated by the ease of cytochrome c extraction by salt from intact mitochondria and by its resistance to extraction in "inside-out" particles. It is also in line with the position of the third loop in the chemiosmotic hypothesis. A direct experimental approach to this problem of localization of the oxidation catalysts with the aid of specific antibodies is now in progress in our laboratory.

It was shown in these experiments[23c] that functional cytochrome c is present on the outer surface (C-side) of the inner membrane in intact mitochondria and on the inner surface of submitochondrial particles. Location on the opposite side (M-side) of the inner membrane was established for F_1. Of special interest is the finding that in contrast to cytochrome c and F_1, cytochrome oxidase could be reached by an antibody prepared against pure preparations of cytochrome oxidase from both sides of the membrane. Ferrocytochrome c oxidation took place in mitochondria as well as in submitochondrial particles and both processes were sensitive to antibody against cytochrome oxidase. An intimate knowledge of the topography of the membrane components may well be of invaluable aid in the design of future experiments on the reconstitution of oxidative phosphorylation from individual components.

XI. CONCLUSIONS

The role of the inner membranes of mitochondria and chloroplasts appears to be the key issue in the current formulations of the mechanisms of the energy generating processes that take place in these organelles. If, as postulated by Mitchell, a translocation of protons across the membrane and a separation of "inside" and "outside" is required for the operation of the ATP generating system, vesicularity and limited permeability to protons are essential features of these membranes. According to Mitchell, it should be impossible to obtain oxidative phosphorylation either in a nonvesicular system or with a membrane freely permeable to protons. Neither has been achieved thus far.

What are the essential constituents of an intact and functional inner membrane and how are they assembled? The question whether a single "structural protein" is responsible for the organization of the membrane is still unanswered. On the other hand, there is a remarkable interplay between the various catalytic components of the membrane (e.g., succinate dehydrogenase and F_1 interact with phospholipids and cytochromes) which opens the alternative possibility of self-assembly without the aid of an "organizational" protein. Indeed, several mitochondrial catalysts have a structural role that can be clearly distinguished from their catalytic role.

Two extreme views can, therefore, be formulated. According to one, there is a single structural protein which organizes all the components of the membrane. According to the other, all proteins of the membrane have a structural as well as a catalytic role and the capabilities of self assembly. A crucial experiment distinguishing between these alternatives would be the reconstitution of a functional membrane from soluble components. This has not as yet been achieved. But, since model membranes have been made by interaction of soluble proteins and phospholipids, further attempts in this direction appear warranted. A third possibility somewhere between the two extremes is that a structural protein, perhaps only a minor component of the membrane, plays an important role in the initiation of membrane formation, followed by the assembly of catalytic components by specific interactions with each other. Such a structural protein would be the logical candidate for a gene product of mitochondrial DNA. Identification and quantitation of such a component represents an important research project which could enhance our understanding of the structural organization as well as of the biogenesis of the inner mitochondrial membrane.

Yet another view could be advanced which includes the presence of multiple structural proteins which participate in the assembly of the inner membrane. Several coupling factors (F_2, F_3, and F_5) which are required for phosphorylation but have not as yet been shown to have a catalytic function, would qualify for such organizational roles. They may function either as coupling factors linking phosphorylation to oxidation, or they may participate in the "sealing" of the membrane, preventing the free permeation of protons or the access of water to sites in the membrane concerned with the preservation of energy during the oxidation process.

It should be apparent to a reader of this review that we know much more about the properties of the individual components of the inner mitochondrial membrane than about the structure of the membrane itself. However, I question whether it is possible to really "see" the proverbial forest without an intimate knowledge of its trees. I believe that we do not have very far to go before a clear image of the function and structure of the inner membrane will emerge.

REFERENCES

1. Arnon, D. I., Tsujimoto, H. Y., and McSwain, B. D. (1967), *Nature* **214,** 5088, 562.
2. Avron, M. (1967), Current Topics in Bioenergetics, Vol. II, D. R. Sanadi, ed., Academic Press, N.Y.
3. Avron, M., Grisario, V., and Sharon, N. (1965), *J. Biol. Chem.* **240,** 1381.
4. Beinert, H., Palmer, G., Cremona, T., and Singer, T. P. (1965), *J. Biol. Chem.* **240,** 475.
5. Bennun, A., and Racker, E. (1969), *J. Biol. Chem.* **244,** 1325.
6. Bernath, P., and Singer, T. P. (1962), *Methods in Enzymology* **5,** 597.
7. Blair, P. V., Oda, T., Green, D. E., and Fernandez-Moran, H. (1963), *Biochemistry* **2,** 756.
8. Boyer, P. D. (1967), Biological Oxidation (Thomas P. Singer, ed.), John Wiley & Sons, New York.
9. Boyer, P. D. (1958), *Proc. Intern. Symposium Enzyme Chem.,* Maruzen, Tokyo.
10. Bruni, A., and Racker, E. (1968), *J. Biol. Chem.* **243,** 964.
11. Bulos, B., and Racker, E. (1968), *J. Biol. Chem.* **243,** 3891, 3901.
12. Butow, R. A., and Racker, E. (1965), "New York Heart Assoc. Symposium in Oxygen," *J. Gen. Physiol.* **49,** 149.
13. Bygrave, F. L., and Lehninger, A. L. (1966), *J. Biol. Chem.* **241,** 3894.
14. Chance, B. (1967), *Fed. Proc.* **26,** 1341.
15. Chance, B., Bonner, W. D., Jr., and Storey, B. T. (1968), "Electron Transport in Respiration," *Ann. Rev. Plant Physiol.* **19,** 295.
16. Chance, B., and Mela, L. (1967), *J. Biol. Chem.* **242,** 830.
16a. Chance, B., Ernster, L., Garland, P. B., Lee, C. P., Light, P. A., Ohnishi, T., Ragan, C. I., and Wong, D. (1967), *Proc. Natl. Acad. Sci.* **57,** 1498.
17. Chance, B., and Williams, G. R. (1956b), *J. Biol. Chem.* **221,** 477.
18. Chang, I. C., and Kahn, J. S. (1966), *Arch. Biochem. Biophys.* **117,** 282.
19. Cockrell, R. S., Harris, E. J., and Pressman, B. C. (1966), *Biochem.* **5,** 2326.
20. Cockrell, R. S., Harris, E. J., and Pressman, B. C. (1967), *Nature* **215,** 1487.
21. Cohn, M., and Drysdale, G. R. (1955), *J. Biol. Chem.* **216,** 831.
22. Conover, T. E., Prairie, R. L., and Racker, E. (1963), *J. Biol. Chem.* **238,** 2831.
23. Criddle, R. S., Bock, R. M., Green, D. E., and Tisdale, H. (1962), *Biochemistry* **1,** 827.
23a. Criddle, R. S., and Bock, R. M. (1959), *Biochem. Biophys. Res. Commun.* **1,** 138.
23b. Davies, R. E. (1961), "Symposium on Membrane Transport and Metabolism," Publishing House of the Czechoslovak Academy of Sciences, Prague.
23c. diJeso, F., Christiansen, R. O., Steensland, H., and Loyter, A. (1969), *Fed. Proc.* **28,** 663.
24. Ernster, C., and Lee, C. P. (1964), *Ann. Rev. Biochem.* **33,** 729.
25. Fernandez-Moran, H. (1962), *Circulation* **26,** 1039.

26. Fessenden-Raden, J. M., Dannenberg, A., and Racker, E. (1966), *Biochem. Biophys. Res. Commun.* **25**, 54 and unpublished experiments.
27. Fessenden-Raden, J. M., and Racker, E. (1968), *Fed. Proc.* **27**, 297.
28. Fessenden-Raden, J. M., and Racker, E. (1966), *J. Biol. Chem.* **241**, 2483.
29. Goldberger, R., Smith, A. L., Tisdale, H., and Bomstein, R. (1961), *J. Biol. Chem.* **236**, 2788.
30. Grant, C. T., and Taborsky, G. (1966), *Biochemistry* **5**, 544.
31. Green, D. E. (1965), "Oxidases and Related Redox Systems," p. 1098 (T. E. King, H. S. Mason, and M. Morrison, eds.), John Wiley & Sons, New York.
32. Guillory, R. J., and Racker, E. (1968), *Biochim. Biophys. Acta* **153**, 490.
33. Haake, P. C., and Westheimer, F. H. (1961), *J. Amer. Chem. Soc.,* **83**, 1102.
34. Halder, D., Freeman, K., and Work, T. S. (1966), *Nature* **212**, 9.
35. Hatefi, Y., and Stempel, K. E. (1967), *Biochem. Biophys. Res. Commun.* **26**, 301.
36. Hinkle, P. C., Butow, R. A., Racker, E., and Chance, B. (1967), *J. Biol. Chem.* **242**, 5169.
37. Hinkle, P. C., Penefsky, H. S., and Racker, E. (1967), *J. Biol. Chem.* **242**, 1788.
38. Hommes, F. A. (1964), *Arch. Biochem. Biophys.* **107**, 78.
39. Jagendorf, A. T. (1967), *Fed. Proc.* **26**, 1361.
40. Kagawa, Y., and Racker, E. (1966), *J. Biol. Chem.* **241**, 2461, 2467, 2475.
41. King, T. E. (1963), *J. Biol. Chem.* **238**, 4037.
42. Kouba, R., and Varner, J. E. (1959), *Biochem. Biophys. Res. Commun.* **1**, 129.
43. Kröger, A., and Klingenberg, M. (1966), *Biochem. Z.* **344**, 317.
44. Lardy, H. A., Johnson, D., and McMurray, W. C. (1958), *Arch. Biochem. Biophys.,* **78**, 587.
45. Lee, C. P., Azzone, G. F., and Ernster, L. (1964), *Nature* **201**, 152.
46. Lee, C. P., and Ernster, L. (1965), *Biochem. Biophys. Res. Commun.* **18**, 523.
47. Lee, C. P., and Ernster, L. (1966), "Regulation of Metabolic Processes in Mitochondria " Vol. 7, p. 218 (J. M. Tager, S. Papa, E. Quagliariello, and E. C. Slater, eds.), Elsevier Publ. Co., Amsterdam.
48. Lehninger, A. L. (1967), *Fed. Proc.* **26**, 1333.
48a. Lehninger, A. L. (1964), "The Mitochondrion," Benjamin, New York.
49. Lester, R. L., and Fleischer, S. (1961), *Biochim. Biophys. Acta* **47**, 358.
50. Livne, A., and Racker, E. (1969), *J. Biol. Chem.* **244**, 1339.
51. Löw, H. Alm, B., and Vallin, I. (1964), *Biochem. Biophys. Res. Commun.* **14**, 347.
52. Löw, H., Vallin, I., and Alm, B. (1963), "Energy-linked Functions in Mitochondria," p. 5 (B. Chance, ed.), Academic Press, New York.
53. Loyter, A., Christiansen, R. O., Steensland, H., Saltzgaber, J., and Racker, E. (1969), *J. Biol. Chem.* (in press).
54. MacLennan, D. H., and Tzagoloff, A. (1968), *Biochemistry* **7**, 1603.
55. McCarty, R. E. (1968), *Biochem. Biophys. Res. Commun.* **32**, 37.

56. McCarty, R. E., and Racker, E. (1966), Brookhaven Symposia in Biology: No. 19, New York.
57. McCarty, R., and Racker, E. (1968), *J. Biol. Chem.* **243**, 129.
58. Mitchell, P. (1966), *Biol. Rev. Cambridge Phil. Soc.* **41**, 445.
59. Mitchell, P. (1967), *Fed. Proc.* **26**, 1370.
60. Mitchell, P. (1961), *Nature* **191**, 144.
61. Mitchell, R. A., Hill, R. D., and Boyer, P. D. (1967), *J. Biol. Chem.* **242**, 1793.
62. Nair, P. M., and Mason, H. S. (1967), *J. Biol. Chem.*, **242**, 1406.
63. Ohnishi, T., Kawaguchi, K., and Hagihara, B. (1966), *J. Biol. Chem.* **241**, 1797.
64. Penefsky, H. S. (1969), Electron and Coupled Energy Transfer (T. King, and M. Klingenberg, eds.) Marcel Dekker, Inc., New York (in press).
65. Penefsky, H. S. (1964), *Fed. Proc.* **23**, 533.
66. Penefsky, H. S. (1967), *J. Biol. Chem.* **242**, 5789.
67. Penefsky, H. S., Pullman, M. E., Datta, A., and Racker, E. (1960), *J. Biol. Chem.* **235**, 3330.
68. Penefsky, H. S., and Warner, R. C. (1965), *J. Biol. Chem.* **240**, 4694.
69. Pullman, M. E., and Monroy, G. C. (1963), *J. Biol. Chem.* **238**, 3762.
70. Pullman, M. E., Penefsky, H. S., Datta, A., and Racker, E. (1960), *J. Biol. Chem.* **235**, 3322.
71. Pullman, M. E., Penefsky, H., and Racker, E. (1958), *Arch. Biochem. Biophys.* **76**, 227.
72. Pullman, M. E., and Racker, E. (1956), *Science* **123**, 1105.
73. Pullman, M. E., and Schatz, G. (1967), *Ann. Rev. Biochem.* **36**, 539.
74. Racker, E. (1963), *Biochem. Biophys. Res. Commun.* **10**, 435.
75. Racker, E. (1964), *Biochem. Biophys. Res. Commun.* **14**, 75.
76. Racker, E. (1967), *Fed. Proc.* **26**, 1335.
77. Racker, E. (1965), "Mechanisms in Bioenergetics," Academic Press, Inc., New York.
78. Racker, E., and Horstman, L. L. (1967), *J. Biol. Chem.* **242**, 2547.
79. Racker, E., and Horstman, L. L. (1967), *7th Intern. Cong. Biochem.*, Tokyo Symposium **6**, 297.
80. Racker, E., and Krimsky, I. (1952), *J. Biol. Chem.* **198**, 731.
81. Racker, E., and Monroy, G. (1964), Abstr. Intern. Cong. Biochem., 6th, New York. (*Intern. Union Biochem.* **32**, 760, X-S13).
82. Racker, E., Tyler, D. D., Estabrook, R. W., Conover, T. E., Parsons, D. F., and Chance, B. (1965), "Oxidases and Related Redox Systems," p. 1077 (T. E. King, H. S. Mason, and M. Morrison, eds.), John Wiley & Sons, New York.
83. Richardson, S. H., Hultin, H. O., and Fleischer, S. (1964), *Arch. Biochem. Biophys.* **105**, 254.
84. San Pietro, A. (1967), "Harvesting the Sun," p. 49, Academic Press, Inc., New York.
85. Schatz, G., Penefsky, H. S., and Racker, E. (1967), *J. Biol. Chem.* **242**, 2552.
86. Schatz, G., and Racker, E. (1966), *Biochem. Biophys. Res. Commun.* **22**, 579.

87. Schatz, G., and Racker, E. (1966), *J. Biol. Chem.* **241,** 1429.
88. Schatz, G., Racker, E., Tyler, D. D., Gonze, J., and Estabrook, R. W. (1966), *Biochem. Biophys. Res. Commun.* **22,** 585.
89. Schroeder, E. A., and Racker, E. (1959), *Fed. Proc.* **18,** 1261.
90. Shavit, N., Skye, G. E., and Boyer, P. D. (1967), *J. Biol. Chem.* **242,** 5125.
91. Slater, E. C. (1966), *Comprehensive Biochem.* **14,** 327 (M. Florkin and E. H. Stotz, eds.), Elsevier Publ. Co., Amsterdam.
92. Slater, E. C. (1967), *European Journ. of Biochemistry* **1,** 317.
93. Slater, E. C. (1953), *Nature* **172,** 975.
94. Slater, E. C., Kaniuga, Z., and Wojtczak, L. (1966), eds. "Biochemistry of Mitochondria," pp. 1–10, Academic Press, Inc., and P. W. N., London and Warsaw.
95. Szarkowska, L. (1966), *Arch. Biochem. Biophys.* **113,** 519.
96. Tzagoloff, A., Byington, K. H., and MacLennan, D. H. (1968), *J. Biol. Chem.* **243,** 2405.
97. Vambutas, V. K., and Racker, E. (1965), *J. Biol. Chem.* **240,** 2660.
98. Wadkins, C. L., and Lehninger, A. L. (1963), *J. Biol. Chem.* **238,** 2555.
99. Yamashita, S., and Racker, E. (1969), *J. Biol. Chem.* **244,** 1220.
100. Zalkin, H., and Racker, E. (1965), *J. Biol. Chem.* **240,** 4017.

5

Outer Membrane
of Mitochondria*

Lars Ernster and Bo Kuylenstierna

Department of Biochemistry

University of Stockholm

Stockholm, Sweden

I. INTRODUCTION

Although the existence of two mitochondrial membranes, an outer limiting membrane and a folded inner membrane giving rise to the cristae, has been established for a long time, it is only recently that methods have been developed for their separation. The purpose of this chapter is to summarize current information concerning the morphological, physical, chemical, and enzymatic properties of isolated mitochondrial inner and outer membranes, and to discuss this information in relation to the functional organization of mitochondria.

The first high-resolution electron micrographs of mitochondria were published by Palade[106] in 1952. They were interpreted to indicate that the mitochondrion is surrounded by a membrane which forms a number of infoldings or ridges protruding into the interior of the mitochondrion, called *cristae mitochondriales* by Palade. The following year, Sjöstrand [137]

* This chapter is partly based on a paper presented at a *Symposium on Mitochondria —Structure and Function,* Fifth Annual Meeting of the Federation of European Biochemical Societies, Prague, 1968.[43] Work quoted from the authors' laboratory has been supported by grants from The Swedish Cancer Society and the Swedish Medical and Natural-Science Research Councils.

discovered that the limiting membrane of mitochondria was double. The inner structures, seen earlier by Palade,[106] were interpreted by Sjöstrand [137] as separate units and called "inner double membranes" or *septae*. The existence of a double limiting membrane was soon confirmed by Palade[107] who maintained, however, that the cristae are infoldings of the inner limiting membrane. According to Palade's definition (cf. Ref. 108), "Two spaces or chambers are outlined by the mitochondrial membranes: an outer chamber contained between the two membranes, and an inner chamber bounded by the inner membrane. The inner chamber is penetrated and, in most cases, incompletely partitioned by laminated structures which are anchored with their bases in the inner membrane and terminate in a free margin after projecting more or less deeply inside the mitochondrion." This definition of the mitochondrial structure has become widely accepted over the past years (cf. Refs. 102, 110, 129), and has also served as the morphological concept underlying most procedures that have recently been devised for the separation of the inner and outer membranes.

  Much information concerning the stuctural relationship of the two mitochondrial membranes has emerged in recent years from morphological observations with isolated mitochondria exposed to various experimental conditions. In particular, four structural states are of interest in the present context: "orthodox," "condensed," "swollen," and "contracted." These are schematically illustrated in Figure 5-1. The terms "orthodox" and "condensed" have been introduced by Hackenbrock[63] to describe reversible structural changes in isolated mitochondria that can be induced by altering the metabolic steady state of the mitochondria. These changes, which are

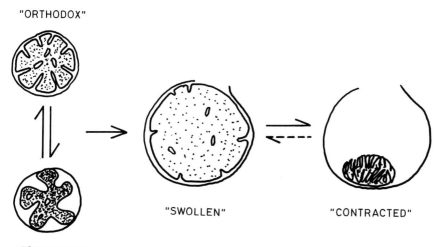

FIGURE 5-1. Conformational states of mitochondria. See text for explanation.

probably related to the phenomenon of "low-amplitude" swelling and contraction first described by Chance and Packer[25] (cf Ref. 63), primarily concern the conformation of the inner membrane with a concomitant shift in the ratio of the matrix space and the space between two membranes. The outer membrane is unaltered. Interestingly, the mitochondria in the condensed state reveal frequent points of attachment of the inner membrane to the outer membrane. Conformational changes of the inner membrane in response to alterations of the metabolic state have also recently been observed by Penniston *et al.*[115] and interpreted to be related to the energy state of the mitochondria.

The "swollen" and "contracted" states of mitochondria, as illustrated in Figure 5-1, refer to the well-known, "large amplitude" structural changes of mitochondria that are observed under a variety of experimental conditions not necessarily related to the metabolic state of the mitochondria. These states probably are reflections of the osmotic properties of the two membranes. Thus, exposure of mitochondria to a hypotonic medium, to low concentrations of certain compounds such as short-chain fatty acids, thyroxine, stilbestrol, etc. (cf Ref. 80), or to an isotonic medium containing readily penetrating cations and anions,[26,28] leads to a swelling of the mitochondria, and subsequent exposure to a hypertonic medium, or to ATP, Mg^{++}, or Mn^{++}, and serum albumin, causes a contraction. As shown by Parsons and associates,[110,159] a characteristic of the swollen mitochondria is a distended and, in most cases, disrupted outer membrane. The inner membrane, however, is largely unfolded, but usually retains its continuity. Significantly, contraction involves a refolding or reaggregation of the inner membrane, without a restoration of the outer membrane which, for most parts, becomes detached from the inner membrane.[97,159] These features have been of great importance as a rationale in elaborating methods for the separation of the two membranes.

Both the inner and outer mitochondrial membranes have been proposed [129] to be built up as classical unit membranes[35] of 50–70 Å thickness. In recent years, however, electron microscopic evidence for the presence of globular substructures in both membranes has been brought forth.[32,48,138–140] Relevant to this point is the observation of Fleischer *et al.*[49] that after extraction of 90% of the mitochondrial phospholipid by acetone, the inner membrane retains its double-layered, unit-membrane structure while the outer membrane is destroyed.

A striking morphological difference between the two membranes is the occurrence of regularly spaced, mushroom-like subunits on the inner surface of the inner membrane, discovered in 1962 by Fernández-Morán[47] in negatively-stained specimens of mitochondria. These subunits are absent from the outer membrane. Projecting subunits have also been observed on the outer surface of the outer membrane,[110,141] but these differ from the inner-membrane subunits in size, shape, and regularity of appearance.

Although the occurrence of the inner-membrane subunits has not yet been proven in intact mitochondria (cf Refs. 140, 157), they appear with great consistency in swollen mitochondria and submitochondrial preparations. They constitute, therefore, a valuable diagnostic tool for distinguishing membrane fractions originating from the inner and outer membranes. It should be pointed out, however, that the subunits may become detached from the inner membrane in the course of subfractionation of mitochondria,[125,147] and their absence or presence thus cannot be used as the sole criterion in identifying a given membrane fraction.

Definitions of the mitochondrial outer membrane, somewhat different from that described in the foregoing, have been proposed and employed by Green and associates in the course of their studies of the chemical and enzymatic organization of the mitochondrion. In 1965, Green[56] visualized the mitochondrion as consisting of an outer double membrane and a row of separate, closed inner membranes—a picture reminiscent of the 1953 model of Sjöstrand.[137] A year later, Green and Perdue[60] adapted Palade's definition[107] of the mitochondrial structure as consisting of a single outer membrane surrounding a continuous, folded inner membrane; the latter was coated on its inner surface by the repeating subunits of Fernández-Morán,[47] and the former on its outer surface by projections described by Parsons.[110] The same year Green[57] proposed a further model, essentially a hybrid between the two earlier ones. According to this model, "the mitochondrion appears to be built up of two membrane systems which closely interlock—an outer membrane system that encloses the mitocondrion and a system of inner membranes that radiate into the interior from the periphery. In the intact mitochondrion the two tubular systems are fused and the space within the tubules of one system is continuous with the space in the tubules of the other." In still further modifications of this model, the area of fusion between the two tubular systems has been reduced to an orifice[115] (similar to that earlier described by, e.g., Whittaker,[157]) and the subunits of the outer membrane have been moved from the outer surface to the space between the "outer" and "inner limiting membrane" components of the outer membrane system.[5] The reasons given for regarding the "inner boundary membrane" as part of the outer membrane, in spite of its recognized continuity with the inner tubular system (i.e., the cristae), were two: its postulated lack of projecting subunits, and its postulated identity in chemical composition with the outer limiting membrane. Both of these postulates, however, lack experimental support, and, indeed, it is difficult to see how a physical separation of the outer and inner mitochondrial membranes would at all be technically feasible if these latest models of Green and associates were correct.

TABLE 5-1
METHODS FOR THE SEPARATION OF THE OUTER AND INNER MITOCHONDRIAL MEMBRANES

For definition of mitochondrial compartments, see Figure 5-2. Abbreviations: OM = outer membrane; IM = inner membrane; IS = intermembrane space; M = matrix; diff = differential; dens gr = density gradient; centr = centrifugation; supnt = supernatant; concn = concentration.

Authors (ref)	Tissue	Rationale	Treatment	Means of separation	Designation of fractions obtained and their proposed contents
Parsons et al.[111]	Rat liver	Rupture of OM by swelling	Hypotonic P_i	Diff and dens gr centr	$B = OM$ $C = OM + IM$ $P = IM$
Parsons et al.[113]	Guinea pig liver	"	"	"	"
Sottocasa et al.[144-146]	Rat liver	Rupture and detachment of OM by swelling and contraction, followed by mechanical fragmentation of OM	Hypotonic P_i, followed by hypertonic sucrose + ATP + Mg^{++} and sonication	Dens gr centr	Heavy = $IM + M$ Light = OM Soluble = $IS + M$
Schnaitman et al.[133]	Rat liver	Rupture of OM by swelling Binding of OM-cholesterol by digitonin	Distilled water Low concns of digitonin	Dens gr centr Diff centr	Small vesicle fraction = OM Ghosts = $IM + M$ P $(9,500g) = IM + M$ P $(40,000g) = IM$ P $(144,000g) = OM$ S $(144,000g) = IS + M$
Schnaitman and Greenawalt[134]	Rat liver	"	"	"	Low speed pellet = $IM + M$ High speed pellet = OM High speed supnt = IS
Lévy et al.[85-87]	Rat liver	Binding of OM-cholesterol by digitonin, followed by mechanical fragmentation	Low concns of digitonin, sonication	Diff and dens gr centr	$C_2 = OM$ $C_4 = IM$ $S_2, S_4 = M$

Reference	Source	Procedure	Treatment	Diff centr	Fractions obtained
Bachmann et al.[9]	Beef heart	Separation of particulate subfractions with and without respiratory-chain enzyme components	Sonication		F_s = *OM* ("composite" fraction) S ($=S_3 + S_4$) = *OM* ("detachable" fraction)
"	"	"	Extraction of acetone powder with dist water	"	S = *OM* ("detachable" fraction)
"	"	"	Phospholipase	"	R_2 = *IM* K = *OM* ("particulate" fraction) S_3 = *OM* ("detachable" fraction)
Allmann and Bachmann[3]	Beef heart	"	"	"	"
Allmann et al.[6]	Beef liver	"	"	"	"
	Beef heart	"	"	"	"
	Beef liver	"	"	"	"
	Beef kidney	"	"	"	"
Green et al.[59]	Beef heart	"	Sonication in the presence of cholate	"	R_2 = *IM* F_c = *OM* ("composite" fraction) S_4 = *OM* ("detachable" fraction)
Allmann et al.[5]	Rat liver	Detachment of "outer boundary membrane" of *OM* by swelling	Oleate	"	Residue = *IM* + *OM* ("particulate" fraction) Supernatant = *OM* ("detachable" fraction)
"	"	Complete detachment of *OM* by swelling and contraction	Oleate, followed by ATP + Mg^{++} + serum albumin	"	Residue = *IM* Supernatant = *OM*
Byington et al.[20,21,142]	Beef heart	Rupture and solubilization of *OM*, and fragmentation of *IM*	Diethylstilbestrol + divalent cations	"	Residue = *IM* Supernatant = *OM*

II. METHODS FOR THE SEPARATION OF THE OUTER AND INNER MITOCHONDRIAL MEMBRANES

Table 5-1 is a survey of various methods described for the subfractionation of mitochondria with the aim of separating the outer and inner membranes. The identification of the subfractions as given in the Table is based on the definition of the four mitochondrial compartments according to Figure 5-2, namely: (a) the outer membrane; (b) the intermembrane

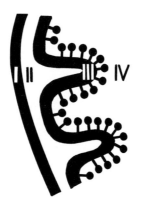

I.   OUTER MEMBRANE

II. INTERMEMBRANE SPACE

III. INNER MEMBRANE

IV. MATRIX

FIGURE 5-2. Definition of mitochondrial compartments. From Ernster and Kuylenstierna.[43]

space, *i.e.*, the space between the outer and inner membranes, including the space bordered by the outer surface of the cristae; (c) the inner membrane, including the cristae and their projecting subunits; and (d) the matrix, i.e., the space within the inner surface of the inner membrane.*

The first successful method for the preparation of what appears to be pure inner and outer membranes was described by Parsons *et al.*[111–113] in 1966. It is based on the observation[110,159] that swelling of liver mitochondria in hypotonic phosphate buffer causes a distension and occasional rupture of the outer membrane, while the inner membrane unfolds without

---

* Occasionally, the terms, "cristate space" and "intracristae space" have been used in the literature to denote what is defined here as the inner membrane and the intermembrane space, respectively. These terms seem to be somewhat unfortunate, since they relate to the infolding portions of the inner membrane and not to the inner membrane as a whole. For example, a swollen mitochondrion may have as much inner membrane as an intact one, although it may be largely devoid of cristae, the latter having been unfolded as a result of the swelling. Also, the term "intracristae space" (i.e. the space *within* the cristae) does not give a clear indication as to whether it refers to the space within the *in*foldings or the *out*foldings of the cristae, or, which would seem to be the most literal interpretation, within the cristal membrane itself.

breaking. Centrifugation of the swollen mitochondria on a suitable gradient separates small but pure specimens of inner and outer membrane from the bulk of the unbroken mitochondria (Figures 5-3, 5-4). A somewhat similar principle was used by Schnaitman et al.[133] who exposed mitochondria to osmotic lysis in distilled water and separated the "ghosts," [22] consisting of inner membranes, from the fragmented outer membranes by means of gradient centrifugation.

A procedure elaborated by Sottocasa et al.[144-146] is based on a swelling of mitochondria in hypotonic phosphate followed by a selective contraction of the inner membrane in the presence of hypertonic sucrose containing ATP and Mg^{++}, and by a gentle sonication to facilitate the detachment of the broken outer membrane. Subsequent density–gradient centrifugation separates outer and inner membranes in a high yield (Figures 5-5–5-9). A modification of this method has been described by Jones and Jones.[70] Another method introduced by Lévy et al.[85-88] and developed further by Schnaitman et al.[133,134] and Hoppel et al.,[67,96] takes advantage of a selective action of low concentrations of digitonin on the outer membrane, and yields, after suitable centrifugal separation, outer membrane fragments and relatively well-preserved inner membranes in good quantities (Figures 5-10– 5-12). Brdiczka et al.[17] have recently employed a procedure based on freezing of mitochondria at liquid-air temperature in a hypotonic medium to detach the outer membrane which results, after density–gradient centrifugation, in inner-membrane structures with well-preserved matrix.

All the above methods have been worked out with liver mitochondria and have been based primarily on morphological criteria in identifying the two membranes. Most of them have received wide use in the past few years in different laboratories engaged in studies of the intramitochondrial distribution of enzymes and other chemical constituents, and the results have shown remarkable agreement among the various procedures and research groups. Moreover, the results are consistent with conclusions drawn from studies with intact mitochondria.

Green and associates[3-6,9,10,20,21,56,58-60,142] have devised a number of procedures involving disruption of the mitochondria by sonication, freezing and thawing, organic solvents, detergents, phospholipase, fatty acids, or diethylstilbestrol, with the purpose of isolating outer and inner membranes from various tissues. These procedures have been based primarily on biochemical criteria for the identification of the two membranes. The conclusions reached by Green and associates regarding the intramitochondrial distribution of enzymes and other constituents differ in several important respects from those arrived at by most other laboratories. In the sections that follow, we shall first summarize these more generally supported conclusions, and subsequently consider possible reasons for the discrepancies between these and the conclusions of Green and associates.

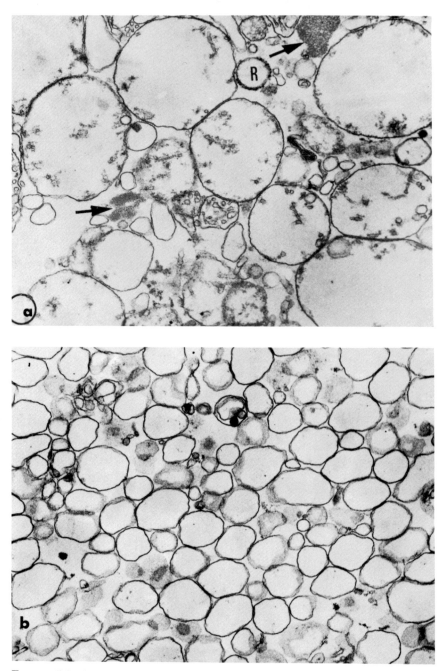

FIGURE 5-3. (a) Inner membrane ("P") fraction, and (b) outer membrane ("B") fraction of guinea-pig liver mitochondria, prepared by the method of Parsons *et al.*[113] Osmium tetroxide-fixed specimens stained with basic lead citrate. Magnification 35,000×. (Reproduced from Parsons *et al.*[113] with kind permission of the authors and publishers.)

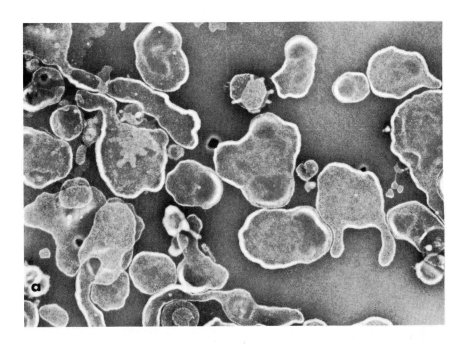

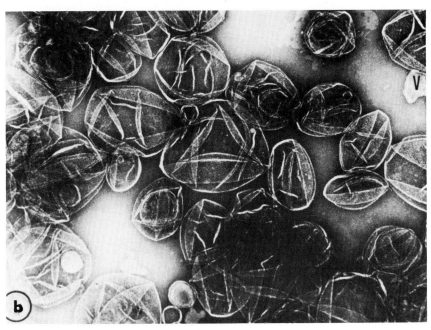

FIGURE 5-4. Outer membrane ("P") fraction of guinea pig liver mitochondria, prepared by the method of Parsons *et al.*[113] Negatively stained specimens, (a) without previous fixation, and (b) after previous fixation with osmium tetroxide. Magnification 35,000×. (Reproduced from Parsons *et al.*[113] with kind permission of the authors and publishers.)

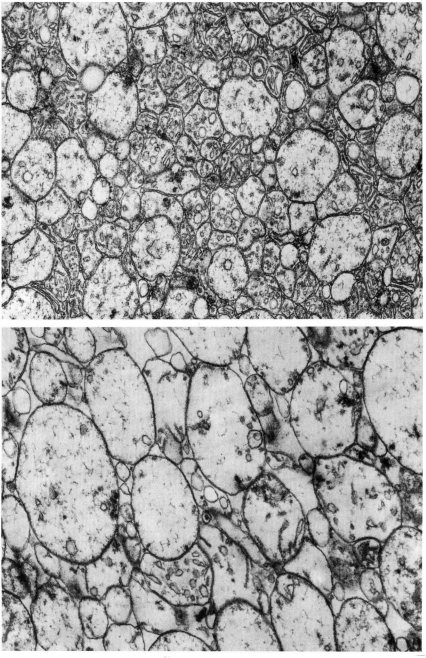

FIGURE 5-5. Inner membrane ("heavy") fraction of rat liver mitochondria, prepared by the method of Sottocasa *et al.*[145] Osmium tetroxide fixation. Magnification 30,000×. Top: Lower part of pellet; Bottom: upper part of pellet. (Reproduced from Sottocasa *et al.*[145] with kind permission of the publishers.)

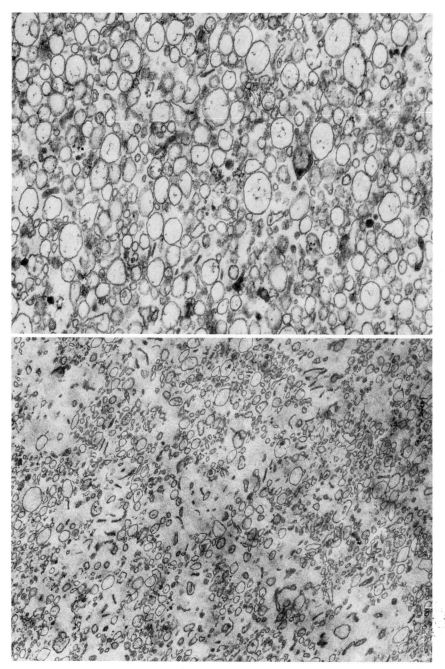

FIGURE 5-6. Outer membrane ("light") fraction of rat liver mitochondria, prepared by the method of Sottocasa et al.[145] Osmium tetroxide fixation. Magnification 30,000×. Top: Lower part of pellet; Bottom: upper part of pellet. (Reproduced from Sottocasa et al.[145] with kind permission of the publishers.)

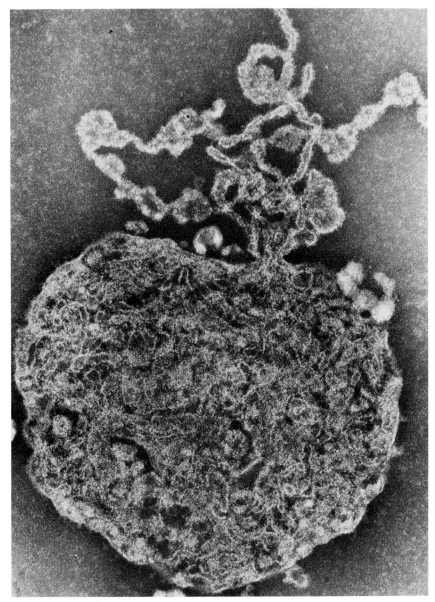

FIGURE 5-7. Negatively stained specimen of inner membrane ("heavy") fraction of rat-liver mitochondria, prepared by the method of Sottocasa *et al.*[145] Magnification 86,000×. (Reproduced from Sottocasa *et al.*[145] with kind permission of the publishers.)

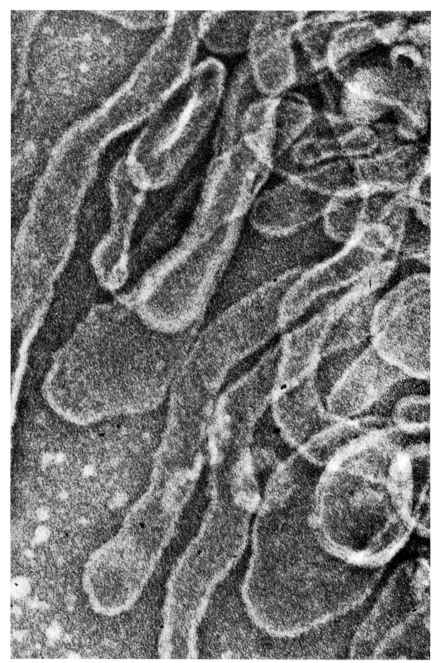

FIGURE 5-8. Negatively stained specimen of outer membrane ("light") fraction of rat liver mitochondria, prepared by the method of Sottocasa *et al.*[145] Magnification 165,000×. (Reproduced from Sottocasa *et al.*[145] with kind permission of the publishers.)

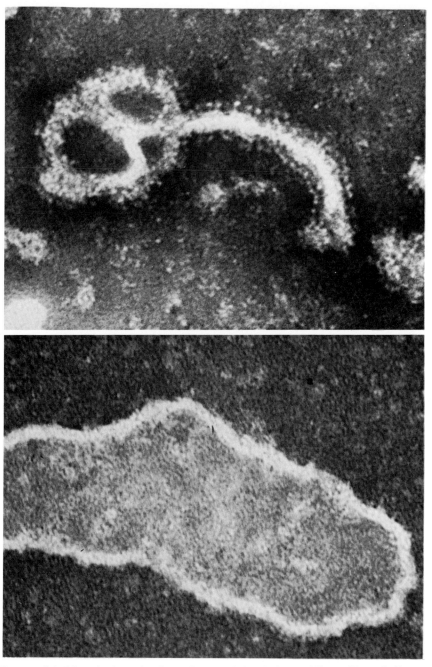

FIGURE 5-9. Negatively stained specimens of (top) inner ("heavy") and (bottom) outer ("light") membrane fractions of rat liver mitochondria, prepared by the method of Sottocasa *et al.*[145] Magnification 240,000× and 278,000×, respectively.

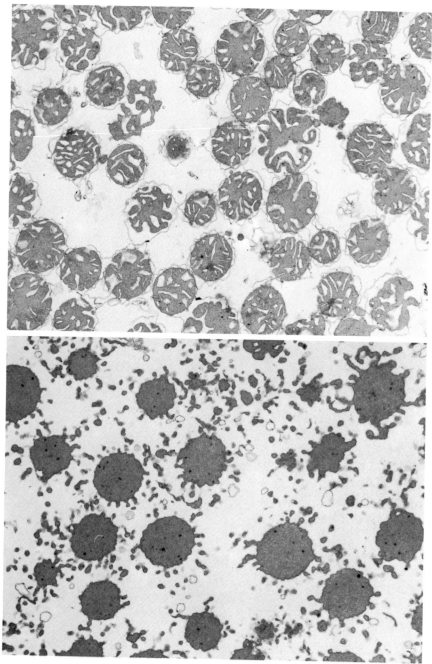

FIGURE 5-10. (Top) Freshly isolated rat liver mitochondria and (bottom) inner membrane-matrix preparation isolated by the improved digitonin fractionation procedure of Schnaitman and Greenawalt.[134] Fixed with glutaraldehyde and osmium tetroxide, and stained with uranyl acetate and lead citrate. Magnification 19,500×. (Reproduced from Schnaitman and Greenawalt[143] with kind permission of the authors and publishers.)

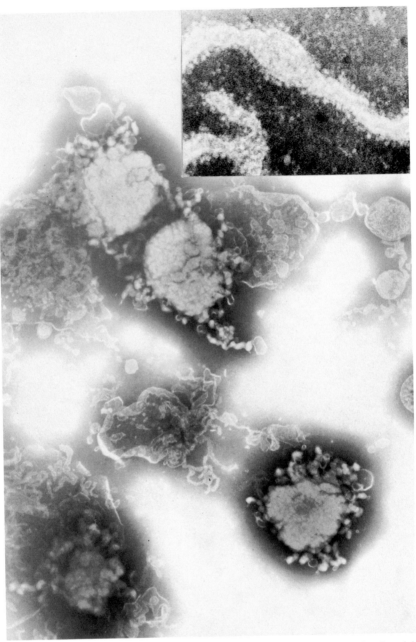

FIGURE 5-11. Inner membrane–matrix preparation of rat liver mitochondria isolated by the improved digitonin fractionation procedure of Schnaitman and Greenawalt.[134] Unfixed preparation negatively stained with phosphotungstic acid. Magnification 33,000×; inset, 187,000×. (Reproduced from Schnaitman and Greenawalt[134] with kind permission of the authors and publishers.)

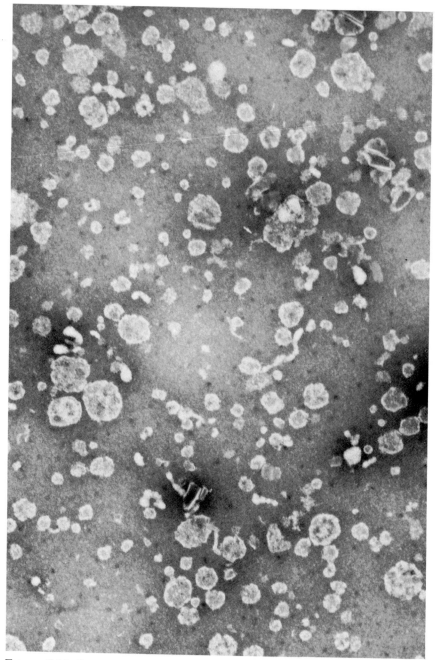

FIGURE 5-12. Outer membrane preparation of rat liver mitochondria isolated by the improved digitonin fractionation procedure of Schnaitman and Greenawalt.[134] Unfixed preparation negatively stained with phosphotungstic acid. Magnification 65,000×. (Reproduced from Schnaitman and Greenawalt[134] with kind permission of the Authors and Publishers.)

III. PHYSICAL PROPERTIES

Very striking differences in physical properties between the two membranes are found in regards to osmotic behavior and permeability. As already mentioned in the previous section, the inner membrane readily unfolds and refolds in response to changes in osmotic pressure. This feature has often been interpreted as suggestive evidence for the occurrence of a contractile protein in this membrane,[80,103] although conclusive evidence for this contention is still lacking (cf Ref. 16). In contrast, the outer membrane shows no reversible response to changes in osmotic pressure, and its distension and rupture during mitochondrial swelling may be interpreted as a passive process following the unfolding of the inner membrane.

Extensive studies,[72,73,117–120] based on measurements of the space occupied by various substances present in the medium in relation to the total water space of the mitochondria, and correlated with morphological observations, have led to the conclusion that the inner mitochondrial membrane possesses only a very limited permeability to most substances except uncharged molecules of a molecular weight not greater than 100–150. The majority of charged molecules of physiological importance pass through the inner membrane by way of specific translocators associated with this membrane (see also p. 198). In contrast, the outer membrane seems to be freely permeable to a wide range of substances, both charged and uncharged, with molecular weights up to about 10,000. An important implication of this concept is that all low-molecular, charged components present in isolated mitochondria, including various nucleotides, are located inside the inner membrane.

Another difference in physical properties between the two membranes concerns their density (approximately 1.13 for the outer membrane and 1.21 for the inner membrane) as estimated by Parsons *et al.*[111] from isopycnic centrifugation data.

X-ray diffraction patterns of the two membranes show a fundamental similarity,[151,152] despite striking differences in physical and chemical properties.

IV. CHEMICAL COMPOSITION

From enzyme distribution data obtained after subfractionation of rat-liver mitochondria with the digitonin method, Schnaitman and Greenawalt[134] arrived at the conclusion that the outer membrane accounts for about 4% and the inner membrane for about 21% of the total mitochondrial protein. About 67% of the mitochondrial protein is accounted for by the matrix, and the remainder by the intermembrane space. Similar values can be deduced from the data of Sottocasa *et al.*[144–146] obtained with the swelling–contraction–sonication procedure.

The most pronounced chemical features for distinguishing the two membranes can be found in their lipid composition.[68,71,84,100,113,114,133,148] Available information concerning the lipid composition of the mitochondrial inner and outer membranes as well as data regarding microsomes are summarized in Tables 5-2 and 5-3. The outer membrane contains, on a protein

TABLE 5-2

PHOSPHOLIPID AND CHOLESTEROL CONTENTS OF GUINEA PIG LIVER MITOCHONDRIA, ISOLATED INNER AND OUTER MITOCHONDRIAL MEMBRANES, AND MICROSOMES*

| | Mitochondria | | | |
	whole	inner membrane	outer membrane	Microsomes
Phospholipid (mg/mg protein)	0.159	0.301	0.878	0.391
Cholesterol (μg/mg protein)	2.28	5.06	30.1	30.2
Cholesterol (μg/mg phospholipid)	14.3	16.8	34.3	77.2

* From Parsons and Yano.[114]

basis, two to three times more total phospholipid than the inner membrane, amounting to a phospholipid content of about 45% of the dry weight.[113] The value is also higher than the phospholipid content of microsomes, and, in general, of any intracellular membrane hitherto investigated. It should be pointed out, however, that both the isolated inner membranes and the microsomes may contain significant amounts of soluble protein (originating from the mitochondrial matrix and from the lumen of the endoplasmic reticulum, respectively) which would lead to an underestimate of the phospholipid–protein ratios for these membranes.

Qualitative differences in phospholipid composition are most striking in the case of cardiolipin, which occurs predominantly, if not exclusively, in the inner membrane, and phosphatidyl inositol, which is found more abundantly in the outer membrane.[113] In both these respects, the mitochondrial outer membrane resembles the microsomes. Among neutral lipids, cholesterol has been reported to be six times more concentrated, on a protein basis, in the outer than in the inner membrane.[114] This explains the preferential action of digitonin on the outer membrane. The high cholesterol content of the outer membrane again is similar to that found with microsomes. Ubiquinone seems to be present only in the inner membrane.[143,146a]

Green and associates (cf Ref. 5) have reported protein contents of the

TABLE 5-3
PHOSPHOLIPID COMPOSITION OF LIVER MITOCHONDRIA, ISOLATED INNER AND OUTER MEMBRANES, AND MICROSOMES
The values are expressed in percentages of total phospholipid.

Authors (ref)	Parsons et al.[113] x				Lévy and Sauner[84]		Stoffel and Schiefer[148]			
Tissue	guinea pig liver				rat liver		rat liver			
		Mitochondria			Mitochondria			Mitochondria		
Phospholipids	whole	inner membr	outer membr	Microsomes	inner membr	outer membr	whole	inner membr	outer membr	Microsomes
Phosphatidyl choline	40.0	44.5	55.2	62.8	39.2	59.0	47	41	49	70
Phosphatidyl ethanolamine	28.4	27.7	25.3	18.3	34.4	20.3	26	35	31	21
Phosphatidyl inositol	7.0	4.2	13.5	13.4	6.6	4.8	} 13	2	17	14
Phosphatidyl serine	N.D.	N.D.	N.D.	4.5	} 3.4	7.5		—	—	5
Sphingomyelin										
Cardiolipin	22.5	21.5	3.2	0.5	} 13.7	6.3	14	21	3	—
Phosphatidic acid	2.3xx	2.2xx	2.5xx	1.1xx						
Phosphatidyl glycerol								1		
Lysophosphatidyl choline					3.1	3.4				

x These figures exclude the alkali stable phospholipid fraction.
xx Unidentified (phosphatidyl glycerol?)
Abbreviation: N.D. = none detected.

outer membrane of liver and heart mitochondria amounting to 30% or higher of the total mitochondrial protein. They have further concluded that the phospholipid contents of the two membranes are essentially similar both quantitatively (on a protein basis) and qualitatively. A possible reason for the discrepancies between these results and those of other laboratories will be considered in a later section (see p. 200).

V. INTRAMITOCHONDRIAL DISTRIBUTION OF ENZYMES

General Considerations

Current concepts regarding the localization of enzymes in the outer and inner mitochondrial membranes are based on results showing a concentration of these enzymes in isolated outer and inner membranes. In the case of enzymes localized in the intermembrane space or in the matrix, the conclusions are based, in addition to fractionation data, on certain assumptions. These involve considerations regarding, on one hand, the possible leakage of enzymes from and through the membranes during the fractionation procedure, and, on the other hand, the permeability properties of the two membranes as established in studies with intact mitochondria. Some pertinent problems are illustrated in Figure 5-13, which indicates the distribution and

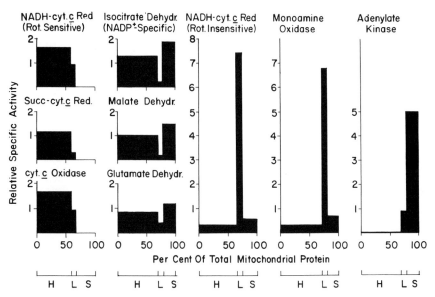

FIGURE 5-13. Distribution of some enzyme activities in the "heavy" (H), "light" (L), and "soluble" (S) fractions of rat liver mitochondria, prepared by the method of Sottocasa *et al.*[146] (Reproduced from Sottocasa *et al.*[146] with kind permission of the publishers.)

recovery of various enzyme activities upon subfractionation of rat-liver mitochondria by the procedure of Sottocasa *et al.*[146] This procedure yields three subfractions: a "heavy," a "light," and a "soluble" subfraction. From electron microscopic examination, the heavy subfraction consists of inner membranes including part of the matrix, and the light subfraction consists of vesicles derived from the outer membrane. The soluble subfraction ought to include the contents of the intermembrane space together with part of the matrix and any material released from the two membranes during the fractionation procedure.

As may be seen in Figure 5-13, the enzymes investigated show four types of distribution pattern. One type, represented by cytochrome oxidase, succinate-cytochrome *c* reductase, and the rotenone-sensitive NADH-cytochrome *c* reductase, is concentrated in the heavy subfraction, and thus constitutes enzymes associated with the inner membrane. A second type, represented by the rotenone-insensitive NADH-cytochrome *c* reductase and monoamine oxidase, is concentrated in the light subfraction, and thus constitutes enzymes associated with the outer membrane. A third type of enzymes, exemplified by adenylate kinase, is concentrated in the soluble subfraction, with little activity found in the light subfraction, and practically none in the heavy one. By the above definition of the origin of the three subfractions, this type of enzyme may originate either from the intermembrane space or from one of the two membranes. It is unlikely to originate from the matrix, since, in that case, part of its activity should have been recovered in the heavy subfraction. Furthermore, the fact that adenylate kinase in the intact mitochondria is insensitive to atractylate as measured with external adenine nucleotides as substrates[27] eliminates its localization in the inner membrane, where the atractylate-sensitive ADP–ATP exchange system is located.[73,158] Thus, adenylate kinase is most probably located in either the intermembrane space, or as a loosely bound enzyme in the outer membrane.

A fourth type of distribution pattern is represented by the enzymes glutamate dehydrogenase, malate dehydrogenase, and the NADP-specific isocitrate dehydrogenase. These enzymes show a bimodal distribution, part of them being recovered in the heavy, and part in the soluble subfraction, with little activity appearing in the light subfraction. Furthermore, the specific activity of the soluble subfraction is higher than that of the heavy subfraction. This pattern of distribution is consistent with the conclusion that these enzymes are located in the matrix. The higher specific activity found in the soluble subfraction as compared with that in the heavy one is readily explained by the fact that the latter is made up to an appreciable portion by inner membrane material. If these dehydrogenases were located in the outer membrane or the intermembrane space, as postulated by Green and associates,[3-6,9,10,20,21,56,58-60,142] one would not expect to find any significant activities in the heavy subfraction obtained by the present procedure.

Localization of Enzymes

Table 5-4 is a survey of current information concerning the intramitochondrial localization of various enzymes.

The enzymes so far found in the outer membrane represent a rather heterogeneous group from the functional point of view, and their localization in the outer membrane could not always be readily predicted from studies with intact mitochondria. For example, although the occurrence of monoamine oxidase in mitochondria has been known for a long time[12,31,54,130] it is only when the outer membrane has been isolated that its localization in this membrane has been recognized.[133] The function of this enzyme in mitochondria is not known. Kynurenine hydroxylase is another example of an enzyme of unexpected localization,[95,104] especially as most other NADPH-linked hydroxylases in the liver are associated with the endoplasmic reticulum (cf Ref. 94). An NAD-specific xylitol dehydrogenase was found by Arsenis et al.[8a] to be localized in the outer membrane or the intermembrane space of guinea pig liver mitochondria. A corresponding NADP-specific enzyme was located in the inner membrane or the matrix. It was pointed out that the two xylitol dehydrogenases may mediate hydrogen transfer between extramitochondrial NAD and intramitochondrial NADP.

A rotenone-, amytal-, and antimycin A-insensitive NADH-cytochrome c reductase,[37,38,41,42,44,79,122] separated from the respiratory chain[41,42,77-79,93] and similar to the NADH-cytochrome b_5 reductase system of microsomes,[92,126-128] has long been known to occur in mitochondria, and its recently-established localization in the outer membrane[144-146] fits logically with its preferential reactivity with external NADH in the intact mitochondria. The function of this enzyme system or of its microsomal counterpart is not yet known, although, in the latter case, its possible involvement in the desaturation of fatty acids has recently been suggested.[105]

The localization of certain ATP- or ADP-involving enzyme reactions in the outer membrane—including an ATP-dependent fatty acyl-CoA synthetase[101,163] (specific for long-chain fatty acids[1,2]), adenylate kinase,[17,89,134,135,146] and nucleoside diphosphokinase[89,134,135] (the latter two possibly being located in the intermembrane space)—is in accordance with the findings that these reactions when assayed in the intact mitochondria with external ATP or ADP as substrate are insensitive to atractylate.[27,53,154,163] Nucleoside monophosphokinase has been concluded to occur in the outer membrane or the intermembrane space, although the bulk of the enzyme is found in the inner membrane or the matrix.[89] A dual localization has also been reported in the case of the mitochondrial fatty acid elongating system, with a preferential activity for C_{14} and C_{16} fatty acids in the outer mem-

TABLE 5-4*
LOCALIZATION OF ENZYMES IN LIVER MITOCHONDRIA

Outer membrane	Intermembrane spece	Inner membrane	Matrix
"Rotenone-insensitive" NADH-cytochrome c reductase[14,86,87,113,134,144-46] [NADH-cytochrome b_5 reductase;[144-146] cytochrome b_5[95,113]]	Adenylate kinase[17,89,134,135,146]	Respiratory chain [cytochromes b, c_1, c, a, a_3;[113] succinate dehydrogenase;[17,133,134] succinate-cytochrome c reductase;[134,144-146] succinate oxidase;[85-87] "rotenone-sensitive" NADH-cytochrome c reductase[86,87,144-146] NADH oxidase;[85-87] choline-cytochrome c reductase;[71] cytochrome c oxidase,[86,111,134,144-146] respiratory chain-linked phosphorylation[23,87,134,135]]	Malate dehydrogenase[17,113,133,134,146]
Monoamine oxidase[14,17,47,133,134,146]	Nucleoside diphosphokinase[89,134,135]		Isocitrate dehydrogenase (NADP-spec)[17,113,134,146]
Kynurenine hydroxylase[95,104,134]	Nucleoside monophosphokinase[89]		Isocitrate dehydrogenase (NAD-spec)[134]
ATP-dependent fatty acyl-CoA synthetase[52,101,163]	Xylitol dehydrogenase (NAD-spec.)?[8a]		Glutamate dehydrogenase[17,101,113,134,146]
Glycerolphosphate-acyl transferase[148,160]		β-Hydroxybutyrate dehydrogenase[101,134]	α-Ketoglutarate dehydrogenase[134] [lipoyl dehydrogenase[134]]
Lysophosphatidate-acyl transferase[148]		Ferrochelatase[70,91]	Citrate synthetase[17]
Lysolecithin-acyl transferase[148]		δ-Aminolevulinic acid synthetase?[91]	Aconitase[113]
Cholinephosphotransferase[148]		Carnitine palmityl-transferase[101]	Fumarase[17,113]
Phosphatidate phosphatase[148]		Fatty acid oxidation system?[13]	Pyruvate carboxylase[109]
Phospholipase A_{II}[155]		Fatty acid elongation system[29,122a]	Phosphopyruvate carboxylase[109]
Nucleoside diphosphokinase[89,134,135]		Xylitol dehydrogenase (NADP-spec)?[8a]	Aspartate aminotransferase[134]
Fatty-acid elongation system[29]			Ornithine-carbamoyl transferase[134]
Xylitol dehydrogenase (NAD-spec.)?[8a]			Fatty acyl-CoA synthetase(s)?[1,2,163]
			Fatty acid oxidation systems?[13] [β-hydroxybutyryl-CoA dehydrogenase[17]]
			Xylitol dehydrogenase (NADP-spec)?[8a]

* The data in this table are based on results obtained in fractionation studies involving separation of the outer and inner mitochondrial membranes. Further information concerning the intramitochondrial localization of enzymes, originating from studies with intact mitochondria, is discussed in the text. For definition of mitochondrial compartments, see Figure 5-2.

brane, and for C_{10} fatty acid in the inner membrane.[29] Quagliariello et al.[122a] have concluded that the enzymes of fatty-acid elongation are located in the inner membrane. The occurrence of a fatty-acid synthetizing system in both the outer and inner membrane of rabbit heart mitochondria has been reported by Whereat et al.[156] Pette[116] and Klingenberg and Pfaff [73] have concluded that creatine kinase in heart mitochondria has a localization similar to that of adenylate kinase in liver mitochondria. Rose and Warms[132] have found that the outer membrane of liver mitochondria contains the majority of the mitochondrial binding sites for hexokinase. In ascites tumor mitochondria, which contain bound hexokinase, the enzyme reaction is insensitive to atractylate with external ATP as substrate, indicating that it is located in the outer membrane.

The outer membrane also contains the enzymes glycerolphosphate-,[148,160] lysophosphatidate-,[148] and lysolecithin-[148] acyl transferase, cholinephosphotransferase,[148] as well as phosphatidate phosphatase[148] and phospholipase A_{II}[155] (the latter being distinct from the phospholipase A_I present in microsomes[153]), suggesting a role of the outer membrane in phospholipid metabolism. Such a function is further indicated by recent reports of Bygrave and associates who have found that the incorporation of ^{14}C-choline[71] and ^{14}C-serine[19] into the phospholipid of isolated rat-liver mitochondria takes place primarily in the outer membrane. This situation is in contrast to that found for the incorporation of amino acids into protein which takes place exclusively in the inner membrane.[15,99,161]

In agreement with early morphological [106,107] and histochemical [11] indications, the inner mitochondrial membrane is the site of the respiratory chain catalysts[17,85–87,111,113,144–146] as well as the associated phosphorylation process;[23,87,134,135] the latter appears to be partly located in the projecting subunits of Fernández-Morán[47] as revealed by studies of Racker and associates.[124,125] A strict orientation of the respiratory chain within the inner membrane is indicated by the fact that NADH [74,79] and succinate[64,123] can reach the chain preferentially from the inside, and cytochrome c preferentially from the outside.[24,75,83] A portion of cytochrome c is readily released from the mitochondria upon treatments leading to a disruption of the outer membrane.[69,69a,73,121] This portion of cytochrome c is unlikely to be localized in the outer membrane or the intermembrane space because of its lack of reactivity with the outer-membrane NADH-cytochrome c reductase system,[145] and is thus probably loosely associated with the inner membrane. Whereas the succinate,[17,144–146] NADH,[17,144–146] choline,[71] and glycerol phosphate[116] dehydrogenases are tightly bound to the membrane, other flavoproteins such as lipoyl dehydrogenase,[65,134] sarcosine dehydrogenase,[50] as well as the flavoenzymes involved in fatty-acid oxidation[51,52] are either easily dissociated or located in the matrix. The nicotinamide nucleotide transhydrogenase[36,76] and β-hydroxybutyrate dehydrogenase[101,134] are firmly associated with the inner membrane.

Of great interest are recent reports concerning the localization of ferrochelatase[70,91] and δ-aminolevulinic acid synthetase[91] in the inner membrane. These findings suggest a role of the inner membrane in heme and porphyrin synthesis, with a possible regulatory function in the ribosomal synthesis of cytochromes.

As already mentioned, the inner membrane is believed to be the site of specific translocators for various charged molecules of physiological significance which cannot readily diffuse through this membrane. Specific anion-translocators described up to now (cf Ref. 26) include those for phosphate (and arsenate), for malate and succinate (and certain nonphysiological dicarboxylic acids, but not fumarate), for α-ketoglutarate, for citrate, isocitrate, and *cis*-aconitate, for glutamate, for aspartate, as well as the atractylate-sensitive ATP–ADP exchange carrier system. From a functional point of view, the enzyme fatty acyl-CoA-carnitine transferase,[101] which is associated with the inner membrane and is responsible for the translocation of long-chain fatty acids, also belongs in this category of catalysts. The inner membrane also contains the binding sites for various divalent cations involved in their energy-linked translocation across the inner mitochondrial membrane[18,61,150] as well as for different ionophoric antibiotics, such as valinomycin or gramicidin, that facilitate the penetration of univalent cations (cf Ref. 82).

Enzymes believed to be localized in the matrix include those involved in the citric acid cycle (except succinate dehydrogenase)[17,73,113,133,134,146] and related processes such as substrate-level phosphorylation, the pyruvate and phosphopyruvate carboxylase reactions,[109] glutamate oxidation,[17,73,101,109,134,146] transamination,[134] citrulline synthesis,[134] a GTP- and an ATP-dependent fatty acyl-CoA synthetase reaction[163] (the latter specific for medium- and short-chain fatty acids),[1,2] and fatty acid oxidation.[13,17] The evidence for the localization of these enzymes, which is in sharp contrast to the conclusions reached by Green and associates (see p. 199), rests, as was already pointed out, on a combination of information from fractionation studies (see Table 5-4) and experience with intact mitochondria. In the intact mitochondria, these enzymes show so-called "latency," i.e., they reveal only limited activity towards externally added substrates or coenzymes which do not readily penetrate the inner mitochondrial membrane or whose penetration, by way of a specific translocator, has been blocked. Upon removal of the outer membrane, part or the whole of these enzymes will remain in the inner-membrane fraction (depending on the degree of damage to the inner membrane and consequent leakage of matrix occurring during the fractionation procedure) and may show latency just as in the intact mitochondria. Subsequent fragmentation of the inner membrane by suitable mechanical or chemical means will abolish the latency, and the bulk of the matrix enzymes will appear in the soluble fraction.

The matrix seems to be the site of mitochondrial DNA, as revealed by electron miscroscopic evidence.[98] Mitochondrial RNA has been found to be localized mainly in the "inner membrane fraction" (probably including part of the matrix) with a minor portion of the RNA being found in the outer membrane.[15] Although the localization of the mitochondrial DNA-dependent RNA polymerase and amino-acid activating enzyme system has not yet been established, the fact, already mentioned, that the amino-acid incorporation into protein catlyzed by isolated mitochondria is recovered exclusively in the inner membrane fraction,[15,99,161] together with the localization of the mitochondrial DNA in the matrix,[98] strongly suggests that the mitochondrial RNA- and protein-synthetizing system as a whole is located in the matrix space.

Are the Citric Acid Cycle and Related Enzymes Associated with the Outer Membranes?

As already pointed out, Green and associates[3-6,9,10,20,21,56,58-60,142] have based their methods for the separation and identification of the outer and inner mitochondrial membranes primarily on biochemical rather than morphological criteria. These, in turn, rested essentially on three postulates: (1) that the inner membrane is the site of the respiratory chain, (2) that little if any mitochondrial protein is not membrane-bound, and (3) that any enzyme found in the soluble form after subfractionation of beef heart or liver mitochondria by various procedures originates from the outer membrane. Based on these postulates, the conclusion was reached that all enzymes involved in the citric acid cycle (with the exception of succinate dehydrogenase), in fatty acid oxidation and elongation, and in substrate-level phosphorylation, are in the intact mitochondrion associated with the outer membrane. It was further concluded that the outer membrane constitutes the permeability barrier for nicotinamide nucleotides as well as the site for the carnitine-mediated transport of fatty acids and the atractylate-sensitive translocation of ADP and ATP.[7,8]

While postulate (1) above is supported by ample evidence from other laboratories,[11,17,85-87,106,111,113,133,134,144-146] postulate (2) remains hypothetical and may, indeed, be difficult to prove experimentally. Regarding postulate (3), the evidence so far provided by Green and associates comes from attempts to recover citric-acid cycle and related enzymes in a particulate submitochondrial fraction devoid of respiratory chain activity. Such a fraction, by definition, would represent the outer membrane. The seemingly strongest evidence of this kind published by Green and associates up to now comes from experiments[59] in which sonication, in the absence or presence of cholate, was used to disrupt beef heart mitochondria, and the resulting subfractions were separated by differential centrifugation. The postulated "outer membrane" subfractions, F_s obtained in the absence of cholate, and F_c in its presence, were reported to exhibit various citric-acid cycle enzyme activities that were two to three times higher, on a protein basis, than that of the sonicated starting material. However, closer examination of the data reveals that the yield of the various enzyme activities recovered was relatively low, ranging between 8 and 23% in the case of F_s, and between 30 and 40% in the case of F_c. Moreover, cytochrome *a*, which was used as a marker for the respiratory chain, was recovered to an extent ranging between 4 and 20% in F_s, and between 23 and 38% in F_c. These data thus show no convincing concentration of citric-acid cycle enzymes over those of the respiratory chain in the postulated "outer membrane" subfractions; however, they do suggest that these subfractions represent inner membrane vesicles including some matrix. The discrepancies regarding the protein and phospholipid contents of the "outer membrane" subfraction of Green and associates as compared to those found by other laboratories (see p. 191), may have a similar explanation.

In another instance, Green and associates, using phospholipase to disrupt mitochondria, described [4] a particulate subfraction devoid of cytochrome *a* but containing 50–60% of the mitochondrial α-ketoglutarate, pyruvate, and β-hydroxybutyrate dehydrogenase activities, while other citric-acid cycle and related enzyme activities were found exclusively in the soluble subfraction. These findings were interpreted by postulating that the two α-keto acid dehydrogenases, which are high-molecular weight enzyme complexes, are firmly associated with the outer membrane, whereas other citric acid cycle enzymes, which represent single proteins of relatively low molecular weight, are easily detachable components of the outer membrane, residing in the postulated inner-surface subunits of the latter. While this remains a hypothetical possibility, the converse assumption, namely, that the two α-keto acid dehydrogenase complexes may become attached to, or cosediment with, the outer membrane during the fractionation procedure, can clearly not be excluded. An even more complicated situation seems to prevail in the case of the β-hydroxybutyrate dehydrogenase where, as Green *et al.*[59] have pointed out, the enzyme is recovered in the outer membrane fraction only with the phospholipase procedure, whereas with the sonication method, it is recovered in the inner membrane fraction. According to Green *et al.*[59] this inconsistency can be explained by an artifactual redistribution of the enzyme during the sonication procedure, owing to a specific "relocation" of β-hydroxybutyrate dehydrogenase from the outer to the inner membrane. In view of the nature of the two procedures, it would appear more logical to assume that the phospholipase treatment, which is a chemical intervention, is more likely to cause such a redistribution of an enzyme (in particular of a phospholipid-dependent enzyme such as β-hydroxybutyrate dehydrogenase[136]) than is mechanical disruption by sonication.

A critical evaluation of the sonication and phospholipase procedures has recently been published by Brdiczka *et al.*[17]

Localization of Nicotinamide Nucleotides

Perhaps the most compelling evidence in favor of the localization of the mitochondrial nicotinamide nucleotide-linked dehydrogenases inside the inner membrane comes from information relating to the localization of the mitochondrial nicotinamide nucleotides themselves. It is well established—and Green has been the pioneer in establishing this[33,55]—that all mitochondrial nicotinamide nucleotide-linked dehydrogenase reactions, including the oxidations of the citric-acid cycle, fatty acids, and glutamate, are catalyzed by isolated, intact mitochondria at maximal rate via endogenous nicotinamide nucleotides without a need for supplementation with externally added coenzymes. Therefore, if the mitochondrial nicotinamide nucleotides were located inside the inner membrane and incapable of rapidly penetrating the latter, then also the nicotinamide nucleotide-linked dehydrogenases ought to be located inside the inner membrane. If the dehydrogenases resided outside the inner membrane, it would be impossible for them to interact with the mitochondrial nicotinamide nucleotides at adequate rates. The following lines of evidence strongly support the conclusion that the mitochondrial nicotinamide nucleotides are, indeed, located inside the inner membrane and do not readily penetrate the latter:

(1) External nicotinamide nucleotides readily penetrate the "sucrose space" of mitochondria (i.e., the intermembrane space),[73] but only very slowly exchange with the intramitochondrial nicotinamide nucleotides.[62]

(2) Incubation of liver mitochondria at 30° in a hypotonic medium containing EDTA results in an almost complete release of adenylate kinase, an enzyme located in the outer membrane or the intermembrane space, in less than 1 min. Under the same conditions, practically no nicotinamide nucleotides are released from the mitochondria even after 3 hrs.[121]

(3) External nicotinamide nucleotides interact only slowly, if at all, with intramitochondrial dehydrogenases, and this interaction is greatly enhanced by agents that cause a swelling or disruption of the mitochondria.[39,45,46,66] Contraction of the mitochondria restores the "latency" of these enzymes.[39,45,46]

(4) Contraction of swollen liver mitochondria, i.e., selective restoration of the inner membrane, results in a restricted accessibility of external nicotinamide nucleotides to the mitochondrial dehydrogenases.[38,40] When present during contraction, NAD is reincorporated into the mitochondria.[69]

(5) "Digitonin particles," which presumably represent derivatives of mitochondria devoid of outer membrane,[85,96,133] contain endogenous NAD, and catalyze the oxidation of β-hydroxybutyrate, but are inaccessible to external nicotinamide nucleotides.[30,74]

(6) Added cytochrome c enhances the aerobic oxidation of external NADH by intact liver mitochondria by a process that is insensitive to

amytal,[42] rotenone,[41] and antimycin A,[37,79,122] and which involves an NADH-cytochrome b_5 reductase system associated with the outer membrane.[144-146] Addition of NAD and cytochrome c fails to restore amytal-inhibited respiration with NAD-linked substrates unless the respective dehyrogenase also is added from the outside.[38]

The above lines of evidence, together with the conclusions already discussed regarding the localization of the various substrate-translocating systems in the inner membrane, constitute the strongest argument presently available at the level of the intact mitochondrion against the localization of the citric-acid cycle and related enzymes in the outer membrane.

VI. SOME PROBLEMS RELATED TO THE LOCALIZATION OF THE ROTENONE-INSENSITIVE NADH-CYTOCHROME C REDUCTASE AND MONOAMINE OXIDASE

Recently, Green and associates[5,58] have seriously criticized the conclusions of other laboratories regarding the localization of the rotenone-insensitive NADH-cytochrome c reductase and monoamine oxidase in the outer membrane of liver mitochondria. They claimed that both of these conclusions were based on experimental artifacts, arising, in one case, from the use of impure preparations of mitochondria, and in the other, from the application of an unsuitable enzyme assay.

The original conclusion[144-146] that the rotenone-insensitive NADH-cytochrome c reductase is a true constituent of the outer membrane of liver mitochondria rather than merely a microsomal contaminant—a conclusion later confirmed in several laboratories[86,113,134]—is supported by the following evidence:

(1) Twice-washed rat-liver mitochondria exhibit about ten times higher activity, on a protein basis, of this enzyme than of other microsomal enzymes such as glucose-6-phosphatase or various NADPH-linked enzyme systems.[144,145] Such a concentration of NADH-cytochrome c reductase in microsomes in relation to other enzymes has not been found by any physical procedure hitherto employed to subfractionate microsomes, including extensive sonication to produce vesicles of a diameter of one-tenth of the original microsomes.[34]

(2) The isolated outer membrane fraction exhibits little or no glucose-6-phosphatase or NADPH-linked enzyme activities.[144,145] Its rotenone-insensitive NADH-cytochrome c reductase activity is higher, on a protein basis, than that of both microsomes and microsomal subfractions obtained after sonication.

(3) The properties of the components of the outer-membrane NADH-cytochrome c reductase differ in several respects from those of the corresponding microsomal enzyme system, including sensitivity of the NADH-cytochrome b_5 reductase to dicoumarol,[126-128,145] the firmness of the association of cytochrome b_5 with the membrane,[126-128,145] the reducibility of the

cytochrome by cysteine,[126–128,145] and the low-temperature spectrum[113] and electrophoretic mobility[36a] of the cytochrome.*

The objection raised by Green et al.[5,58] against the above conclusion was based on the finding that further washing of twice-washed liver mitochondria, prepared and washed at relatively high centrifugal forces (10–15,000xg), removed the rotenone-insensitive NADH-cytochrome c reductase activity along with that of glucose-6-phosphatase, i.e., with the removal of residual microsomal contamination. Ernster and Kuylenstierna[43] have recently carried out similar experiments, using both the centrifugal force employed in their earlier work (6000xg),[145,146] and the high centrifugal force employed by Green et al.[5,58] With the 6000xg preparations, the rotenone-insensitive NADH-cytochrome c reductase activity remained virtually constant between zero and four washes, while the glucose-6-phosphatase activity decreased exponentially. Similar results were recently reported by Beattie.[14] After two to four washes, the rotenone-insensitive NADH-cytochrome c reductase activity of the mitochondria was 40 to 50%, and the glucose-6-phosphatase activity 4–5% of that of the microsomes. This was in good agreement with the values reported earlier[144,145] for the twice-washed mitochondria (37 and 4%, respectively). Furthermore, the NADPH-cytochrome c reductase activity of the 6000xg preparations decreased to practically nil after only one wash. This finding suggests that contaminating microsomes are readily removed from the mitochondria by a single wash, and that the residual glucose-6-phosphatase activity may originate from nonmicrosomal—possibly lysosomal—material, exhibiting presumably nonspecific phosphatase activity. The 15,000xg preparations exhibited considerably higher rotenone-insensitive NADH-cytochrome c reductase, glucose-6-phosphatase, and NADPH-cytochrome c reductase activities than did those obtained at 6000xg, and these were not efficiently removed even after four washes. These preparations, which evidently were heavily contaminated with both microsomes and nonmicrosomal material, are apparently not suitable for studies requiring pure cell fractions, and any conclusions or criticisms based on the use of such preparations may be of doubtful validity.

The conclusion that monoamine oxidase is located in the outer membrane originates from studies of Schnaitman et al.[133] who have used the

---

* Note added in the proofs. A further difference between the mitochondrial, rotenone-insensitive, and the microsomal NADH-cytochrome c reductase systems is the much greater sensitivity of the mitochondrial system to trypsin (B. Kuylenstierna, D. G. Nicholls, S. Hovmöller and L. Ernster, unpublished results). Adenylate kinase also is readily inactivated by trypsin, but only provided the mitochondria are exposed to hypotonic treatment for a short period of time; under the same conditions, the respiratory chain and various dehydrogenase activities remain unaffected by trypsin. These findings provide additional support to the conclusions that (1) the mitochondrial rotenone-insensitive NADH-cytochrome c reductase system is a true outer-membrane constituent rather than an artifact arising from microsomal contamination; (2) adenylate kinase is located in the intermembrane space.

method of Tabor *et al.*[149] to assay the enzyme in submitochondrial fractions obtained by the digitonin procedure. This method is based on the spectrophotometric measurement of benzaldehyde formed from benzylamine in the course of the monoamine oxidase reaction. The conclusion of Schnaitman *et al.*[133] has been confirmed by Sottocasa *et al.*[146] using the same enzyme assay and submitochondrial fractions obtained by the swelling–contraction–sonication procedure. Green *et al.*[5,58] have pointed out that the enzyme assay based on the determination of an aldehyde may give misleading results in the present connection, owing to the presence of aldehyde dehydrogenase in the inner-membrane subfraction which would interfere with the demonstration of any monoamine oxidase activity associated with this fraction. By employing another enzyme assay, devised by McCaman *et al.*[90] and based on the use of ^{14}C-tyramine as the substrate for monoamine oxidase, they arrived at the conclusion that the enzyme is concentrated in the inner-membrane subfraction of liver mitochondria obtained by either phospholipase treatment or oleate-induced swelling and subsequent contraction in the presence of ATP.

Clearly, the criticism of Green *et al.*[5,58] regarding the use of the enzyme assay based on aldehyde determination is perfectly valid, and this assay *would* lead to false conclusions *if* the monoamine oxidase were located in the inner membrane. However, the data reported by Green *et al.*,[5,58] from which they conclude that the latter indeed is the case, are, in turn, of doubtful validity, because of the lack of convincing documentation of their procedures employed for the subfractionation of mitochondria and the identity and purity of the subfractions. Indeed, using the fractionation procedure of Sottocasa *et al.*[146] and an enzyme assay similar to that employed by Green *et al.*,[5,58] involving the use of ^{14}C-tryptamine,[162] Ernster and Kuylenstierna[43] have recently obtained clear evidence for the concentration of monoamine oxidase in the outer membrane of rat-liver mitochondria, in agreement with the earlier conclusion of Schnaitman *et al.*[133] Independently, Beattie[14] and Schnaitman and Greenawalt[134] have reported similar evidence, based on the ^{14}C-tyramine method. In addition, the latter authors have shown that chloral hydrate, in a concentration that inhibits aldehyde dehydrogenase to an extent of 80–90%, did not alter the distribution pattern of monoamine oxidase as determined by the benzaldehyde assay.

The criticism raised by Green *et al.*[5,58] against the conclusions of other laboratories regarding the localization of rotenone-intensitive NADH-cytochrome *c* reductase and of monoamine oxidase thus seems to lack valid experimental ground.

VII. CONCLUSION

For a long time, our knowledge of the chemical and enzymatic organization of mitochondria has been based almost exclusively on information from studies with intact mitochondria. Although submitochondrial particle preparations of various kinds, exhibiting respiratory enzyme activity and identified with fragments of the inner mitochondrial membrane, have been studied extensively over the past years, relatively little attention has been paid to the question as to the chemical and enzymatic composition of the rest of the mitochondrion, namely, the outer membrane and the spaces surrounded by the two membranes.

Methods developed in recent years for the separation of the mitochondrial inner and outer membranes have made it possible to approach this problem. In fact, progress has been amazingly rapid in this field, and, as a result, there is now information available on the intramitochondrial distribution of almost all enzymes and other chemical constituents so far known to be present in mitochondria. Insofar as the methods employed have been adequately documented morphologically as well as biochemically, they have yielded results of remarkable reproducibility and agreement among various laboratories. Moreover, the results of the majority of fractionation studies are in excellent agreement with observations with intact mitochondria.

There is still some uncertainty regarding the localization of those enzymes which are recovered in the soluble form after the separation of the two membranes, and it is difficult in some cases to decide whether a given enzyme is present in the intact mitochondrion in the truly soluble form or as loosely attached to one of the membranes. On the other hand, the available information, in most cases, allows a clear distinction between enzymes located in the outer membrane or the intermembrane space, and those located in the inner membrane or the matrix.

The inner membrane and the matrix constitute the mitochondrion in the classical functional sense, in that this "inner compartment" embodies the enzymic complements of substrate oxidation, respiration, and energy conservation. It also constitutes the mitochondrion as a metabolic entity with its permeability barriers and specific ion-translocating systems, as well as an organelle with its own machinery for DNA, RNA, and protein synthesis, endowing it with a certain extent of genetic autonomy. It is now quite obvious that ideas put forward in recent years concerning the possible bacterial origin of mitochondria (cf Refs. 81, 131) are pertinent to this inner mitochondrial compartment only.

The available information concerning the enzymic composition of the "outer compartment" of mitochondria, i.e., the outer membrane and the intermembrane space, suggests a rather diversified function which, in most cases, lacks an obvious metabolic relationship to the function of the inner compartment. In fact, the ready permeability of the outer membrane to substances of low-molecular weight suggests that the outer compartment of the mitochondrion in the intact cell is in equilibrium with the rest of the cytoplasm with respect to free metabolites.

Several authors have pointed out certain similarities in enzyme composition between the outer mitochondrial membrane and the endoplasmic reticulum[133,146,148] and discussed these in relation to a possible phylogenetic and/or ontogenetic relationship between the two membranes.[113,146] It would be interesting in the future to extend these studies to a comparison of the chemical and enzymic properties of the outer mitochondrial membrane with other intracellular membranes as well, such as the nuclear, lysosomal, and Golgi membranes, and with the plasma membrane.

REFERENCES

1. Aas, M. (1968), *Abstr. 5th FEBS Meeting,* Prague.
2. Aas, M., and Bremer, J. (1968), *Biochim. Biophys. Acta* **164,** 157.
3. Allmann, D. W., and Bachmann, E. (1967), *Methods Enzymol.* **10,** 438.
4. Allmann, D. W., Bachmann, E., and Green, D. E. (1966), *Arch. Biochem. Biophys.* **115,** 165.
5. Allmann, D. W., Bachmann, E., Orme-Johnson, N., Tan, W. C., and Green, D. E. (1968), *Arch. Biochem. Biophys.* **125,** 981.
6. Allmann, D. W., Galzinga, L., McCaman, R. E., and Green, D. E. (1966), *Arch. Biochem. Biophys.* **117,** 413.
7. Allmann, D. W., Harris, R. A., and Green, D. E. (1967), *Arch. Biochem. Biophys.* **120,** 693.
8. Allmann, D. W., Harris, R. A., and Green, D. E. (1967), *Arch. Biochem. Biophys.* **122,** 766.
8a. Arsenis, C., Maniatis, T., and Touster, O. (1968), *J. Biol. Chem.* **243,** 4396.
9. Bachmann, E., Allmann, D. W., and Green, D. E. (1966), *Arch. Biochem. Biophys.* **115,** 153.
10. Bachmann, E., Lenaz, G., Perdue, J. F., Orme-Johnson, N., and Green, D. E. (1967), *Arch. Biochem. Biophys.* **121,** 73.
11. Barrnett, R. J. (1962), "Enzyme Histochemistry," p. 537, M. S. Burstone, ed. Academic Press, Inc., New York.
12. Baudhuin, P., Beaufay, H., Rahman-Li, Y., Sellinger, O., Wattiaux, R., Jacques, P., and de Duve, C. (1963), *Biochem. J.* **92,** 179.
13. Beattie, D. S. (1968), *Biochim. Biophys. Res. Commun.* **30,** 57.
14. Beattie, D. S. (1968), *Biochim. Biophys. Res. Commun.* **31,** 901.
15. Beattie, D. S., Basford, R. E., and Koritz, S. B. (1967), *Biochemistry* **6,** 3099.
16. Bemis, J. A., Bryant, G. M., Arcos, J. C., and Argus, M. F. (1968), *J. Mol. Biol.* **33,** 299.
17. Brdiczka, D., Pette, D., Brunner, G., and Miller, F. (1968), *European J. Biochem.* **5,** 294.
18. Brierley, G. P., and Slautterback, D. B. (1964), *Biochim. Biophys. Acta* **82,** 183.
19. Bygrave, F. L., and Bücher, T. (1968), *Abstr. 5th FEBS Meeting,* Prague.
20. Byington, K. H., Morey, A. V., and Smoly, J. (1968), *Fed. Proc.* **27,** 461.

21. Byington, K. H., Smoly, J. M., Morey, A. V., and Green, D. E. (1968), *Arch. Biochem. Biophys.* **128**, 762.
22. Caplan, A. I., and Greenawalt, J. W. (1966), *J. Cell Biol.* **31**, 455.
23. Caplan, A. I., and Greenawalt, J. W. (1968), *J. Cell Biol.* **36**, 15.
24. Carafoli, E., and Muscatello, U. (1968), *Abstr. 5th FEBS Meeting*, Prague.
25. Chance, B., and Packer, L. (1958), *Biochem. J.* **68**, 295.
26. Chappell, J. B. (1968), *Brit. Med. Bull.* **24**, 150.
27. Chappell, J. B., and Crofts, A. R. (1965), *Biochem. J.* **95**, 707.
28. Chappell, J. B., and Crofts, A. R. (1966), "Regulation of Metabolic Processes in Mitochondria," B.B.A. Library, Vol. 7, p. 293; Tager, J. M., Papa, S., Quagliariello, E., and Slater, E. C., eds., Elsevier Publ. Co., Amsterdam.
29. Colli, W., Hinkle, P., and Pullman, M. E. (1968), *Fed. Proc.* **27**, 648.
30. Cooper, C., and Lehninger, A. L. (1956), *J. Biol. Chem.* **219**, 489.
31. Cotzias, G., and Dole, V. (1951), *Proc. Exptl. Biol. Med.* **78**, 157.
32. Crane, F. L., Stiles, J. W., Prezbindowski, K. S., Ruzicka, F. J., and Sun, F. F. (1968), "Regulatory Functions of Biological Membranes," B.B.A. Library, Vol. 11, p. 21, Järnefelt, J., ed., Elsevier Publ. Co., Amsterdam.
33. Cross, R. J., Taggart, J. V., Covo, A. G., and Green, D. E. (1950) *J. Biol. Chem.* **177**, 655.
34. Dallman, P. R., Dallner, G., Bergstrand, A., and Ernster, L. (1969), *J. Cell Biol.* **41**, 357.
35. Danielli, J. F., and Davson, H. (1935), *J. Cell Comp. Physiol.* **5**, 495.
36. Danielson, L., and Ernster, L. (1963), *Biochem. Z.* **338**, 188.
36a. Davis, K. A., and Kreil, G. (1968), *Biochim. Biophys. Acta* **162**, 627.
37. de Duve, C., Pressman, B. C., Gianetto, R., Wattiaux, R., and Appelmans, F. (1955), *Biochem. J.* **60**, 604.
38. Ernster, L. (1956), *Exptl. Cell Res.* **10**, 721.
39. Ernster, L. (1959), *Biochem. Soc. Symp.* **16**, 54.
40. Ernster, L. (1967), "Mitochondrial Structure and Compartmentation," p. 341, Quagliariello, E., Papa, S., Slater, E. C., and Tager, J. M., eds., Adriatica Editrice, Bari.
41. Ernster, L., Dallner, G., and Azzone, G. F. (1963), *J. Biol. Chem.* **238**, 1124.
42. Ernster, L., Jalling, O., Löw, H., and Lindberg, O. (1955), *Exptl. Cell Res. Suppl.* **3**, 124.
43. Ernster, L., and Kuylenstierna, B. (1968), "Mitochondria—Structure and Function," Ernster, L., and Drahota, Z., eds., Proc. 5th FEBS Meeting, Prague, Publ. Czechosl. Acad. Sci., Prague, and Academic Press, London, in press.
44. Ernster, L., Löw, H., and Lindberg, O. (1955), *Acta Chem. Scand.* **9**, 200.
45. Ernster, L., and Navazio, F. (1956), *Acta Chem. Scand.* **10**, 1038.
46. Ernster, L., and Navazio, F. (1956), *Exptl. Cell Res.* **11**, 483.
47. Fernández-Morán, H. (1962), *Circulation* **26**, 1039.
48. Fernández-Morán, H., Oda, T., Blair, P. V., and Green, D. E. (1964), *J. Cell Biol.* **22**, 63.
49. Fleischer, S., Fleischer, B., and Stoeckenius, W. (1967), *J. Cell Biol.* **32**, 193.

50. Frisell, W. R., Patwardhan, M. V., and Mackenzie, C. M. (1965), *J. Biol. Chem.* **240**, 1829.
51. Garland, P. B., Chance, B., Ernster, L., Lee, C. P., and Wong, D. (1967) *Proc. Natl. Acad. Sci.* **58**, 1696.
52. Garland, P. B., Haddock, B. A., and Yates, D. W. (1968), "Mitochondria —Structure and Function," Ernster, L., and Drahota, Z., eds., Proc. 5th FEBS Meeting, Prague, Publ. Czechosl. Acad. Sci., Prague, and Academic Press, London, in press.
53. Garland, P. B., and Yates, D. W. (1967), *Round Table Discussion on Mitochondrial Structure and Compartmentation*, p. 385 Quagliariello, E., Papa, S., Slater, E. C., and Tager, J. M., eds., Adriatica Editrice, Bari.
54. Gorkin, V. (1966), *Pharmacol. Rev.* **18**, 115.
55. Green, D. E. (1951), *Biol. Rev.* **26**, 410.
56. Green, D. E. (1965), "Oxidases and Related Redox Systems" Vol. II, p. 1061, King, T. E., Mason, H. S., and Morrison, M., eds., John Wiley & Sons, Inc., New York.
57. Green, D. E. (1966), "Comprehensive Biochemistry," Vol. 14, p. 309, Florkin, M., and Stotz, E., eds., Elsevier Publ. Co., Amsterdam.
58. Green, D. E., Allmann, D. W., Harris, R. A., and Tan, W. C. (1968), *Biochem. Biophys. Res. Commun.* **31**, 368.
59. Green, D. E., Bachmann, E., Allmann, D. W., and Perdue, J. F. (1966), *Arch. Biochem. Biophys.* **115**, 172.
60. Green, D. E., and Perdue, J. F. (1966), *Ann. N.Y. Acad. Sci.* **137**, 667.
61. Greenawalt, J. W., Rossi, C. S., and Lehninger, A. L. (1964), *J. Cell Biol.* **23**, 21.
62. Greenspan, M. D., and Purvis, J. L. (1965), *Biochim. Biophys. Acta* **99**, 191.
63. Hackenbrock, C. R. (1966), *J. Cell Biol.* **30**, 269.
64. Harris, E. J., van Dam, K., and Pressman, B. C. (1967), *Nature* **213**, 1126.
65. Hassinen, I., and Chance, B. (1968), *Biochem. Biophys. Res. Commun.* **31**, 895.
66. Hogeboom, G. H., and Schneider, W. C. (1953), *J. Biol. Chem.* **204**, 233.
67. Hoppel, C., and Cooper, C. (1968), *Biochem. J.* **107**, 367.
68. Huet, C., Lévy, M., and Pascaud, M. (1968), *Biochim. Biophys. Acta* **150**, 521.
69. Hunter, F. E., Malison, R., Bridgers, W. F., Schultz, B., and Atchison, A. (1959), *J. Biol. Chem.* **234**, 693.
69a. Jacobs, E. E., and Sanadi, D. R. (1960), *J. Biol. Chem.* **235**, 531.
70. Jones, M. S., and Jones, O. T. G. (1968), *Abstr. 5th FEBS Meeting*, Prague, 1968, *Biochem. Biophys. Res. Commun.* **31**, 977.
71. Kaiser, W., and Bygrave, F. L. (1968), *European J. Biochem.* **4**, 582.
72. Klingenberg, M. (1963), "Energy-linked Functions of Mitochondria," p. 381, Chance, B., ed., Academic Press, Inc., New York.
73. Klingenberg, M., and Pfaff, E. (1966), "Regulation of Metabolic Processes in Mitochondria," B.A.A. Library, Vol. 7, p. 180, Tager, J. M., Papa, S., Quagliariello, E., and Slater, E. C., eds., Elsevier Publ. Co., Amsterdam.
74. Lee, C. P. (1963), *Fed. Proc.* **22**, 527.

75. Lee, C. P., and Carlson, K. (1968), *Fed. Proc.* **27**, 828.
76. Lee, C. P., and Ernster, L. (1966), "Regulation of Metabolic Processes in Mitochondria," B.A.A. Library, Vol. 7, p. 218, Tager, J. M., Papa, S., Quagliariello, E., and Slater, E. C., eds., Elsevier Publ. Co., Amsterdam.
77. Lehninger, A. L. (1951), *J. Biol. Chem.* **190**, 345.
78. Lehninger, A. L. (1951), *Phosphorus Metab.* **1**, 344.
79. Lehninger, A. L. (1955), *Harvey Lectures* **49**, 176.
80. Lehninger, A. L. (1962), *Physiol. Rev.* **42**, 467.
81. Lehninger, A. L. (1965), "The Mitochondrion," W. A. Benjamin, Inc., New York.
82. Lehninger, A. L., Carafoli, E., and Rossi, C. S. (1967), *Adv. Enzymol.* **29**, 259.
83. Lenaz, G., and MacLennan, D. H. (1966), *J. Biol. Chem.* **241**, 5260.
84. Lévy, M., and Sauner, M.-T. (1967), *Compt. Rend. Soc. Biol.* **161**, 277.
85. Lévy, M., Toury, R., and André, J. (1966), *C.R. Acad. Sci. Paris, Sér. D* **262**, 1593.
86. Lévy, M., Toury, R., and André, J. (1966), *C.R. Acad. Sci. Paris, Sér. D* **263**, 1766.
87. Lévy, M., Toury, R., and André, J. (1967), *Biochim. Biophys. Acta* **135**, 599.
88. Lévy, M., Toury, R., Sauner, M.-T., and André, J. "Mitochondria—Structure and Function," Ernster, L., and Drahota, Z., eds., Proc. 5th FEBS Meeting, Prague, 1968, Publ. Czechosl. Acad. Sci., Prague, and Academic Press, London, in press.
89. Lima, M. S., Nachbaur, G., and Vignais, P. (1968), *C.R. Acad. Sci. Paris, Sér. D* **266**, 739.
90. McCaman, R. E., McCaman, M. W., Hunt, J. M., and Smith, M. S. (1965), *J. Neurochem.* **12**, 15.
91. McKay, R., Druyan, R., and Rabinowitz, M. (1968), *Fed. Proc.* **27**, 774.
92. Mahler, H. R., Raw, I., Molinari, R., and Ferreira do Amaral, D. (1958), *J. Biol. Chem.* **233**, 230.
93. Maley, G. F. (1957), *J. Biol. Chem.* **224**, 1029.
94. Mason, H. S. (1965), *Ann. Rev. Biochem.* **34**, 595.
95. Mayer, G., Ullrich, V., and Staudinger, H. (1968), *Hoppe Seyler's Z. Physiol. Chem.* **349**, 459.
96. Morton, D. J., Hoppel, C., and Cooper, C. (1968), *Biochem. J.* **107**, 377.
97. Munn, E. A., and Blair, P. V. (1967), *Z. Zellfrsch. Mikroskop. Anat.* **80**, 205.
98. Nass, M. M. K., and Nass, S. (1963), *J. Cell Biol.* **19**, 593.
99. Neupert, W., Brdiczka, D., and Bücher, T. (1967), *Biochem. Biophys. Res. Commun.* **27**, 488.
100. Newman, H. A. I., Gordesky, S. E., Hoppel, C., and Cooper, C. (1968), *Biochem. J.* **107**, 381.
101. Norum, K., Farstad, M., and Bremer, J. (1966), *Biochem. Biophys. Res. Commun.* **24**, 797.
102. Novikoff, A. B. (1961), "The Cell" Vol. II, p. 299, Brachet, J., and Mirsky, A. E., eds., Academic Press, New York.

103. Ohnishi, T., and Ohnishi, T. (1962), *J. Biochem.* (Japan), **51,** 380; **52,** 230.
104. Okamoto, H., Yamamoto, S., Nozaki, M., and Hayaishi, O. (1967), *Biochem. Biophys. Res. Commun.* **26,** 309.
105. Oshino, N., Imai, Y., and Sato, R. (1967), *Abstr. 7th Intern. Congr. Biochem.,* Tokyo.
106. Palade, G. E. (1952), *Anat. Record* **114,** 427.
107. Palade, G. E. (1953), *J. Histochem. Cytochem.* **1,** 188.
108. Palade, G. E. (1956), "Enzymes: Units of Biological Structure and Function," p. 185, Gaebler, O. H., ed., Academic Press, Inc., New York.
109. Papa, S., Landriscina, C., Lofrumento, N. E., and Quagliariello, E. (1968), *Abstr. 5th FEBS Meeting,* Prague.
110. Parsons, D. F. (1965), *Intern. Rev. Exptl. Pathol.* **4,** 1.
111. Parsons, D. F., Williams, G. R., and Chance, B. (1966), *Ann. N.Y. Acad. Sci.* **137,** 643.
112. Parsons, D. F., and Williams, G. R. (1967), *Meth. Enzymol.* **10,** 443.
113. Parsons, D. F., Williams, G. R., Thompson, W., Wilson, D., and Chance, B. (1967), "Round Table Discussion on Mitochondrial Structure and Compartmentation," p. 29, Quagliariello, E., Papa, S., Slater, E. C., and Tager, J. M., eds., Adriatica Editrice, Bari.
114. Parsons, D. F., and Yano, Y. (1967), *Biochim. Biophys. Acta* **135,** 362.
115. Penniston, J. T., Harris, R. A., Asai, J., and Green, D. E. (1968), *Proc. Natl. Acad. Sci.* **59,** 624.
116. Pette, D. (1966), "Regulation of Metabolic Processes in Mitochondria," B.B.A. Library, Vol. 7, p. 28, Tager, J. M., Papa, S., Quagliariello, E., and Slater, E. C., eds., Elsevier Publ. Co., Amsterdam.
117. Pfaff, E. (1965), *Unspezifische Permeabilität und spezifischer Austausch der Adeninnucleotide als Beispiel mitochondrialer Compartmentierung.* Thesis, Phillips-Universität, Marburg.
118. Pfaff, E. (1967), "Round Table Discussion on Mitochondrial Structure and Compartmentation," p. 165, Quagliariello, E., Papa, S., Slater, E. C., and Tager, J. M., eds., Adriatica Editrice, Bari.
119. Pfaff, E., Klingenberg, M., and Heldt, H. W. (1965), *Biochim. Biophys. Acta* **104,** 312.
120. Pfaff, E., Klingenberg, M., Ritt, E., and Vogell, W. (1968), *European J. Biochem.* **5,** 222.
121. Pfaff, E., and Schwalbach, K. (1967), "Round Table Discussion on Mitochondrial Structure and Compartmentation," p. 346, Quagliariello, E., Papa, S., Slater, E. C., and Tager, J. M., eds., Adriatica Editrice, Bari.
122. Pressman, B. C., and de Duve, C. (1954), *Arch. Intern. Physiol.* **62,** 306.
122a. Quagliariello, E., Landriscina, C., and Coratelli, P. (1968), *Biochim. Biophys. Acta* **164,** 12.
123. Quagliariello, E., and Palmieri, F. (1968), *European J. Biochem.* **4,** 20.
124. Racker, E., and Horstman, L. L. (1967), *J. Biol. Chem.* **242,** 2547.
125. Racker, E., Tyler, D. D., Estabrook, R. W., Conover, T. E., Parsons, D. F., and Chance, B. (1965), "Oxidases and Related Redox Systems," Vol. II, p. 1077, King, T. E., Mason, H. S., and Morrison, M., eds., John Wiley & Sons, Inc., New York.

126. Raw, I., and Mahler, H. R. (1959), *J. Biol. Chem.* **234,** 1867.
127. Raw, I., Molinari, R., Ferreira do Amaral, D., and Mahler, H. R. (1958), *J. Biol. Chem.* **233,** 225.
128. Raw, I., Petragnani, N., and Camargo-Nogueira, O. (1960), *J. Biol. Chem.* **235,** 1517.
129. Robertson, J. D. (1959), *Biochem. Soc. Symp.* **16,** 3.
130. Rodriguez de Lores Arnaiz, G., and de Robertis, E. (1962), *J. Neurochem.* **9,** 503.
131. Roodyn, D. B., and Wilkie, D. (1968), "The Biogenesis of Mitochondria," Methuen, London.
132. Rose, I. A., and Warms, J. V. B. (1967), *J. Biol. Chem.* **242,** 1635.
133. Schnaitman, C., Erwin, V. G., and Greenwalt, J. W. (1967), *J. Cell Biol.* **32,** 719.
134. Schnaitman, C., and Greenawalt, J. W. (1968), *J. Cell Biol.* **38,** 158.
135. Schnaitman, C. A., and Pedersen, P. L. (1968), *Biochem. Biophys. Res. Commun.* **30,** 428.
136. Sekuzu, I., Jurtshuk, P., and Green, D. E. (1963), *J. Biol. Chem.* **238,** 975.
137. Sjöstrand, F. S. (1953), *Nature* **171,** 30.
138. Sjöstrand, F. S. (1963), *Nature* **199,** 1262.
139. Sjöstrand, F. S. (1964), *J. Ultrastruct. Res.* **10,** 263.
140. Sjöstrand, F. S. (1968), "Regulatory Functions of Biological Membranes," B.B.A. Library, Vol. 11, p. 1, Järnefelt, J., ed., Elsevier Publ. Co., Amsterdam.
141. Smith, D. S. (1963), *J. Cell Biol.* **19,** 115.
142. Smoly, J. M., Byington, K. H., Tan, C. W., and Green, D. E. (1968), *Arch. Biochem. Biophys.* **128,** 774.
143. Sottocasa, G. L. (1968), Abstr., 4th FEBS Meeting, Oslo, 1967; *Biochem. J.* **105,** 1 p.
144. Sottocasa, G. L., Ernster, L., Kuylenstierna, B., and Bergstrand, A. (1967), "Round Table Discussion on Mitochondrial Structure and Compartmentation," p. 74, Quagliariello, E., Papa, S., Slater, E. C., and Tager, J. M., eds., Adriatica Editrice, Bari.
145. Sottocasa, G. L., Kuylenstierna, B., Ernster, L., and Bergstrand, A. (1967), *J. Cell Biol.* **32,** 415.
146. Sottocasa, G. L., Kuylenstierna, B., Ernster, L., and Bergstrand, A. (1967), *Meth. Enzymol.* **10,** 448.
146a. Sottocasa, G. L., and Sandri, G. (1968), *Italian J. Biochem.* **17,** 17.
147. Stasny, J. T., and Crane, F. L. (1964), *J. Cell Biol.* **22,** 49.
148. Stoffel, W., and Schiefer, H.-G. (1968), *Hoppe Seyler's Z. Physiol. Chem.* **349,** 1017.
149. Tabor, C. W., Tabor, H., and Rosenthal, S. M. (1954), *J. Biol. Chem.* **208,** 645.
150. Thomas, R. S., and Greenawalt, J. W. (1968), *J. Cell Biol.* **39,** 55.
151. Thompson, J. E., Coleman, R., and Finean, J. B. (1967), *Biochim. Biophys. Acta* **135,** 1074.
152. Thompson, J. E., Coleman, R., and Finean, J. B. (1968), *Biochim Biophys. Acta* **150,** 405.

153. van Deenen, L. M., van den Bosch, H., van Golde, L. M. G., Scherphof, G. L., and Waite, B. M. (1968), "Symp. on Cellular Compartmentalization and Control of Fatty Acid Metabolism, Proc. 4th FEBS Meeting, Oslo, 1967," p. 89 (Gran, F. C., ed.), Universitetsforlaget, Oslo, and Academic Press, Inc., London and New York.

154. van den Bergh, S. G. (1967), "Round Table Discussion on Mitochondrial Structure and Compartmentation" p. 400, Quagliariello, E., Papa, S., Slater, E. C., and Tager, J. M., eds., Adriatica Editrice, Bari.

155. Vignais, P. M., Nachbaur, J., André, J., and Vignais, P. V. "Mitochondria —Structure and Function" Ernster, L., and Drahota, Z., eds., Proc. 5th FEBS Meeting, Prague, 1968, Publ. Czechosl. Acad. Sci., Prague, and Academic Press, London, in press.

156. Whereat, A. F., Orishimo, M. W., and Nelson, J. B. (1968), *Fed. Proc.* **27**, 362.

157. Whittaker, V. P. (1966), "Regulation of Metabolic Processes in Mitochondria," B. B. A. Library, Vol. 7, p. 1, Tager, J. M., Papa, S., Quagliariello, E., and Slater, E. C., eds., Elsevier Publ. Co., Amsterdam.

158. Winkler, H. H., Bygrave, F. L., and Lehninger, A. L. (1968), *J. Biol. Chem.* **243**, 20.

159. Wlodawer, P., Parsons, D. F., Williams, G. R., and Wojtczak, L. (1966), *Biochim. Biophys. Acta* **128**, 34.

160. Wojtczak, L., and Zborowski, J. (1967), *Abstr. 4th FEBS Meeting*, Oslo.

161. Work, T. S. (1968), "Biochemical Aspects of the Biogenesis of Mitochondria," p. 367, Slater, E. C., Tager, J. M., Papa, S., and Quagliariello, E., eds., Adriatica Editrice, Bari.

162. Wurtman, R. J., and Axelrod, J. (1963), *Biochem. Pharmacol.* **12**, 1439.

163. Yates, D. W., and Garland, P. B. (1967), *Biochem. J.* **102**, 40 p.

6

Energy-Linked Transport in Mitochondria

Berton C. Pressman*

Johnson Foundation

University of Pennsylvania

Philadelphia, Pennsylvania

I. INTRODUCTION

The ability of mitochondria to carry out the energy-dependent transport of ions now seems hardly less remarkable than their earlier recognized capacity for the energy-dependent synthesis of ATP. Both these processes can meet their energy requirement by the oxidation of substrates via a sequence of reactions which are shared, at least, in part. A detailed discussion of possible reaction mechanisms of energy transfer has been covered in Chapter 4. This chapter will be confined to the characteristics of energy-linked transport in mitochondria and their general implications concerning mitochondrial metabolism. The reader's attention is also directed to other recent reviews on mitochondrial ion transport.[72,103]

The first ion whose energy-linked accumulation by mitochondria was recognized is K^+.[9,108,178,179] Slater and Cleland also reported the binding of Ca^{++} to heart mitochondria at an early date, but regarded the process as nonenergetic, since under their conditions it even occurred at $0°$.[177] The energy-linked interaction of mitochondria with Ca^{++} was, however, implied by earlier observations that Ca^{++} stimulated latent ATPase[51] and

* This chapter was written during the tenure of a Career Development Award (K3-GM-3626) from the National Institute of Health. Present address: Papanicolaou Cancer Research Institute, Miami, Florida.

213

inhibited phosphorylation.[101,130] Ca^{++} was also known to stimulate the respiration of mitochondria in the absence of an energy acceptor system.[106,174] Chance noted that, with small additions of Ca^{++}, this respiratory stimulation was transient and proportional to the quantity of added Ca^{++}.[28] Eventually, the work of Saris,[162,163] DeLuca and Engstrom,[50] and Vasington and Murphy[185] established that the translocation of Ca^{++} *per se* was responsible for the energy dissipation which gave rise to the increase in respiratory activity.

The rate and extent at which mitochondria can accumulate Ca^{++} as well as other divalent ions, such as Sr^{++} and Mn^{++}, greatly exceed those reported earlier for K^{+}. However, the work of Pressman and collaborators has shown that extremely low concentrations of certain antibiotics accelerate energy-dependent K^{+} transport so that it compares favorably with that of divalent ions.[119,136] Since the complexing properties of divalent ions with various functional groups complicate determination of their intramitochondrial activity, it is difficult to establish their precise transmitochondrial gradient and, hence, to calculate the work requirement for their intramitochondrial accumulation. Fortunately, K^{+} is less capable of forming complexes, hence, its movements are much more amenable to thermodynamic analysis. It is relatively simple to establish that the mitochondrial transport of K^{+} is not merely energy-linked but unequivocally requires energy at the prevailing gradients. As will be seen, several of the various ions susceptible to mitochondrial transport offer technical advantages for the design of experiments for studying transport phenomena.

II. RELATIONSHIP OF ION TRANSPORT TO MITOCHONDRIAL ENERGY TRANSFER

The concept of purifying bulk preparations of subcellular organelles and studying them by biochemical techniques stems chiefly from the work of Claude, who characterized mitochondria as the cellular "power plants."[44] The power plant functions by accepting reducing equivalents from a variety of substrates and transferring them stepwise through a sequence of respiratory carriers to oxygen. At discrete regions of the respiratory chain, energy may be tapped off and converted into a common hypothetical energized intermediate (represented in Figure 6-1 as "$\sim$"), which can then be directed to drive various endergonic reactions including ATP synthesis and ion transport.

For our operational purposes it is not necessary to know the precise chemical or physical nature of $\sim$, for these points are discussed in greater depth in Chapter 4. The principal point brought out by the scheme of Figure 6-1 is that a three way reversible communication between respiration, ATP synthesis, and ion transport is established through the common intermediate, $\sim$. For example, the sequence leading to ATP synthesis can be experimentally reversed, i.e., ATP hydrolysis can produce $\sim$

which can in turn either support energy-linked ion transport,[47,119] or else drive reducing equivalents back up the respiratory chain, e.g., from succinate to pyridine nucleotide.[29] Other sequences which have been demonstrated experimentally are reversed electron transport[45,46] and the syn-

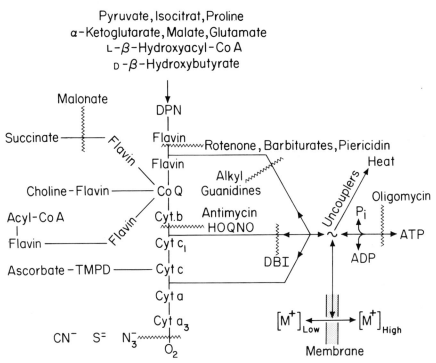

FIGURE 6-1. Relationship of electron transport to mitochondrial energy transfer. The fluted lines represent loci of action of various inhibitory agents.

thesis of ATP driven by the valinomycin-induced "downhill" release of K^+ from mitochondria.[46,48]

With fresh, properly prepared mitochondria, the coupling of reactions through $\sim$ is so tight that the rate of a reaction yielding $\sim$ may be controlled by the availability of a $\sim$ utilizing system. Thus, the flow of reducing equivalents to O_2, i.e., respiration, can be made dependent on the concentrations of ADP and Pi available to form ATP[98] or, alternatively, on the availability of ions suitable for energy-linked transport.[136,185] The metabolic state in which mitochondrial respiration is not limited by $\sim$ dissipation has been termed "State 3."[33] In the absence of $\sim$ dissipating systems, i.e., "State 4,"[33] reactions yielding $\sim$ will not halt completely, because of the dissipation of $\sim$ due to either the presence of endogenous transportable ions or because ~, or some other energized intermediate in equilibrium with it, breaks down spontaneously. Thus, the State 4 rate of intact mitochondria usually ranges from 12–30% of the State 3 rate.

When designing experiments, advantage is often taken of various agents which affect specific reactions related to energy production and utilization. One such group of agents are the uncouplers which act catalytically to de-energize ~, releasing the energy primarily in the form of heat. Some of the more frequently employed members of this group are nitro- and halophenols (e.g., DNP * and pentachlorophenol) and substituted carbonylcyanide phenylhydrazones, e.g., m-ClCCP. Such compounds permit ~ producing reactions, e.g., substrate oxidation[98] or ATP hydrolysis,[99] to proceed at maximal rates in the absence of other ~ consuming reactions. Conversely, uncouplers can depress the level of available ~, thereby preventing energy-linked ion transport, or even reversing its direction.[119,136,185] A state resembling that produced by uncoupling agents occurs spontaneously on aging of mitochondria,[90] or by disrupting them with surface active agents[148] or mechanical means.[89] All these processes can be assumed to lead to an accelerated spontaneous discharge of ~.

A second category of agents used as tools for studying mitochondria are the energy transfer blocking agents. The first recognized member of this category was guanidine, which appears to block the first energy transfer conservation site in the flavin region of the respiratory chain.[80] It has much less effect on the corresponding branch leading to the second energy conservation site. Later, various alkylguanidines were found which were more effective and more site specific than the parent compound [134,135] while certain related biguanides, notably the phenethyl derivative, DBI, have a predilection to produce an analogous block at the second energy conservation site in the cytochrome *b* region.[135] The antibiotics oligomycin,[97] rutamycin,[100] and aureovertin[94] block strongly between ~ and the terminal ATP synthesis reaction.

A third category of agents block specific loci in the electron transfer chain. Rotenone[53a] and piericidin[66a] block electron flow between NADH-dehydrogenase and cytochrome *b*; amytal does likewise, but at much higher concentrations.[55] Antimycin A and HOQNO block the chain between the second and third energy conservation sites,[131] while cyanide, sulfide, and azide, inhibit cytochrome oxidase. Malonate serves as a highly specific competitive inhibitor of succinate dehydrogenase, hence, it is particularly effective in blocking the oxidation of either endogenous succinate or that which arises metabolically during the course of an experiment.

Recently, a fourth general category of agents has emerged which func-

---

* The following abbreviations have been employed: Pi, inorganic phosphate; ADP and ATP, adenosine di- and triphosphates respectively; m-ClCCP, carbonylcyano-m-chlorophenylhydrazone; DNP, 2,4-dinitrophenol; MICA, 5-methoxyindole-2-carboxylic acid; DBI, phenethylbiguamide; HOQNO, 2-*n*-heptyl-4-hydroxyguinoline-N-oxide.

tion by blocking the transport of metabolites across the mitochondrial membrane. Examples of this group are atractyloside, which blocks the transport of adenine nucleotides,[93] and 2-butyl malonate, which prevents certain oxidizable substrates, e.g., malate from entering the mitochondria.[153] Inhibitors of this category will be discussed later.

III. TRANSPORT OF CALCIUM

The energy-linked accumulation of Ca^{++} by mitochondria has been studied more extensively than that of any other ion, partly because of the rapidity of the process and partly because animal mitochondria are intrinsically highly permeable to this ion, presumably because of a Ca^{++}-carrying system present in the membrane.[163] The earlier observed stimulation of State 4 respiration by Ca^{++} appeared to resemble closely the effects of such uncoupling agents as DNP.[174] However, Chance observed that low concentrations of Ca^{++} caused a transient burst of respiration proportional to the amount of added Ca^{++} [28] which was quite distinct from the earlier observed sustained stimulation of respiration caused by slightly larger amounts.[106,174] This observation was particularly significant in that it revealed a stoichiometric rather than a catalytic interaction of Ca^{++} with mitochondria.[28]

The true energy-linked nature of this interaction was clarified principally by Vasington and Murphy, who found that the uptake of radioactive Ca^{++} by kidney mitochondria was dependent on ATP and augmented by oxidizable substrates, Mg^{++} and Pi. Ca^{++}-accumulation was inhibited by the uncouplers DNP and dicumarol, as well as the respiratory chain inhibitors, cyanide, azide, and antimycin. Indeed, DNP or antimycin even caused the release of endogenous Ca^{++} present in the mitochondria.[185]

In his comprehensive thesis, "The Calcium Pump in Mitochondria," Saris[163] found that with liver mitochondria exposed to low levels of Ca^{++}, either ATP or oxidizable substrate alone was fully capable of supporting cycles of Ca^{++}-accumulation. Higher Ca^{++} levels produced irreversible changes indicative of mitochondrial damage. Since either ATP or oxidizable substrate ought to be equally capable of generating $\sim$, according to the scheme of Figure 6-1, the role of ATP in sustaining Ca^{++}-uptake in respiring kidney mitochondria[185] could possibly be to protect the particles against damage by excess Ca^{++}.

Saris also correlated the kinetics of several parameters affected during the transport of Ca^{++} and other divalent ions. Among these were:

(1) pH of medium (glass electrode). This normally decreased transiently, concomitantly with Ca^{++}-uptake. If a threshold value of Ca^{++} was exceeded, the membranes were damaged, and the pH reverted to its original value. The same effect was observed when the membranes were treated with detergent, e.g., Triton.

(2) Light scattering (transmission). This increased in the presence of

Pi, although subsequent experiments in the presence of an alternate permeant anion, acetate, led to a marked scattering decrease indicative of mitochondrial swelling (cf Ref. 151).

(3) ATPase (measured colorimetrically). This rose transiently during the uptake of small amounts of Ca^{++}, but excess Ca^{++} induced a sustained ATPase correlated with damage to the mitochondrial membrane.

(4) Fluorescence. This decreased transiently with the uptake of small doses of Ca^{++}, corresponding to the oxidation of pyridine nucleotide, which remained permanently oxidized when the damaging threshold-level of Ca^{++} was exceeded.

(5) Respiration (polarographic O_2 electrode). This increased transiently with small doses of Ca^{++}, but remained permanently elevated with larger Ca^{++} additions.

(6) Direct measurement of $^{45}Ca^{++}$ uptake (by rapid centrifugation).

The subsequent work of Chance extended these multiparameter measurements by developing instrumentation for simultaneous information retrieval, adding the parameter of electron-carrier redox changes as assayed spectrophotometrically.[30]

Chance and Mela[31] concluded that spectrophotometrically measured changes in the mitochondrially-bound pH indicator bromthymol blue yielded information about the pH of either the matrix or localized regions within the membrane. These experiments indicated an internal alkalinization of as much as one pH unit during Ca^{++} uptake, reflecting the extramitochondrial acidification. Inclusion of permeant anions in the system quenched the Ca^{++}-induced pH response. The validity of this approach has been challenged, recently, on the grounds that alterations in the membrane potential could cause translocations of the indicator.[118] Although similar values for mitochondrial pH gradients were also calculated[1] on the basis of transmembrane distribution of the weak acid, 5,5-dimethyl-2,4-oxazolidimedione (DMO), the distribution of this indicator may also be affected by changes in membrane potential.

In early experiments for which conditions of optimal cation uptake were selected, Pi was included in the reaction medium.[50,185] It was subsequently recognized that Pi accumulated along with the Ca^{++} [104] as had earlier been observed during the uptake of Mg^{++} by heart mitochondria.[15] This could obviously be related to the retention of intramitochondrial Ca^{++} by precipitating it as an insoluble phosphate. This precipitate could account for the increase in light scattering in such systems,[163] the opposite of what occurs when comparable amounts of Ca^{++} are taken up with acetate replacing Pi in the medium.[30] Attempts have been made to identify the species of precipitated calcium phosphate as one of the classical crystalline forms such as hydroxyapatite.[155] However, freshly precipitated calcium phosphate is a stable, highly absorptive gel, and apparently exists in a similar state within the mitochondria associated with protein and anions, such as hydroxyl, in a relatively ill-defined form. Such amorphous granules have been isolated from mitochondria by Weinbach.[186a]

A rather voluminous literature exists on the stoichiometry of the various parameters affected by Ca^{++} transport. Under certain conditions, notably high pH and salt concentrations, the earlier determined $Ca^{++}/\sim$ value of approximately two[28,30,155,156] can be exceeded considerably,[23,24,154,157] bringing it closer to the higher values observed for the $K^{+}/\sim$ ratio.[47] A more detailed treatment of these quantitative relationships will not be attempted here, as it has been covered extensively in a previous review.[103]

IV. TRANSPORT OF STRONTIUM

The uptake of Sr^{++} by liver mitochondria was briefly reported by Mraz,[120] Chappell *et al.*,[36] and Saris.[163] It was examined in greater detail by Carafoli and co-workers[22,25] and by Wenner.[187] In general, Sr^{++} accumulation resembles that of Ca^{++}; it can be driven by either ATP or a variety of oxidizable substrates. Relatively large quantities of Sr^{++} (i.e., 370 μM per g protein) has been shown to accumulate in the absence of permeant anions. However, Sr^{++} transport is somewhat slower than that of Ca^{++}, and, under appropriate conditions, competition between Sr^{++} and Ca^{++} can be demonstrated. Higher concentrations of Sr^{++} are less injurious to mitochondria than equivalent concentrations of Ca^{++}, and can protect mitochondria against swelling induced by Ca^{++}, Pi, thyroxin, and oleate.[21a]

V. TRANSPORT OF MAGNESIUM

Brierley *et al.* observed that, in the presence of Pi, beef heart mitochondria can accumulate massive amounts of Mg^{++} i.e., over 1 mM per g protein.[13,15,16] While the uptake is not rapid enough to produce observable bursts of respiration when supported by the oxidation of substrate, it can be sustained for ten minutes or longer. Mg^{++} uptake may also be supported by ATP hydrolysis. Recognition that only the ATP-driven transport of Mg^{++}, but not the substrate-driven uptake, can be inhibited by oligomycin was instrumental in developing the concepts expressed in Figure 6-1, i.e., that the flow of energy from oxidizable substrate to ion transport need not pass directly through ATP.[13,15,16] This is significant in that nonmitochondrial-membrane transport systems, devoid of a means of deriving energy directly from substrate oxidation, are presumably obligatorily dependent on ATP as a source of energy. For a general review of the biochemistry of nonmitochondrial-membrane transport systems, see Ref. 175.

Although a marked parallelism exists between the uptake of Ca^{++} by heart, kidney, and liver mitochondria, the latter are (ordinarily) extremely sluggish in accumulating Mg^{++}. It is interesting that the relatively low level of accumulated Mg^{++} which saturates liver mitochondria, ca 25 μM per g protein, can be displaced by monovalent guanidine on a mole for mole basis.[149] Since guanidine does not form an insoluble sesquiphosphate, this implies that Mg^{++} need not form an insoluble sesquiphosphate in rat-liver mitochondria, but possibly does in heart mitochondria during massive accumulation. Rasmussen and his collaborators found that the accumulation of Mg^{++} by rat-liver mitochondria could be greatly enhanced by relatively high levels (ca 10^{-6} M) of parathormone in the presence of appropriate anions, e.g., Pi or acetate.[152,160] Under these conditions, a Mg^{++}-dependent respiratory burst could be obtained. It was inferred that this property might be related to the physiological action of this hormone. However, its effects in enhancing Mg^{++} permeability of liver mitochondria are mimicked by other basic proteins of low molecular weight such as histones,[69] and recent work indicates that the hormone more likely mediates its true physiological effects via AMP cyclase which it stimulates at extremely low concentrations (10^{-8} M).[41]

VI. TRANSPORT OF MANGANESE

The fact that radioactive Mn^{++} administered to rats accumulates primarily in mitochondria,[109] led to the observation that isolated liver mitochondria also take up Mn^{++}.[8] Since this process is inhibited as $0°$ or by anaerobiosis, it was concluded that Mn^{++} accumulation is energy-linked. The observed acidification of the medium during Mn^{++} uptake has since become recognized as an important general characteristic of mitochondrial cation accumulation. Since the mitochondria appeared to shrink during Mn^{++} uptake under the conditions employed, it was concluded that the ion accumulated in a form making little osmosic contribution to the intramitochondrial fluid, i.e., that it was bound.[8]

Chappell et al.[39] observed a transient burst of respiration during Mn^{++} uptake in the presence of Pi. By recording the pH shift with a glass electrode, they measured a H^+/Mn^{++} stoichiometry close to unity. They further established that the pH shift represented a translocation of protons across the mitochondrial membrane, since damage of the membrane with detergent restored the initial pH, i.e., there was no net chemical production of protons in the system.

Subsequently, Chappell et al. utilized a rapid filtration technique for determining the uptake of $^{54}Mn^{++}$ directly.[36] Pi enhanced the extent of Mn^{++} uptake, and the resultant Pi/Mn^{++} and H^+/Mn^{++} ratios observed were consistent with the intramitochondrial precipitation of $Mn_3(PO_4)_2$. In the absence of Pi, extensive energy-linked Mn^{++} uptake occurred at slightly higher H^+/Mn^{++} ratios. They concluded that under these conditions Mn^{++} was bound to some preexisting component of the mitochondria, such as phospholipid, a suggestion similar to that advanced earlier to explain Ca^{++} binding.[177]

The ability of Mn^{++} to enhance the relaxation of H_2O protons as measured by a pulsed nuclear magnetic resonance technique was applied to the mitochondrial transport of Mn^{++}. This property of paramagnetic Mn^{++} is unique among the ions known to be actively accumulated by mitochondria. The data obtained suggested different states of mitochondrially bound Mn^{++} as a function of time, depending on whether or not Pi was present in the medium.[39]

An interesting synergism has been observed between Ca^{++} and Mn^{++} uptake in which the former greatly stimulates the rate of uptake of the latter.[32] A complicated relationship of Mg^{++} to the $Ca^{++}-Mn^{++}$ synergism has also been reported.[54]

VII. TRANSPORT OF GUANIDINE

Hollunger discovered that guanidine and several of its derivatives block the transfer of energy between mitochondrial electron transport and the synthesis of ATP.[80] The inhibition could be prevented or reversed by uncouplers. These studies were extended by Pressman, who found that the inhibitory potency of a series of alkylguanidine homologues was proportional to the ability of mitochondria to accumulate them.[134] Pressman also observed that guanidine inhibition led to extensive Pi uptake.[133] Each equivalent of guanidine or alkylguanidine accumulated was accompanied by two thirds of an equivalent of Pi and the accumulation process was energy-linked.

The poor ability of valinomycin and gramicidin to release the inhibition of mitochondrial respiration by guanidine derivatives was the first indication that these antibiotics are not classical uncoupling agents.[135] This observation eventually led to the discovery of the ionophorous properties of these antibiotics.[143] Whether or not a direct mechanistic link exists between the inhibition of the mitochondrial energy-transfer sequence (cf Ref. Figure 6-1) and guanidine transport has not been established.

VIII. TRANSPORT OF POTASSIUM AND OTHER ALKALI CATIONS

The general phenomenon of energy-linked ion transport in mitochondria was first revealed by studies of the metabolism of K^+ by these organelles.[9,108,178,179] Other observations also indicated a related dependence on K^+ of mitochondrial metabolism and energy transfer,[146,147] and independently suggested the maintenance of mitochondrial K^+ reserves by active metabolism.[146] Water movements and attendant light scattering changes were also associated with the translocation of K^+.[150] These observations were extended by Gamble, who found that even digitonin particles exhibit energy-dependent K^+ transport.[61] Several subsequent studies have been made of the slow energy-dependent accumulation of K^+ by mitochondria from both liver[42,62,107,159,165,184] and heart.[19]

Evidence was also presented on the basis of isotopic exchange with $^{42}K^+$ for the existence of slowly and rapidly exchanging K^+ pools in a mitochondrial pellet.[3,4,62,82] However, the rapidly exchanging pool probably represents primarily the extramitochondrial water space of the pellet. On treating mitochondria with small doses of valinomycin, we have failed to observe evidence for the existence of more than a single intramitochondrial $^{42}K^+$ pool.[86]

The rate of energy-dependent accumulation of K^+ by liver mitochondria can be raised considerably by first depleting the endogenous K^+ level by incubation in isotonic sucrose at 37°.[43] This accelerated ion transport differs from that stimulated by ionophorous agents in that, under comparable conditions, a given K^+ uptake is accompanied by less H^+ ejection.[70]

The spontaneously released K^+ can, furthermore, be replaced by other alkali cations, e.g., Li^+. The spontaneous alkali ion uptake of acetate salts by heart mitochondria also differs from ioniphore-induced ion uptake in that Na^+ is preferred to K^+.[19]

The rapid reaccumulation rate of K^+ by K^+-depleted liver mitochondria is perhaps aided by the reduction of the K^+ concentration gradient, but it also implies that the treatment raises the intrinsically low permeability of fresh, intact mitochondria to this ion. Analogous effects have been obtained by treating mitochondria with various heavy metals such as Hg^{++},[63] organic mercurials[18,164] and Zn^{++}.[14,17] An explanation for these effects within the framework of a general scheme for the regulation of mitochondrial ion transport will be offered later.

The ultimate capacity of mitochondria for transporting alkali ions became apparent only with the introduction of the "ionophorous" antibiotics.[143] These agents increase the inherently low permeability of mitochondria to alkali ions,[3,4,70,82] and permit rates and extents of K^+ transport[47] which can exceed even those of Ca^{++}, the most rapidly transported divalent cation.

Before going into a detailed analysis of the general aspects and mechanisms of mitochondrial transport, it is of interest to consider certain inherent advantages the study of monovalent ion transport offers over divalent ion transport.

Detection.   Moore and Pressman introduced the use of specific ion glass electrodes[52] for monitoring the kinetics of alkali ion uptake by mitochondria.[119] The effective time constant of a complete monitoring system, allowing for the electrode amplifier and mixing, is between 0.5–1 second, and, therefore, is not limiting in tracking the movements of K^+ in mitochondria, which usually extend over a period of a half minute or longer. Liquid ion exchange electrodes for divalent cations are also available; however, their time response is slower and too limited for tracking rapid pulses of Ca^{++} uptake of short duration.[85] For reliability, both the glass and liquid exchanger electrodes must equilibrate with the complete reaction medium several minutes before initiation of ion movements. This can be ideally accomplished by initiating monovalent cation transport with small quantities of ionophorous antibiotics after the electrodes have equili-brated; no equivalent means of initiating divalent transport is currently available. Although glass electrodes for monovalent cations are widely used for kinetic studies with mitochondria,[137] the use of divalent ion electrodes is primarily limited to measuring the completeness rather than the kinetics of Ca^{++} uptake.[35]

Interpretation.   Reductions in *activity* of monovalent ions in biological systems due to binding is minimal. Experiments with penetrating microelec-trodes of ion-specific glass indicate that even within cells the activity of K^+ is in reasonable agreement with that calculated from its chemical concentra-tion; by the same criteria, some evidence for binding of Na^+ has been ob-tained.[79,105] On the other hand, the divalent cations have relatively high chemical affinities for such groups as basic nitrogens, phosphate, and car-boxylate. This can cause a large discrepancy between chemical concentra-tion and activity (ion specific electrodes of all types respond to chemical activity, not concentration). Some idea of the precautions necessary to hold a reasonably constant relationship between Ca^{++} concentration and the Ca^{++} activity detected by an electrode is given in a study of ATP-driven Ca^{++} transport by muscle microsomes.[85]

IX. IONOPHOROUS ANTIBIOTICS

The ionophorous (ion-carrying) antibiotics are of particular significance, since they establish a common link between the specialized phenomenon of mitochondrial ion transport and the general phenomenon of transport across various biological and artificial membranes.[140] The first compound of this class to be recognized was gramicidin, which appeared to be a powerful un-coupler of oxidative phosphorylation in mitochondria (cf. Ref. 96). A simi-lar finding was later reported for valinomycin.[111] However, it was noted that both these antibiotics differed from the true uncoupling agents in that they evoke an energy-dependent proton ejection from mitochondria[135] remi-niscent of that seen during divalent cation transport (cf. Ref. 163). The ejection of protons was found to require the presence of K^+, and the con-clusion that K^+ was moving counter to the protons was confirmed by means of the K^+-selective glass electrode.[119]

Ionophorous activity was found in a variety of other compounds illustrated in Figure 6-2. Valinomycin consists of a ring of alternating hydroxy and amino acids. Metabolites of this type have been generically termed *cyclic depsipeptides*,[169] and usually contain unnatural or D configurations. In val-inomycin, the sequence D-valine-D-hydroxyisovalerate-L-valine-L-lactate re-peats three times, resulting in a ring of 36 atoms.[170] Using the K^+-depend-ent stimulation of the respiration in phosphate acceptor-free mitochondria as an assay, it was found that inverting the configuration of one or more asymmetric centers, or changing the ring size by inserting or withdrawing a repeating sequence, i.e., changing the ring size to 48 or 24 atoms, virtually

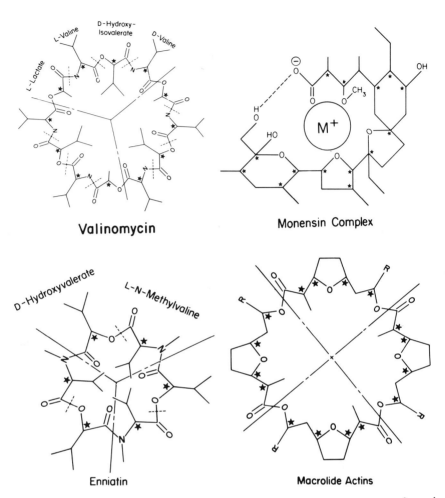

FIGURE 6-2. Structures of representative ionophores. These structures are formal representations and not intended to indicate the actual three-dimensional conformations where known.[140]

abolished activity.[136] Yet the ring size *per se* cannot be critical, since a second group of depsipeptides, the enniatins (Figure 6-2), again are active only with the number of repeating sequences found in nature, corresponding to 18 ring atoms.

One of the most remarkable aspects of ionophore-induced mitochondrial transport is its extreme ion selectivity.[119,136,141,142] Although the ionophores facilitate the transport of all species of alkali ions,[140,141,144] their ability to discriminate sharply between K^+ and Na^+ is of profound biological significance. The extreme preference for K^+ over Na^+ exhibited by valinomycin in mitochondrial transport systems is about 10,000:1.[136,142]

Several groups of investigators have attempted to explain this selectivity on the basis of the fit between the hole in the larger cyclic ionophores (cf Figure 6-2) and the radius of the hydrated alkali cations.[37,121,181] However, the preferential effect on K^+ transport over Na^+ transport falls as the ring *exterior* is altered during the homologous transition from nonactin (4R = H, cf Figure 6-2) to trinactin (1R = H; 3R = CH_3).[66] Furthermore, the ionic preference series for the smaller enniatins, as measured by mitochondrial transport[136,142] or electrometrically with model membranes[121] closely resembles that of the valinomycin-type ionophores. This suggests a common basis for the molecular interaction between cations and this entire ionophore group which cannot be related directly to the molecular dimensions of planar ionophore rings. Significantly, the 18 atom enniatin rings are sufficiently small so that, to apply the postulate of molecular fit, interacting alkali ions must be completely stripped of their hydration shells.[96a,121]

Gramicidins are *linear* polypeptides[161] and, thus, constitute another series of naturally occurring ionophorus antibiotics.[136] These compounds must not be confused with the cyclic gramicidin S [53] which, like the related tyrocidins, is an uncoupling agent[96] without demonstrable ionophorous activity towards mitochondria. The precise conformation of the gramicidins other than "S" is not known but by analogy with the cyclic ionophores, their secondary structure could cyclize when they are complexed to cations.[96a] The gramicidins show little discrimination between Na^+ and K^+, either with mitochondria or model systems, possibly owing to multiple cyclizing options to accommodate ions of various radii.

Because of the steric structural requirements for ionophorous activity of valinomycin and the enniatins, it was first suggested that they could interact with a preexisting receptor which controlled the mitochondrial ion pump.[119] Chappell and Crofts observed, however, that gramicidin (presumably a commercial sample of mixed fractions A, B, and C) and not valinomycin could produce a K^+-for-H^+ or a K^+-for-Na^+ exchange across an erythrocyte membrane, and that valinomycin and gramicidin S accelerated the efflux of $^{42}K^+$ from artificially produced phospholipid micelles.[38] This they interpreted as supporting their earlier conclusion that the ionophorus antibiotics acted by rendering membranes in general permeable to alkali ions by introducing "flaws" or "pores" in the membrane.

Mueller and Rudin discovered that the ionophorus antibiotics increased the alkali ion permeability of aqueously immersed artificial "black" lipid membranes of various composition.[121] Since, in this system, both sides of the membrane are accessible to electrodes, the selectivity of the induced permeability can be conveniently measured quantitatively by electrometric techniques analogous to those used in electrophysiology. Thus, permeability may be measured either by determining the alkali ion-dependent current carried through the membrane, i.e., its ohmic resistance, or by the potential developed when the cation diffusion forces on both sides of the membrane are unequal. These data show a remarkably close correlation between the activity and selectivity of a wide variety of ionophorus antibiotics in both artificial membranes and mitochondrial systems. It might be added that antibiotic activities of the test substances also parallel their effects on the mitochondrial and model systems,[173] and, indeed, they can be shown to affect ion transport in susceptible microorganisms.[67,138] Several mechanisms have been proposed for the transport-mediating and ion-selectivity properties of valinomycin and related antibiotics:

(1) Triggering of a specific receptor site.

(2) Introduction of a "flaw" or "pore" in a membrane by surface distortion.

(3) Formation of a channel in a membrane by several antibiotic molecules.

(4) Acting as a mobile carrier via formation of a complex.

The first possibility, originating from early observations of structure–activity relationships and the extremely low concentrations at which the antibiotics work,[119] has been rendered unlikely since both optical antipodes of enniatin B have equal activity in the microbial [161] and mitochondrial [141] test systems. A natural specific biological receptor site would undoubtedly possess enough asymmetry to discriminate between optical antipodes. Therefore, the required asymmetry arises from requirements intrinsic to the ionophorous antibiotics themselves, and would be equivalent in the two antipodes but different for all other disstereoisomers. The second and third possibilities are both related, and assumed to fit between cations and the molecular dimensions of the hole in the center of the ring. Molecular models of the ionophorus antibiotics and the hydrated and nonhydrated cations offer no obvious explanation for why two antibiotics of such radically different ring sizes as the enniatins and macrolide actins have such closely parallel activities, or why increasing the number of methyl groups at a locus distal to the ring center alters ionic selectivity of the actins (cf Figure 6-2). It should be noted that the progression of ionic radii increases with the atomic weight for the anhydrous alkali ions, but decreases for the hydrated ions, hydrated Li^+ being considerably larger than hydrated Cs^+. The fact that bulk phases of CCl_4 containing ionophorous agents can develop ion-dependent interphase potentials also renders the formation of channels by stacking antibiotic molecules unlikely.[143,180]

We believe that a more satisfactory mechanism for the transport mediating agents is suggested by the properties of a second group of antibiotics which differ from the valinomycin type in possessing a free carboxyl group. This group, typified by nigericin, reverses the valinomycin-induced, energy-linked mitochondrial ion transport. Lardy and his collaborators[65,96a] suggested that the alkali ion complexes formed by these agents were inhibiting the mitochondrial cation pump. However, Pressman *et al.*[143] showed that nigericin actually increases the energy-dependent turnover of K^+, provided substrate depletion is prevented. Furthermore, nigericin produces a K^+-for-H^+ exchange in such diverse systems as de-energized mitochondria,[143] erythrocytes,[74] microsomes,[143] chromatophores,[167] and chloroplasts[125,166,168] indicating that this ionophore is not specific for the energy-linked mitochondrial ion pump, but rather increases the ion permeability of membranes in general, as does the valinomycin group.

If the permeability effects of the nigericin group result from permitting uncomplexed charged ions to pass across membranes through pores or channels, one would expect to be able to measure the same electrical effects seen with the valinomycin group of antibiotics. No such electrometric phenomena can be detected either with model membranes[143] or with bulk phase systems.[83] The negative findings suggest that nigericin travels with the translocated ion as a neutral charge paired complex. Experiments with radioactive Rb^+ verify that nigericin does indeed transport this cation across bulk phases of CCl_4.[143] Nigericin would also be expected to carry protons across a lipid barrier in its neutral, undissociated carboxylic acid form, thus explaining its ability to promote a K^+-for-H^+ exchange.

The valinomycin group of antibiotics also forms alkali ion complexes, as first described with the macrotetralide actins,[129] suggesting that these complexes are also capable of carrying ions across a lipid barrier. Since the valinomycin class of antibiotics are devoid of charged functional groups, their complexes acquire the charge of the ion complexed. Accordingly, movement of the complex across a lipid barrier translocates charges, producing electrochemical phenomena which the nigericin group cannot.

Therefore, it appears that both valinomycin and nigericin translocate ions by functioning as mobile ion carriers. Thus, it seems appropriate to classify them as "ionophorus," i.e., "ion-bearing" agents.

Model systems consisting of lipid solutions of ionophorus agents are able to manifest the extreme degree of alkali cation selectivity[139,140,141,144] heretofore encountered only in biological systems. The origin of the high degree of alkali ion selectivity of biological systems has long been a challenging problem. The correspondingly high selectivity of models systems containing lipophilic solutions of ionophorus agents accordingly offers a model which is amenable to study at the molecular level.

Enniatin, the smallest antibiotic of the valinomycin type, contains an 18 atom ring which has the approximate dimensions of the anhydrous alkali ions. Among the synthetic "crown" polyethers,[128] the one containing 18 ring atoms is also the best alkali ion complex former. The simplicity of the latter compound, which is capable of inducing K^+ uptake in mitochondria, indicates that the basis of complex formation is the interaction of the alkali ions with the electronegative ring oxygens.[128] This has also been confirmed by infrared spectra for the enniatins[172] as well as for the crown polyethers.[128] The planar ring representation of the macrolide actins has the oxygens spread too far apart to focus on a dehydrated alkali cation; however, x-ray crystallographic analysis of the KCNS complex of nonactin indicates that the antibiotic assumes the conformation of a tennis ball seam, thereby bringing eight cubically deployed oxygens into contact with the K^+, which is encaged in three dimensions[91] in what would be most properly termed a clatherate. The complete structure of nigericin has not yet been reported; however, x-ray crystallographic analysis of a smaller analogue, monensin (Figure 6-2) indicates that its structure[2] conforms to the cage pattern established for the macrolide actins. Optical rotatory dispersion measurements have indicated that a conformational change occurs[172] even when a compact enniatin forms a cation complex.

The energetic factors determining the ability of an aqueous alkali ion to complex with an ionophorus agent within a lipophilic phase are: the energy necessary to dehydrate the ion, the energy of association with the antibiotic, and the energy necessary to change the conformation of the ionophorus agent to one favorable for complex formation (cf. Figure 3). All of the antibiotics of the valinomycin group qualitatively exhibit the same preference pattern for the various cations, although the degree of selectivity of the enniatins and macrotetralide actins is moderate and, in the case of the gramicidins, only slight.[142] The ion selectivity observed with mitochondria occasionally shows slight displacements from the above order. This discrepancy, however, may be ascribed to an interplay between the selectivity of the antibiotic and the selectivity of the mitochondria. The nigericin group is not nearly as homogeneous in its ionic selectivity pattern.[140,141]

The demonstration that relatively simple molecules can carry ions across lipid barriers, and, thereby, mimic many of the properties of permselective biological membranes, gives strong impetus to the concept of biological transport carriers (cf Ref. 189). It appears unlikely that animal systems are able to produce carriers containing D configurations such as those produced by the Streptomyces, but perhaps the unnatural optical configurations are only a highly efficient, although not obligatory, means of fabricating a compact ionophorus agent. Thus, alametricin, which is synthesized by the organism *Trichoderma viride*,[114] contains L configurations exclusively, yet shows ionophorus activity in model membranes,[122] in two-phase partitions,[144] and in mitochondrial systems.[145] We have been able to induce K^+ transport in mitochondria with an even simpler all L cyclic decapeptide, antamanide, extracted from the toxic mushroom *Amanita phalloides*.[188]

The picture of alkali ion carriers which emerges from the study of ionophorus agents could well be extended to most biological carriers. Carriers could function in general by engulfing lipophobic substrates and encasing them in a lipophilic shield which renders them capable of traversing membranous barriers. Electrogenic forces could be generated if the translocated species is charged. The concept in principle would apply equally well to positively or negatively charged, or to uncharged carrier complexes. Additional attributes which could prove advantageous for a natural carrier include an anchor to restrict its locus within a membrane, and an allosteric center which would, when perturbed energetically, alter the conformation of the ionophorous region and change its binding affinity. This would provide a means for feeding energy into the carrier and enable it to transport ions against an electrochemical gradient. Since biological control can be viewed as regulation of the passage of metabolites between compartments across a barrier membrane, the relevance of the above picture to this important area of current biological research is evident.

X. EFFECTS OF OTHER CATIONS ON TRANSPORT

Although La^{+++} does not appear to be transported by mitochondria, it inhibits Ca^{++} transport at concentrations as low as 10^{-7} M.[111a] This inhibitory effect of La^{+++} shared by other lanthanides[112] appears to be noncompetitive and compatible with its binding to a specific Ca^{++} carrier. This binding could be related to the strong chemical affinity of La^{+++} for phosphate groups, which might be an essential component of a hypothetical Ca^{+++} carrier. These effects of La^{+++} are reminiscent of the inhibition of the glucose permease activity in yeast by UO_2^{++} which also binds phosphate.[158] La^{+++} is more specific than UO_2^{++} as a high affinity phosphate binder, since it has less tendency to react with sulfhydryl groups. Thus, La^{+++} appears to be a transport inhibitor, the fourth inhibitor category outlined in the introduction.

Although the widely used biological buffer tris is usually regarded as being relatively inert, it can be taken up by heart mitochondria in an energy-dependent process.[19] This factor should be considered in studying mitochondrial ion transport.

The local anesthetics, which are themselves cationic, inhibit cation transport of a wide variety of ions in both mitochondria[6,87] and model membranes.[7,124] This could be due to their tendency to lodge in the membrane, thereby impeding the passage of other cations by coulombic repulsion.

On the other hand, Zn^{++} [14,17] Hg^{++},[61] and organic mercurials[18,164] increase the turnover and, under carefully controlled conditions, the net energy-linked transport of K^+ and Mg^{++}. These active sulfhydryl inhibitors in some manner appear capable of increasing the permeability of the mitochondrial membrane.

XI. MECHANISM OF TRANSPORT—GENERAL CONSIDERATIONS

To a large extent, the chemiosmotic hypothesis of Mitchell has been responsible for introducing the classical concepts of electrophysiology into the field of mitochondrial transport.[115,116,117] This hypothesis applies the earlier concepts of Davies and Ogston,[49] linking the establishment of a membrane potential via an anisotropic separation of the elements of water, H^+, and OH^-, to the mitochondrial synthesis of ATP. According to Mitchell, the individual electron transport carriers are arranged across a "coupling" membrane so that the passage of electrons from substrate to oxygen establishes an electrochemical, or "chemiosmotic," proton gradient, the interior becoming negatively charged. This gradient in turn can supply the driving force for abstracting the elements of water from ADP and Pi to form ATP. In keeping with the earlier model,[49] the energized "intermediate" for the synthesis of ATP, namely the electrochemical proton gradient, should also be able to drive the transport of various ions across the coupling membrane, which is usually identified with the inner mitochondrial membrane.

Accordng to the equations of Nernst and Gibbs, the molar-free energy change for the ion translocation of a single ion species across the mitochondrial membrane reduces to the expression:

$$\Delta F = 1320 \left(\log \frac{C_{in}}{C_{ex}} - \frac{n \Delta E}{59} \right) \qquad (1)$$

where C_{in} and C_{ex} are the intra- and extramitochondrial ion activities, ΔE the transmembrane potential (exterior potential minus, interior positive for cations), and n the charge of the ion. Thus, if the electropotential term exceeds the concentration term, ions complexed to a neutral valinomycin-like carrier can move spontaneously against a concentration gradient, since this would be energetically down the composite electro-osmotic gradient. If, on the other hand, one ion species moves in exchange for another of identical charge, i.e., via an exchange diffusion carrier analogous to nigericin, the resulting movements would not be influenced by any existing membrane potential, and the free energy change would be:

$$\Delta F = 1320 \left(\log \frac{C_{2\,in}}{C_{1\,ex}} \cdot \frac{C_{2\,ex}}{C_{2\,in}} \right) \qquad (2)$$

Thus, the movement of ion species C_1 against a chemical gradient can be driven by the movement of a second species C_2 moving down an even steeper gradient (cf Ref. 74).

Chappell and Crofts provided a detailed explanation for the accumulation of ions by mitochondria consistent with the chemiosmotic hypothesis, which features an outward directed, metabolically driven proton pump.[37,38] The accumulation of Ca^{++} in the presence of acetate is accompanied by a decrease in mitochondrial light scattering equivalent to an increase in volume, i.e., an uptake of water. This has been interpreted in terms of an increase of osmotically active ions within the mitochondria as the proton pump drives the uptake of both anions and cations.

It may seem somewhat paradoxical that the energy-linked movement of a single ion species, H^+, could drive the uptake of both anions and cations. The hydrogen ion gradient, however, is also equivalent to an anion, i.e., OH^-, gradient. The membrane potential produced by the expulsion of protons would pull in cations if they could move as a charged species through the mitochondria on a neutral carrier. The concomitantly formed hydroxyl gradient could drive anion uptake, assuming that the anions move in an uncharged form on an anionic carrier in exchange for OH^- through the membrane.

The mechanism suggested, by which acetate and other monocarboxylic acids support cation movement, is that, being weak acids, they move through the mitochondrial membrane as the undissociated species, even at neutral pH.[38] Once exposed to the higher intramitochondrial pH resulting from proton expulsion, the weak acids would dissociate and neutralize the excess alkalinity:

$$\text{Membrane}$$

$$A^- \xrightleftharpoons[\text{H}_2\text{O} \quad \text{OH}^-]{} HA \xleftrightarrow{} HA \xrightleftharpoons[\text{OH}^- \quad \text{H}_2\text{O}]{} A^- \qquad (3)$$

This would reduce the concentration of intramitochondrial OH^- which would, in turn, reduce the net electro-osmotic H^+ gradient facilitating further proton expulsion, and enable ion transport to continue until sizable quantities of both anions and cations were accumulated.

The fact that anions of relatively strong, not readily associable, acids such as Pi ($pK_1 = 2.1$), mono- ($pK = 2.9$), and dichloroacetate ($pK = 1.3$), can, like acetate (pK, 4.8), also facilitate ion accumulation, casts some doubt on the above proposed mechanism.[70] Furthermore, the more lipophilic fatty acids of longer chain length, are less effective than acetate in supplying co-ions for cation transport.[34]

An alternative mechanism is that the anions enter in the charged condition via an appropriate carrier. As pointed out previously, a pre-requisite condition for the anion carrier to move anions against an unfavorable membrane potential is that the anion-carrier complex carry no net charge, i.e., the carrier must bear a positive charge to offset that of the transported anion. Thus, if the carrier had a single site which could bind either carboxylate ions or hydroxyl, it could catalyze an hydroxyl for anion exchange until equilibrium is established, as governed by Equation 6-2 rather than Equation 6-1. Thermodynamically, Equation 6-2 would apply to the anion movement regardless of whether the carboxylate ion moves on a positively charged carrier, or diffuses through the mitochondrial membrane as an undissociated acid followed by the reaction shown in Equation 6-3.[38]

The fact that inorganic anions such as Cl^- or HCO_3^-, are unable to permeate the mitochondrial membrane[38,40] implies the existence of anion carriers with high ionic discrimination. We have observed that even the relatively lipophilic CNS^- is incapable of increasing the K^+ uptake by valinomycin-treated mitochondria in contrast to permeant ions such as Pi or acetate.

It has been pointed out that the chemiosmotic hypothesis cannot account for the rather large number of equivalents of K^+ (as many as seven) which may be transported for every $\sim$ expended.[47] The hypothesis also fails to explain the apparent multiple energy transfer pathways between the electron transport chain and $\sim$ as implied by the selective inhibition of the upper energy transfer pathway (cf Figure 6-1) by the alkylguanidines while the diguanides, such as DBI, preferentially inhibit the middle energy conservation pathway.[135] If the energized intermediate for driving ATP synthesis were truly a proton gradient, all pathways between the electron transport chain and $\sim$ ought to be equivalent. Furthermore, equilibration of K^+ with a proposed membrane potential of 130 mV [118] (a minimum value if such a potential is to drive the synthesis of ATP)[47] should induce mitochondria to accumulate virtually all of the intracellular K^+ *in vivo,* and this obviously does not occur. From thermodynamic considerations, a requirement for transmembrane potentials as high as 350 mV can be calculated,[47] which is considerably in excess of any biological precedent, and is of sufficient magnitude to threaten dielectric breakdown of the mitochondrial membrane. A more comprehensive critique of the chemiosmotic hypothesis has been compiled by Slater.[176]

An alternative conclusion is that the capability of mitochondria for transporting cations is directly linked to their vital requirement for translocating anionic metabolites. Such a coupling could be expected, since, regardless of mechanism, appreciable quantities of a single ionic species could not be transported across membranes without producing prohibitively large coulombic forces opposing further transport.

Rasmussen *et al.* account for the interrelationship of electron transport, ATP synthesis, and cation transport by suggesting that Ca^{++} reacts directly with an energized electron-transport carrier moving across the membrane and is released upon discharge of the energized state.[30,151] This explanation does not account for the observed stoichiometries of two Ca^{++} transported per equivalent $\sim$ discharged. Each energized carrier unit with the potentiality of generating a single ATP would be required to transport two Ca^{++}, equivalent to four charges, in its ion carrying mode. This problem becomes even greater in attempting to accommodate as many as seven K^+ on an energized intermediate which would have the potentiality of driving the formation of a single ATP.[47] It also raises questions as to why the equivalent intermediate in plant mitochondria is so chemically different that it interacts with and transports Ca^{++} to a negligible extent,[12] or why phosphorylating submitochondrial particles, in which the carrier ought to be even more accessible, fail to respond to Ca^{++} or K^+, even in the presence of valinomycin.[71] Although the active transport system might well be more sensitive to mitochondrial disruption than the phosphorylation system, there has been a recent report of energy-linked Ca^{++} accumulation by a submitochondrial particle.[42]

XII. PHOSPHATE

Early studies of divalent cation uptake by mitochondria, especially Mg^{++} and Ca^{++}, attempted to define conditions of optimal ion uptake irrespective of their true metabolic relevance.[15,50,104,185] By this criterion, Pi was found to be a favorable component of such systems, leading to the conclusion that its uptake was primarily associated with the precipitation of insoluble sequiphosphates within the mitochondria. It had earlier been observed, however, that Pi accumulation is also promoted by guanidine[133] or its alkyl derivatives in the molar ratio of 2:3,[134] roughly the same Pi: cation ratio as that observed in the case of divalent ions. Since the guanidines do not form insoluble phosphate precipitates, the Pi uptake could best be explained by its being a preferred anion for preventing significant charge imbalances during cation accumulation; the similarity of Pi:cation stoichiometries during singly and doubly charged cation accumulation are compensated for by the expulsion of protons from the mitochondria in the case of divalent cations.

The facultative nature of Pi uptake during cation uptake was confirmed by the ability of mitochondria to accumulate modest amounts of divalent ions (Mn^{++}, Ca^{++}) in its absence.[36,163] Under these conditions, the ejected protons are insufficient to counterbalance the cation uptake, indicating that other ions must also traverse the mitochondrial membrane. Tracer studies have indicated that mitochondria are normally impermeable to the dominant anion species in such systems, i.e., Cl^-.[62] The charge balance could then be accounted for by substrate anion translocation.

XIII. FATTY ACIDS

While investigating the effect of parathyroid hormone on the transport of Mg^{++} and K^+ by mitochondria, Rasmussen *et al.* observed that acetate could substitute for Pi as a "permeant" anion.[152] Acetate can also substitute for Pi in promoting the transport of Ca^{++},[151] and the valinomycin-induced uptake of K^+.[70] Among the homologous fatty acids, propionate works about as well as acetate as a permeant anion, while formate and the higher fatty-acid anions are measurably less effective.[34] For the reasons cited previously,[70] the shorter chain fatty acids probably do not penetrate the membrane in their undissociated form, but rather are transported as anions by a carrier system, either in exchange for OH^-[40] (Transport Mode 4, Table 6-1), electrophoretically (Mode 3), or simultaneously with cations (Mode 5).

A rather specialized mitochondrial transport mechanism also exists, principally for the uptake of longer chain fatty acids.[59,182] At the outer edge of the inner membrane, the acid is condensed with *L*-carnitine in a reaction driven by ATP to form an acylcarnitine derivative. It was assumed earlier that the resulting zwitterion is the actual species traversing the

TABLE 6-1

**POSSIBLE TRANSPORT MODES MEDIATED
BY A CATION-BINDING CARRIER**

1. $M^{n+} \leftrightarrow M^{n+}$	Electrogenic or electrophoretic cation movement
2. $M_1^{n+} \leftrightarrow M_1^{n+}$ $M_2^{n'+} \leftleftarrows M_2^{n'+}$	Cation for cation exchange
3. $A^{n-} \leftrightarrow A^{n-}$	Electrogenic or electrophoretic anion movement
4. $A_1^{n-} \leftrightarrow A_1^{n-}$ $A_2^{n'-} \leftleftarrows A_2^{n'-}$	Anion for anion exchange
5. $M^{n+}A^{n'-} \leftrightarrow M^{n+}A^{n'-}$	Ion pair movement

membrane with the carnitine itself acting as a mobile carrier for acyl groups.[59] More recently, it has been suggested that the acylcarnitine donates its acyl group to an enzyme which transports the acyl group across the membrane.[182] On the other side of the membrane, the acyl group is donated to CoA to form the corresponding acyl-CoA. This transport system can be specifically inhibited by acyl derivatives of the unnatural D form of carnitine,[60] or α-bromoacyl-CoA.[182]

XIV. OTHER CARBOXYLATE ANIONS

Chappell and his collaborators[38,40] utilized a criterion previously applied to erythrocytes[81] as a general test for permeant anions. Whereas the salts of nonpermeant ions (e.g., K^+, Cl^-) can counterbalance the osmotic gradient across the mitochondrial membrane, isotonic solutions consisting of anions and cations, which are both permeant, cannot. Addition of such salts to isolated mitochondria induce rapid swelling which can be conveniently detected as a decrease in light scattering.

Ammonium, a permeant cation, can presumably cross the lipid barrier of the mitochondrial membrane in the form of undissociated ammonia:

$$NH_4^+ \underset{H^+}{\leftrightarrow} NH_3 \leftrightarrow NH_3 \underset{H^+}{\leftrightarrow} NH_4^+ \qquad (4)$$

The procedure, used to establish a given anion as permeant is to dilute a sucrose suspension of mitochondria with an isotonic solution of its ammonium salt. The ensuing swelling indicates that the anion is permeant. To avoid complications by metabolic transformations, the inhibitor rotenone is included.

These experiments indicated that Cl^-, Br^-, SO_4^-, and HCO_3^-, as well as the metabolite fumarate are impermeant, while Pi, arsenate, and the fatty acid anions are permeant. A second category of anions, including malonate, D and L-malate, and succinate, are by themselves impermeant, but become permeant in the presence of low concentrations (2 mM) of Pi.[40] A third category of anions, consisting of the tricarboxylate metabolites, citrate, *iso*-citrate, and *cis*-aconitate, as well as D and L-tartrate, require low concentrations of both Pi and L-malate (3 mM) for penetration.[40] Tartronate and *iso*-malate can partially substitute for L-malate. α-Ketoglutarate resembles the third category in that its permeation also requires Pi and malate; however, it may be distinguished by the fact that tartronate and *iso*-malate are ineffective substitutes for malate.[153]

On the basis of these data, a series of exchange-diffusion carriers were postulated for permitting entry of the various anions. One for Pi and arsenate, a second for malate and succinate, a third for tricarboxylate anions, and a fourth for α-ketoglutarate.[40,153] No distinction was suggested between the Pi and a possible monocarboxylate carrier. Fonyo, however, reported that p-chlormercuribenzoate inhibits the antimycin-induced release of Pi which had been accumulated together with K^+ in mitochondria treated with gramicidin. The release of accumulated acetate was not inhibited by the mercurial. This was interpreted as evidence for separate permeation mechanisms for Pi and monocarboxylate anions.[58]

It may be questioned whether the above criteria for permeation are valid for determining the metabolic availability of a given anionic species. Thus, succinate may be oxidized at maximal rates by mitochondria in the presence of rotenone plus uncoupler, even in the absence of Pi. Low concentrations of many isotopically tagged metabolites rapidly enter and concentrate within mitochondria,[68,77] even in the absence of permeation "activators" as presented by Chappell and co-workers. This apparent discrepancy has been attributed to the ability of mitochondria to accumulate moderate levels of radioactive substrates against considerable gradients by exchange with endogenous anions such as malate and Pi.[68] The permeation of citrate in the absence of Pi and divalent carboxylate ions has also been reported with cyanide-inhibited mitochondria.[64] Accumulation of anionic metabolites is retarded by lowering the mitochondrial energy reserves with appropriate inhibitors and enhanced by increasing the mitochondrial anion reserves by inducing energy-linked K^+ accumulation with valinomycin. After depletion of mitochondrial anion reserves by transient uncoupling, citrate accumulation develops a relatively unspecific requirement for a dicarboxylic acid and an energy source.[68] Thus, it appears that the requirements for net anion uptake are distinctly different from those of anion exchange. The turnover of low levels of anionic metabolites is presumably more significant for the *in vivo* function of mitochondria than the massive net accumulation of anions.

Besides the anion permeability critera cited, swelling with high concentrations of ammonium salt and direct measurement of the entry of isotopically labeled substrates into mitochondria, a third criterion may be employed, namely the ability of a substrate to undergo metabolic transformation by intramitochondrial enzyme systems.[113] Thus, the stimulation of the reaction of exogenous aspartate with mitochondrial aspartate aminotransferase by glutamate, hydroxyglutamate, or aminoadipate indicates a carrier which exchanges glutamate and its analogues for aspartate.[5] This observation depends on pretreating mitochondria with uncoupler to deplete their anion reserves.

One must also bear in mind that the permeability of mitochondria can

TABLE 6-2
MITOCHONDRIAL ANION-TRANSPORT SYSTEMS

System	Substrates	Counterions	Inhibitors	References
I	Pi, Arsenate	OH^- (Co-ions, Ca^{++}, K^+, etc.)	p-Chlormer-curibenzoate	38, 40, 58
	(Azide?)			126
II	$CH_3(CH_2)_nCOOH$ $n < 7$ (Azide?)	OH^- (Co-ions, Ca^{++}, K^+, etc.)		126
III	$CH_3(CH_2)_nCOOH$ $n \geqslant 0$	$\left\{\begin{array}{l} \text{L-Carnitine} \\ \text{CoA} \\ \text{ATP} \end{array}\right\}$ *	Acyl-D-Carnitine α-Bromoacyl-CoA	60 182
IV	Succinate D, L-Malate Malonate	Pi, Arsenate	Chlorosuccinate 2-Butylmalonate	76 40
V	D, L-Tartrate Citrate iso-Citrate cis-Aconitate	Pi + L-Malate, Tartronate, iso-Malate	2-Butylmalonate	40, 153
VI	α-Ketoglutarate	Pi + L-Malate, Succinate Malonate meso-Tartrate	2-Butylmalonate MICA	40, 153 95, 190
VII	L-Glutamate	Pi (slightly)	2-Aminoadipate 4-Hydroxy-glutamate	5
VIII	L-Aspartate	Glutamate 2-Aminoadipate 4-Hydroxy-glutarate		5
IX	ADP, ATP	ATP, ADP	Atractyloside	78, 186
X	α-Glycero-phosphate	Dihydroxy-acetonephosphate		

* These are components of the fatty acid transferring system discussed in the text, not actual counterions.

vary with the source from which they are isolated. Thus, it has been found that housefly mitochondria are impermeable to succinate, malate, citrate, and isocitrate according to the criterion of swelling in the presence of ammonium ions.[183] Since the components of the α-glycerophosphate-dihydroxyacetone phosphate shuttle pass across the insect mitochondrial barrier with extreme rapidity,[21,56] this suggests the presence of a highly selective exchange diffusion carrier for these metabolites.

Permeation specificities can be inferred from studies with inhibitors, the effects of which depend on mitochondrial integrity. Thus, the inhibition by hydroxyglutamate or 2-aminoadipate of the reduction of intramitochondrial pyridine nucleotides by glutamate in intact mitochondria, but not in mitochondrial extracts, has been taken as evidence for a specific glutamate carrier.[5] In analogous fashion, chlorosuccinate, a substrate for succinic dehydrogenase in submitochondrial particles, is not metabolized by intact mitochondria, but does inhibit the utilization of succinate competitively.[76]

The inhibitor MICA produces a general inhibiton of liver metabolism[10] which has been attributed to a general interference with the transport of mitochondrial substrates.[95] By the criterion of mitochondrial swelling in the presence of ammonium ions, MICA appears to be particularly effective in blocking the uptake of α-ketoglutarate.[190]

By means of the various criteria outlined, several mitochondrial anion-transport systems have been operationally defined. These are summarized in Table 6-2.

XV. ADENINE NUCLEOTIDES

It had long been tacitly assumed that the penetration of a biological membrane by so highly charged a molecule as ATP would be unlikely. However, this assumption became untenable when it was observed that ATP preparations specifically labeled in either the adenine moiety or in the α, β or γ phosphate exchanged rapidly with intramitochondrial ATP without extensive relocation of the label.[132] These observations were subsequently confirmed and extended by Klingenberg and Pfaff.[93] Thus, there is little doubt that intact ATP can, indeed, penetrate the mitochondrial membrane.

Further support for this conclusion was obtained in experiments with the inhibitor atractyloside. Superficially, the effects of this agent resemble those of oligomycin. It prevents the phosphorylation of exogenous ADP and the hydrolysis of exogenous ATP induced by uncoupling agents, but does not prevent the release of respiratory control by uncoupling agents in the absence of a phosphate acceptor system.[20] A distinction between atractyloside and oligomycin was observed, in that the latter, but not the former, prevents the labeling of endogenous ATP by ^{32}Pi. Rather than inhibiting a metabolic process, atractyloside, therefore, blocks the passage of ADP and ATP across the mitochondrial membrane.[78,186] This effect is rather specific for adenine nucleotides and supports the existence of a specific adenine nucleotide carrier uniquely sensitive to the inhibitor atractyloside.

XVI. TRANSPORT OF OTHER ANIONS

Although $SO_4^=$ is impermeant according to the criterion of mitochondrial swelling in the presence of ammonium ions,[40] its uptake by mitochondria against considerable concentration gradients has been observed. The transport of $SO_4^=$ was also shown to be energy-linked and stimulated by K^+.[191]

Palmieri and Klingenberg found that the classical cytochrome oxidase inhibitor azide (N_3^-)[88] (cf Figure 6-1) is also actively accumulated by mitochondria.[126] Accumulation was measured with ferricyanide as the terminal electron acceptor in order to bypass the azide block in the electron transfer chain. The uptake of azide, like that of succinate,[77] is inhibited by uncouplers and enhanced by valinomycin-induced K^+ transport.[126] This presumably accounts for some of the earlier reports of direct inhibitory effects of azide on energy transfer which have been summarized in Ref. 126.

Although pyridine nucleotides penetrate the mitochondrial membrane extremely slowly, this process could be shown to be energy linked.[110]

XVII. WATER

The swelling of mitochondria, i.e., water uptake, is usually associated with the accumulation of ions and, thus, qualitatively consistent with a simple osmotic mechanism. The water uptake, however, appears to exceed that accountable for by osmotic forces alone.[69] The mitochondria may possess a mechanochemical system which could augment the osmotically-induced water movements.[102] The activity of such a hypothetical system might be regulated by the mitochondrial energy supply and by transport systems, thereby explaining the observed qualitative correlation between ion and water uptake. The energy-dependent, histone-induced swelling of mitochondria which is associated with the *release* of K^+ down a concentration gradient[84] indicates that significant discrepancies between the usual relationship of ion and water movements do exist.[69] At the moment, the existence of nonosmotic factors controlling mitochondrial water uptake is still an open question.

XVIII. INTERDEPENDENCE OF THE TRANSPORT OF CATIONS AND ANIONS IN MITOCHONDRIA

The *in vitro* concentration gradients of both cations and anions which energized mitochondria can establish and maintain are dependent on metabolic processes occurring *in vivo* within the cell. The translocation of an ion across a membrane against a concentration gradient requires either coupling to even greater concentration gradients of other ions via exchange diffusion carriers, or else electrophoretic propulsion by electropotential gradients. Once a primary electrochemical gradient is established, these mechanisms can, in turn, generate secondary gradients. Accordingly, the energetic requirements for mitochondrial ion translocation are ultimately reduced to the generation of a primary gradient or gradients.

As stressed in the introductory discussion of mitochondrial energetics, the chemiosmotic hypothesis has the virtue of explaining both oxidative phosphorylation as well as the generation of primary electropotential and hydrogen ion gradients. Despite a certain degree of popular acceptance, as outlined earlier in this review, it is difficult to reconcile this hypothesis with several experimental observations, and its ultimate validity must presently be regarded as uncertain.

A seemingly more attractive mechanism for mitochondrial transport is that of primary translocating pumps utilizing mobile ion carriers. The discovery of the ionophorous agents and, in particular, their detailed conformational behavior during complex formation provides a new impetus for earlier concepts of carrier-based mechanisms of active transport which were largely based on mathematical analyses of experimental transport data.[189] The existence of a multitude of ionophorous agents with structural features inherent in proteins, i.e., multiple, regularly spaced, carbonyl groups, suggests that it should be possible for organisms to fabricate highly selective cation carriers from proteins. Such an ionophore–cation complex could, in turn, function as an anion carrier with the cation functioning as a cofactor providing a lipid-compatible positive charge for pulling the anion across the lipoidal region of the membrane (Figure 6-3).[141a]

As previously pointed out, a lipophilic cation-complexing carrier could carry out several modes of transport (Table 6-2). These processes could either be energy driven or spontaneous. Transport Modes 1 and 3 would be electrophoretic (if the ions move down an electropotential gradient) or electrogenic, i.e., potential generating, if the ions move against a prevailing electrical potential. Transport Modes 2, 4, and 5 would be electrically neutral if n and n' are equal, in which case they would be neither helped nor hindered by any existing membrane potential. These neutral modes would, nevertheless, proceed either spontaneously or would require energy, depending on the direction of the prevailing concentration gradients.

Ion movements may be coupled in various ways. The potential set up by an energy driven, i.e., electrogenic transport, of ion species M_1^+ could drive the electrophoretic countermovement of M_2^+ via a second carrier system. The resultant M_1^+ for M_2^+ exchange differs from that of Mode 2, in that the reaction would be mediated through, and influenced by, the membrane potential, while the coupled exchange process of Mode 2 would not be. Electrogenic transport of Mode 1 could also drive an electrophoretic anion movement in the same direction via transport Mode 3.

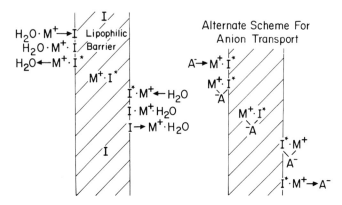

FIGURE 6-3. Resolution of processes involved in ionophore-mediated transport. The left sequence depicts cation translocation catalyzed by a valinomycin type ionophore, i.e., Transport Mode 1 of Table 6-2. The ionophore "I" forms a loose complex with the hydrated cation $M^+ \cdot H_2O$. It then undergoes a conformational change to I^* as it displaces water from $M^+ \cdot H_2O$ to form the lipophilic complex $M^+ \cdot I^*$ which diffuses across the barrier. At the opposite interphase, the first reaction sequence is reversed, releasing $M^+ \cdot H_2O$, and freeing I to diffuse back to the original interphase.

A varient reaction sequence for anion transport is indicated on the right. Entry of the $M^+ \cdot I^*$ complex into the barrier is facilitated by co-migration with the anion A^- to offset the charge of the cation M^+. The anion is released at the opposite interphase. This would constitute Mode 3 transport if the $M^+ \cdot I^*$ complex returns to the starting membrane interphase. If the $M^- \cdot I^*$ complex requires a replacement anion to recross the membrane, this would produce Mode 4 transport.[141a]

Such a net movement of ion pairs differs from that of Mode 5 in being membrane potential mediated.

While carriers are sufficient to explain most transport phenomena, other model systems suggest the permeation of ions across membranes via organized channel systems. Such conclusions are based on the high order of concentration dependence of induced membrane conductance and the lack of ionic selectivity of the induced permeability. Agents apparently capable of forming such channels include amphotericin B,[57] alamethicin,[122] and monazomycin.[122,123] Artificial membranes treated with the latter agents display a remarkable potential-controlled permeability which, under appropriate conditions, can simulate the wave form of biological action-potentials.[122]

The conformational options of ionophore-based carrier systems would lend themselves well to the altered affinity states required by the active transport model of Wilbrandt and Rosenberg.[189] On the other hand, the capability of the channel-based systems to respond to metabolic factors such as the transmembrane potential make them particularly suitable for operating a rapidly responding feedback control system for metabolic regulation.

Certain characteristics of mitochondrial transport seem best explained by an ionophore-based pump in series with a barrier region as shown in Figure 6-4. Mitochondria are known to undergo transient cation perme-

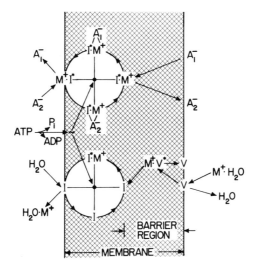

FIGURE 6-4. Diagrammatical representation of transport system in mitochondrial membrane. The upper mode depicts the pump operating as an energy-driven anion exchange system. The lower system illustrates how the addition of valinomycin (V) to the mitochondrial system would transform the anion exchange pump into an energy-driven cation pump. Other symbols are the same as employed in Figure 6-3.

ability changes in response to uncouplers.[26,27] K^+ flux is reduced considerably by uncouplers; the effect of further reducing the energy reserves by the addition of respiratory inhibitors is synergistic.[69] Treatment of mitochondria with sulfhydryl reagents such as mercurials[18,61,164] or Zn^{++} [14,17] unaccountably increases mitochondrial permeability. Although such responses cannot yet be rationalized completely, they suggest the existence of a mitochondrial control system containing functional SH groups which can either increase or decrease permeability.

Any comprehensive mechanism for energy-driven transport must take several factors into consideration. The increased intramitochondrial concentration of anionic substrates, such as succinate, during valinomycin-induced K^+ uptake and their decreased concentration upon treatment with uncoupling agents suggest a close link between mitochondrial anion and cation permeability.[68,77] However, according to isotope flux experiments, untreated mitochondria are normally highly permeable to anionic substrates, although they are but poorly permeable to K^+.[68,77] Furthermore, any scheme for mitochondrial transport and permeability must explain the high *in vitro* capacity of mitochondria for accumulating K^+ which would serve little purpose *per se, in vivo,* where a small K^+ gradient exists between the mitochondria and the cytosol.

According to the scheme of Figure 6-4 the most important role of the mitochondrial ionophore-based pump is inferred to be an energy-driven anion exchange, i.e., Mode 4 transport.[144] The relatively slow turnover of cation is explained by the relatively low cation permeability of the barrier region resulting in each K^+ remaining bound to the carrier for several cycles of anion turnover. Thus, anion turnover rates may exceed cation turnover rates even though both depend on a common pump mechanism.

According to the lower portion of Figure 6-4, the addition of valinomycin to the mitochondria permits the ionophore to carry K^+ across the lipoidal barrier giving it free access to the pump. This provides a low resistance pathway for K^+ which now completely breaches the membrane. When the pump is energized, this can lead to the rapid accumulation of K^+ linked with anion transport either via transport Mode 5 or transport Mode 4 operating in an electrophoretic fashion.

Although the selectivities of the transport systems cataloged in Table 6-2 support the existence of a large variety of distinct carrier systems, the inclusion of several unnatural transport substrates raises questions as to the ultimate significance of these selectivities. Such dubious transport metabolites as malonate, D, L, and *meso*tartrate, $SO_4^=$ and azide, suggest that classification of various systems listed in Table 6-2 is largely operational, and may simply reflect coincidental molecular properties and biologically functional groupings.

Chappell and co-workers have postulated that the uptake of citrate is mediated by a specific permease system which is dependent on malate and Pi.[40] However, freshly isolated mitochondria which have undepleted anion reserves are rapidly penetrated by isotopically labeled citrate.[68] Since there is every reason to believe that a similar situation prevails within the cell, the significance of the malate requirement observed *in vitro* is difficult to assess.

Thus, although several mitochondrial permeation systems may exist, the number of such systems may be fewer than suggested by Table 6-2.

One possible explanation for how a given ionic pathway may exhibit different environmental requirements for the movement of different ionic species is clarified by Table 6-1. Various hybrid transport modes may arise if a carrier operating in an exchange mode operates on ions of unequal charges. Thus, if the charge n of A_1 does not equal n' of A_2 in Mode 4 transport, the resulting exchange will become a hybrid of Mode 3 and 4 transport, i.e., it would become a function of the membrane potential as well as the concentration gradients. Alternatively, the pure Mode 4 electrically neutral transport could be preserved if other ions were carried simultaneously to preserve electroneutrality, e.g., citrate taken up in exchange for malate and phosphate. Then, too, other metabolic factors may exert a regulatory effect on transport by modifying the conformation of a given carrier allosterically. The observation that only acetate, β-hydroxy-butrate, glutamate, $SO_4^=$, and HCO_3^- are accumulated during Ca^{++} uptake but that valinomycin-induced K^+ accumulation will promote the uptake of all Krebs cycle intermediates indicates that even the cation species serving as a carrier co-factor may help determine anion specificity.[92] These results appear to be at variance with those derived from swelling studies that indicate $SO_4^=$ and HCO_3^- to be impermeant.[38,40]

The possibility must also be considered that the cation pump operates electrogenically in Mode 1 to increase the positive charge on the interior of the mitochondrial membrane which then serves to drive in anions electrophoretically via Mode 3. This would require a membrane potential of the opposite sign than that suggested by the chemiosmotic hypothesis, and one which is consistent with the observed relationships between the gradients established by several anionic species of charges ranging from minus one to minus three.[75]

In addition to the extensive list of specific transport inhibitors included in Table 6-1, evidence exists that the natural mitochondrial substrates themselves compete mutually for entry into the mitochondria. Such competitions could exert a considerable control on metabolism. Furthermore, the apparent K_m values for these substrates, exhibited by intact mitochondria, could be quite different from the inherent values for the appropriate dehydrogenases. Thus, the intramitochondrial substrate levels are probably controlled by a complex interplay between the extramitochondrial anionic environment, the energy-linked transport systems, and the rates of intramitochondrial substrate utilization.[73]

Two of the "substrates" listed in Table 6-1, malonate and azide, are actually inhibitors of specific electron-transfer reactions (cf Figure 6-1). Thus, the various K_i values reported in the literature for these agents are, likewise, not strictly inherent to the mitochondrial enzyme sites, but are also functions of the mitochondrial transport systems and the energetic state of the mitochondria.

XIX. CONCLUSION

The capacity of mitochondria for the energy-linked transport of divalent ions decreases according to the series $Ca^{++} > Sr^{++} > Mn^{++} > Mg^{++}$. Certain complexing agents termed "ionophores" enable mitochondria to accumulate alkali ions as well at rapid rates. Mitochondrial cation transport may arise from an energy-linked pump intended to drive metabolite anions across the mitochondrial membrane. Hypotheses both for the mechanism of action of ionophores and the linkage of mitochondrial cation and anion transport have been presented in detail.

We may speculate that the next phase in the study of mitochondrial transport could be the isolation and purification of elements of the mitochondrial transport system so that their affinities and other properties may be studied in the same detail as is currently being applied to the chemically pure ionophores. Analogous substances involved in bacterial transport have already been isolated, e.g., the crystalline sulfate transferring protein from Salmonella.[127] Mueller and Rudin have found that certain crude fractions of mitochondria exhibit ionophore-like electrometric effects on artificial membranes.[124] Hopefully, the general principles emerging from the study of mitochondrial transport and the chemically pure ionophores will extend to nonmitochondrial and nonelectrolyte transport systems as well.

REFERENCES

1. Addanki, S., Cahill, F. D., and Sotos, J. F. (1968), *J. Biol. Chem.* **243**, 2337.
2. Agtarap, A., Chamberlin, J. W., Pinkerton, M., and Steinrauf, I. (1967), *J. Amer. Chem. Soc.* **89**, 5737.
3. Amoore, J. E. (1960), *Biochem. J.* **76**, 438.
4. Amoore, J. E., and Bartley, W. (1958), *Biochem. J.* **69**, 223.
5. Azzi, A., Chappell, J. B., and Robinson, B. H. (1968), *Biochem. Biophys. Res. Commun.* **32**, 624.
6. Azzi, A., and Scarpa, A. (1967), *Biochim. Biophys. Acta* **135**, 1087.
7. Bangham, A. D., Standish, M. M., and Miller, N. (1965), *Nature* **208**, 1295.
8. Bartley, W., and Amoore, J. E. (1958), *Biochem. J.* **69**, 348.
9. Bartley, W., and Davies, R. E. (1954), *Biochem. J.* **57**, 37.
10. Bauman, N., and Hill, C. J. (1968), *Biochemistry* **7**, 1322.
11. Beck, J., Gerlach, H., Prelog, V., and Voser, W. (1962), *Helv. Chim. Acta* **45**, 620.
12. Bonner, W. D., and Pressman, B. C. (1965), *Plant Physiol.* **40**, lv-lvi.
13. Brierley, G. P. (1963), "Energy-Linked Functions of Mitochondria," p. 237 (B. Chance, ed.), Academic Press, Inc., New York.

14. Brierley, G. P. (1967), *J. Biol. Chem.* **242,** 1115.
15. Brierley, G. P., Bachmann, E., and Green, D. E. (1962), *Proc. Natl. Acad. Sci.* **48,** 1928.
16. Brierley, G. P., Murer, E., Bachmann, E., and Green, D. E. (1963), *J. Biol. Chem.* **238,** 3482.
17. Brierley, G. P., and Settlemire, C. T. (1967), *J. Biol. Chem.* **242,** 4324.
18. Brierley, G. P., Settlemire, C. T., and Knight, V. A. (1967), *Biochem. Biophys. Res. Commun.* **28,** 420.
19. Brierley, G. P., Settlemire, C. T., and Knight, V. A. (1968), *Arch. Biochem. Biophys.* **126,** 276.
20. Bruni, A., Bistocchi, M., and Contessa, A. R. (1958), *Boll. Soc. Ital. Biol. Sper.* **34,** 459.
21. Bücher, T., and Klingenberg, M. (1958), *Angew. Chem.* **70,** 552.
21a. Caplan, A. I., and Carafoli E. (1965), *Biochim. Biophys. Acta* **104,** 317.
22. Carafoli, E. (1965), *Biochim. Biophys. Acta* **97,** 99, 107.
23. Carafoli, E., Gamble, R. L., and Lehninger, A. L. (1965), *Biochem. Biophys. Res. Commun.* **21,** 215.
24. Carafoli, E., Gamble, R. L., Rossi, C. S., and Lehninger, A. L. (1967), *J. Biol. Chem.* **242,** 1199.
25. Carafoli, E., Weiland, S., and Lehninger, A. L. (1965), *Biochim. Biophys. Acta* **97,** 88.
26. Casewell, A. H., J. *Membrane Biol.* (in press).
27. Casewell, A. H., and Pressman, B. C. (1968), *Biochem. Biophys. Res. Commun.* **30,** 637.
28. Chance, B. (1956), *Proc. 3rd Intern. Cong. Biochem.* p. 300.
29. Chance, B. (1961), *J. Biol. Chem.* **236,** 1544.
30. Chance, B. (1965), *J. Biol. Chem.* **240,** 2729.
31. Chance, B., and Mela, L. (1966), *J. Biol. Chem.* **241,** 4588.
32. Chance, B., and Mela, L. (1966), *Biochemistry* **5,** 3220.
33. Chance, B., and Williams, G. R. (1956), *Advan. Enzymol.* **17,** 65.
34. Chance, B., and Yoshioka, T. (1965), *Fed. Proc.* **24,** 425.
35. Chance, B., and Yoshioka, T. (1966), *Biochemistry* **5,** 3224.
36. Chappell, J. B., Cohn, M., and Greville, G. D. (1963), "Energy-Linked Functions of Mitochondria," p. 219 (B. Chance, ed.), Academic Press, Inc., New York.
37. Chappell, J. B., and Crofts, A. R. (1965), *Biochem. J.* **95,** 393.
38. Chappell, J. B., and Crofts, A. R. (1966), "Regulation of Metabolic Processes in Mitochondria," Vol. 7, p. 293 (J. M. Tager, S. Papa, E. Quagliariello, and E. C. Slater, eds.), Biochim. Biophys. Acta Library, Amsterdam.
39. Chappell, J. B., Greville, G. D., and Bicknell, K. E. (1962), *Biochem. J.* **84,** 61.
40. Chappell, J. B., and Haarhoff, K. N. (1967), "Biochemistry of Mitochondria," p. 75 (E. C. Slater, Z. Kaniuga and L. Wojtczak, eds.), Academic Press Ltd., London.
41. Chase, L. R., and Aurbach, G. D. (1968), *Science* **159,** 545.
42. Christiansen, R. O., and Loyter, A. (1968), *Fed. Proc.* **27,** 527.

43. Christie, G. S., Ahmed, K., McLean, A. E. M., and Judah, J. D. (1965), *Biochim. Biophys. Acta* **94**, 432.
44. Claude, A. (1950), "The Harvey Lectures," Series XLIII, pp. 121–164.
45. Cockrell, R. S. (1968). *Fed. Proc.* **27**, 528.
46. Cockrell, R. S. (1968), "Relationship of K$^+$ Transport to Oxidative Phosphorylation," Doctoral Thesis, Univ. of Pennsylvania.
47. Cockrell, R. S., Harris, E. J., and Pressman, B. C. (1966), *Biochemistry* **5**, 2326.
48. Cockrell, R. S., Harris, E. J., and Pressman, B. C. (1967), *Nature* **215**, 1487.
49. Davies, R. E., and Ogston, A. G. (1950), *Biochem. J.* **46**, 324.
50. DeLuca, H. F., and Engstrom, G. W. (1961), *Proc. Natl. Acad. Sci.* **47**, 1744.
51. DuBois, K. P., and Potter, V. R. (1943), *J. Biol. Chem.* **150**, 185.
52. Eisenman, G. (1967), "Glass Electrodes for Hydrogen and Other Cations," p. 268 (G. Eisenman, ed.), Marcel Dekker, Inc., New York.
53. Erlanger, B. F., and Goode, L. (1954), *Nature* **174**, 840.
54. Ernster, L. "The Energy Level and Metabolic Control in Mitochondria," (J. M. Tager, S. Papa, E. Quagliariello, and E. C. Slater, eds.), Adriatica Editrice, Bari, in press.
55. Ernster, L., Dallner, G., and Azzone, F. (1963), *J. Biol. Chem.* **238**, 1124.
56. Estabrook, R. W., and Sacktor, B. (1958), *J. Biol. Chem.* **233**, 1014.
57. Finkelstein, A., and Cass, A. (1968), *J. Gen. Phys.* **52**, 145 S.
58. Fonyo, A. (1968), *Biochem. Biophys. Res. Commun.* **32**, 624.
59. Fritz, I. B. (1963), *Advan. Lipid Res.* **1**, 285.
60. Fritz, I. B., and Delisle, G. (1967), *Proc. Natl. Acad. Sci.*, **58**, 790.
61. Gamble, J. L. (1957), *J. Biol. Chem.* **228**, 955.
62. Gamble, J. L., Jr. (1963), *Biochim. Biophys. Acta* **66**, 158.
63. Gamble, J. L., Jr. (1963), *Proc. Soc. Exptl. Biol. Med.* **113**, 375.
64. Gamble, J. L. (1965), *J. Biol. Chem.* **240**, 2668.
65. Graven, S. N., Estrada-O, S., and Lardy, H. A. (1966), *Proc. Natl. Acad. Sci.* **56**, 654.
66. Graven, S. N., Lardy, H. A., Johnson, D., and Rutter, A. (1966), *Biochemistry* **5**, 1729.
66a. Hall, C., Wu, M., Crane, F. L., Takahashi, H., Tamura, S., and Folkers, K. (1966), *Biochem. Biophys. Res. Commun.* **25**, 373.
67. Harold, F. M., and Baarda, J. (1967), *Bacteriology* **94**, 53.
68. Harris, E. J. (1968), *Biochem. J.* **109**, 247.
69. Harris, E. J., Catlin, G., and Pressman, B. C. (1967), *Biochemistry* **6**, 1360.
70. Harris, E. J., Cockrell, R. S., and Pressman, B. C. (1966), *Biochem. J.* **99**, 200.
71. Harris, E. J., Hoeffer, M. P., and Pressman, B. C. (1967), *Biochemistry* **6**, 1348.
72. Harris, E. J., Judah, J. D., and Ahmed, K. (1966), *Current Topics in Bioenergetics,* **1**, 255.
73. Harris, E. J., and Manger, J. R. (1968), *Biochem. J.* **109**, 239.
74. Harris, E. J., and Pressman, B. C. (1967), *Nature,* **216**, 918.

75. Harris, E. J., and Pressman, B. C. (1969), *Biochim. Biophys. Acta* **172**, 66.
76. Harris, E. J., and van Dam, K., unpublished results.
77. Harris, E. J., van Dam, K., and Pressman, B. C. (1967), *Nature* **213**, 1126.
78. Heldt, H. W., Jacobs, H., and Klingenberg, M. (1965), *Biochem. Biophys. Res. Commun.,* **18**, 174.
79. Hinke, J. A. M. (1961), *J. Physiol. (London)*, **156**, 314.
80. Hollunger, G. (1955), *Acta Pharmacol. Toxicol. II,* Suppl. 1.
81. Jacobs, M. H., and Stewart, D. R. (1932), *J. Cell. Comp. Physiol.* **1**, 71.
82. Jackson, K. L., and Pace, N. (1956), *J. Gen. Physiol.* **40**, 47.
83. Jagger, W. S., and Pressman, B. C., unpublished experiments.
84. Johnson, C. L., Mauritzen, C. M., Starbuck, W. C., and Schwartz, A. (1967), *Biochemistry* **6**, 1121.
85. Johnson, J. H., and Pressman, B. C. (1968), *Biochim. Biophys. Acta* **153**, 500.
86. Johnson, J. H., and Pressman, B. C., *Arch. Biochem. Biophys.,* **132**, 139.
87. Judah, J. D., McLean, A. E. M., Ahmed, K., and Christie, G. S. (1965), *Biochim. Biophys. Acta* **94**, 441.
88. Keilin, D. (1936), *Proc. Roy. Soc. London,* **121B**, 165.
89. Kielley, W. W., and Bronk, J. R. (1958), *J. Biol. Chem.* **230**, 521.
90. Kielley, W. W., and Kielley, R. K. (1951), *Fed. Proc.* **10**, 207.
91. Kilbourn, B. T., Dunitz, J. D., Pioda, L. A. R., and Simon, W. (1967), *J. Mol. Biol.* **30**, 559.
92. Kimmich, G., and Rasmussen, H. (1968), *Fed. Proc.,* **27**, 528.
93. Klingenberg, M., and Pfaff, E. (1966), in "Regulation of Metabolic Processes in Mitochondria," p. 180 (J. M. Tager, S. Papa, E. Quagliariello, and E. C. Slater, eds.), Elsevier Publ. Co., Amsterdam.
94. Lardy, H. A. (1962), *First IUB/IUBS Symposium, Biological Structure and Function,* Stockholm, Academic Press, Inc., New York.
95. Lardy, H. A. (1969), "The Energy Level and Metabolic Control in Mitochondria," (S. Papa, J. M. Tager, E. Quagliariello, and E. C. Slater, eds.), Adriatica Editrice, Bari.
96. Lardy, H. A., and Elvejhem (1945), *Ann. Rev. Biochem.* **14**, 1.
96a. Lardy, H. A., Graven, S. N., and Estrada, O. S. (1967), *Fed. Proc.* **26**, 1355.
97. Lardy, H. A., Johnson, D., and McMurray, W. C. (1958), *Arch. Biochem. Biophys.* **78**, 587.
98. Lardy, H. A., and Wellman, H. (1952), *J. Biol. Chem.* **195**, 215.
99. Lardy, H. A., and Wellman, H. (1953), *J. Biol. Chem.* **201**, 357.
100. Lardy, H. A., Witonsky, P., and Johnson, D. (1965), *Biochemistry* **4**, 552.
101. Lehninger, A. L. (1949), *J. Biol. Chem.* **178**, 625.
102. Lehninger, A. L. (1962), *Physiol. Rev.* **42**, 467.
103. Lehninger, A. L., Carafoli, E., and Rossi, C. S. (1968), *Advan. Enzymol.* **29**, 259.
104. Lehninger, A. L., Rossi, C. S., and Greenawalt, J. W. (1963), *Biochem. Biophys. Res. Commun.* **10**, 444.
105. Lev, A. A. (1964), *Nature* **201**, 1132.
106. Lindberg, O., and Ernster, L. (1954), *Nature* **173**, 1038.

107. Lynn, W. S., and Brown, R. H. (1966), *Biochim. Biophys. Acta* **110**, 459.
108. MacFarlane, M. G., and Spencer, A. G. (1953), *Biochem. J.* **54**, 569.
109. Maynard, L., and Cotzias, G. (1955), *J. Biol. Chem.* **214**, 489.
110. Max, S. R., and Purvis, J. L. (1965), *Biochem. Biophys. Res. Commun.* **21**, 587.
111. McMurray, W. and Begg, R. W. (1959), *Arch. Biochem. Biophys.* **84**, 546.
111a. Mela, L. (1968), *Arch. Biochem. Biophys.* **123**, 286.
112. Mela, L. (1968), "Symposium on Ion Transport and Intramitochondrial pH," *Ann. N.Y. Acad. Sci.*, in press.
113. Meijer, A. J., De Haan, E. J., and Tager, J. M. (1967), "Mitochondrial Structures and Compartmentation," p. 207 (E. Quagliariello, S. Papa, E. C. Slater, and J. M. Tager, eds.), Adriatica Editrice, Bari, Italy.
114. Meyer, C. E., and Reusser, F. (1967), *Experientia* **23**, 85.
115. Mitchell, P. (1961), *Nature* **191**, 144.
116. Mitchell, P. (1966), "Chemiosmotic Coupling in Oxidative and Photosynthetic Phosphorylation," Glynn Research Ltd., Bodmin, Cornwall, England.
117. Mitchell, P. (1966), *Biol. Rev.* **41**, 445.
118. Mitchell, P., Moyle, J., and Smith, L. (1968), *European J. Biochem.* **4**, 9.
119. Moore, C., and Pressman, B. C. (1964), *Biochem. Biophys. Res. Commun.* **15**, 562.
120. Mraz, F. R. (1962), *Proc. Soc. Exptl. Biol. Med.* **111**, 429.
121. Mueller, P., and Rudin, D. O. (1967), *Biochem. Biophys. Res. Commun.* **26**, 398.
122. Mueller, P., and Rudin, D. O. (1968), *Nature* **217**, 713.
123. Mueller, P., and Rudin, D. O. "Current Topics in Bioenergetics," 3 (R. Sanadi, ed.), Academic Press, Inc., New York, in press.
124. Mueller, P., and Rudin, D. O., personal communication.
125. Packer, L. (1967), *Biochem. Biophys. Res. Commun.* **28**, 1022.
126. Palmieri, F., and Klingenberg, M. (1967), *Europ. J. Biochem.* **1**, 439.
127. Pardee, A. B., *Ann. N.Y. Acad. Sci.*, in press.
128. Pedersen, C. J. (1967), *J. Amer. Chem. Soc.* **89**, 7017.
129. Pioda, L. A. R., Wacter, H. A., Dohner, R. E., and Simon, W. (1967), *Helv. Chim. Acta* **50**, 1373.
130. Potter, V. R. (1947), *J. Biol. Chem.* **169**, 17.
131. Potter, V. R., and Reif, A. E. (1952), *J. Biol. Chem.* **194**, 287.
132. Pressman, B. C. (1958), *Fed. Proc.* **17**, 291.
133. Pressman, B. C. (1958), *J. Biol. Chem.* **232**, 967.
134. Pressman, B. C. (1963), *J. Biol. Chem.* **238**, 401.
135. Pressman, B. C. (1963), "Energy-Linked Functions of Mitochondria," p. 181 (B. Chance, ed.), Academic Press, Inc., New York.
136. Pressman, B. C. (1965), *Proc. Natl. Acad. Sci.* **53**, 1076.
137. Pressman, B. C. (1967), *Methods Enzymol.* **10**, 714.
138. Pressman, B. C. (1967), "Proceedings of International Symposium on Mechanisms of Action of Fungicides and Antibiotics," Castle Reinhardsbrunn, D.D.R., May 1966, Akademie-Verlag, Berlin.

139. Pressman, B. C. "Symposium on Ion Transport and Intramitochondrial pH" (S. Adenki, and J. F. Sotos, eds.), Annals N.Y. Acad. Sci., in press.

140. Pressman, B. C. (1968), in Symposium on Biological and Artificial Membranes, Am. Physiol. Soc. Society Meeting, *Fed. Proc.* **27**, 1283.

141. Pressman, B. C. (1968), *Proc. 5th FEBS Meeting*, Prague. Symposium on Mitochondria–Structure and Function (L. Ernster, and Z. Drahota, eds.), Academic Press Ltd., London and Acad. Sci. Pub., Prague, Czech. in press.

141a. Pressman, B. C. (1969), in "The Energy Level and Metabolic Control in Mitochondria," S. Papa, J. M. Tager, E. Quagliariello, and E. C. Slater, Eds., p. 87, Bari, Adriatica Editrice.

142. Pressman, B. C., and Harris, E. J. (1967), *Abstr. 7th Intern. Congr. Biochem.* (Tokyo) **5**, 900.

143. Pressman, B. C., Harris, E. J., Jagger, W. S., and Johnson, J. H. (1967), *Proc. Natl. Acad. Sci.* **58**, 1949.

144. Pressman, B. C., and Haynes, D. H. (1968), "Symposium on Molecular Basis of Membrane Function," Prentice Hall, New York.

145. Pressman, B. C., and Fortes, P. A. G., unpublished observations.

146. Pressman, B. C., and Lardy, H. A. (1952), *J. Biol. Chem.* **197**, 547.

147. Pressman, B. C., and Lardy, H. A. (1955), *Biochim. Biophys. Acta* **18**, 482.

148. Pressman, B. C., and Lardy, H. A. (1956), *Biochim. Biophys. Acta* **21**, 458.

149. Pressman, B. C., and Park, J. K. (1963), *Biochem. Biophys. Res. Commun.* **11**, 182.

150. Price, C. A., Fonnesu, A., and Davies, R. E. (1956), *Biochem. J.* **64**, 754.

151. Rasmussen, H., Chance, B., and Ogata, E. (1965), *Proc. Natl. Acad. Sci.* **53**, 1069.

152. Rasmussen, H., Fischer, J., and Arnaud, C. (1964), *Proc. Natl. Acad. Sci.* **52**, 1198.

153. Robinson, B. H., and Chappell, J. B. (1967), *Biochem. Biophys. Res. Commun.* **28**, 249.

154. Rossi, C., and Azzone, G. F. (1965), *Biochim. Biophys. Acta* **110**, 434.

155. Rossi, C. S., and Lehninger, A. L. (1963), *Biochem. Z.* **338**, 698.

156. Rossi, C. S., and Lehninger, A. L. (1964), *J. Biol. Chem.* **239**, 3971.

157. Rossi, E., and Azzone, G. F. (1968), *J. Biol. Chem.* **243**, 1514.

158. Rothstein, A., and Van-Steveninck, J. (1966), "Biological Membranes: Recent Progress," *Ann. N.Y. Acad. Sci.* **137**, 606–623.

159. Rottenberg, H., and Solomon, A. K. (1965), *Biochim. Biophys. Res. Commun.* **20**, 85.

160. Sallis, J. D., DeLuca, H. F., and Rasmussen, H. (1963), *J. Biol. Chem.* **238**, 4098.

161. Sarges, R., and Witkop, B. (1965), *J. Amer. Chem. Soc.* **87**, 2011.

162. Saris, N. E. (1959), *Finska Kemistamfundets Medd.* **68**, 98.

163. Saris, N. E. (1963), *Soc. Sci. Fennica, Commentationes Phys.-Math.* **28**, 11.

164. Scott, R. L., and Gamble, J. L., Jr. (1961), *J. Biol. Chem.* **236**, 570.

165. Share, L. (1958), *Amer. J. Physiol.* **194**, 47.

166. Shavit, N., Dilley, R. A., and San Pietro, A. (1968), *Biochemistry* **7**, 2356.
167. Shavit, N., and San Pietro, A. (1967), *Biochem. Biophys. Res. Commun.* **28**, 277.
168. Shavit, N., Thore, A., Keister, D. L., and San Pietro, A. (1968), *Proc. Natl. Acad. Sci.* **59**, 917.
169. Shemyakin, M. M. (1960), *Angew. Chem.* **72**, 342.
170. Shemyakin, M. M., Aldanova, N. A., Vinogradova, E. I., and Feigina, M. U. (1963), *Tetrahedron Letters, 1921*.
171. Shemyakin, M. M., Oychinnikov, Yu. A., Ivanov, V. T., and Evstratov, A. V. (1967), *Nature* **213**, 412.
172. Shemyakin, M. M., Oychinnikov, Yu. A., Ivanov, V. T., Antonov, V. K., Shkrob, A. M., Mikhaleva, I. I., Evstratov, A. V., and Malenkov, G. G. (1967), *Biochem. Biophys. Res. Commun.* **29**, 834.
173. Shemyakin, M. M., Vinogradova, E. I., Feigina, M., Yu, A., Aldanova, N. A., Loginova, N. F., Ryabova, I. E., and Pavlenko, I. A. (1965), *Experientia* **28**, 548.
174. Siekevitz, P., and Potter, V. R. (1953), *J. Biol. Chem.* **201**, 1.
175. Skou, J. C. (1965), *Physiol. Rev.* **45**, 596.
176. Slater, E. C. (1967), *European J. Biochem.* **1**, 317.
177. Slater, E. C., and Cleland, K. W. (1953), *Biochem. J.* **55**, 566.
178. Spector, W. G. (1953), *Proc. Roy. Soc. (London)*, **141B**, 268.
179. Stanbury, S. W., and Mudge, G. H. (1953), *Proc. Soc. Exptl. Biol. Med.* **82**, 675.
180. Stephanac, Z., and Simon, W. (1967), *Microchem. J.* **12**, 125.
181. Tosteson, D. C., Cook, P., Andreoli, T., and Tieffenberg, M. (1967), *J. Gen. Physiol.* **50**, 2514.
182. Tubbs, P. K., and Garland, P. B. (1968), *Brit. Med. Bul.* **24**, 158.
183. Van Den Bergh, S. G. (1967), "Mitochondrial Structure and Compartmentation," p. 207 (E. Quagliariello, S. Papa, E. C. Slater and J. M. Tager, eds.), Adriatica Editrice, Bari, Italy.
184. Ulrich, F. (1960), *Amer. J. Physiol.* **198**, 847.
185. Vasington, F. D., and Murphy, J. V. (1962), *J. Biol. Chem.* **237**, 2670.
186. Vignais, P. V., and Duce, E. D. (1966), *Bull. Soc. Chim. Biol.* **48**, 1169.
186a. Weinbach, E. C., and von Brand, Theodor (1965), *Biochem. Biophys. Res. Commun.* **9**, 133.
187. Wenner, C. E. (1966), *J. Biol. Chem.* **241**, 2810.
188. Wieland, T., Lueben, G., Ohenheym, H., de Vries, J. X., Prox, A., and Schmid, J. (1968), *Angew. Chem. Intern. Edit.* **7**, 204.
189. Wilbrandt, W., and Rosenberg, T. (1961), *Pharmacol. Rev.* **13**, 109.
190. Williamson, J. R., and Fukami, M., personal communication.
191. Winters, R. W., Delluva, A. M., Deyrup, I. J., and Davies, R. E. (1962), *J. Gen. Physiol.* **45**, 757.

7

Biogenesis of Mitochondria*

Gottfried Schatz
Institut für Biochemie†
University of Vienna, Austria

I. INTRODUCTION

The mechanism of mitochondrial formation has been under discussion for almost eighty years[4] and is, thus, one of the old problems of cell biology. Initially, the mitochondrion was mainly viewed as a morphological entity, and interest in its biogenesis was, therefore, largely restricted to cytologists. Some twenty years ago, however, this question also attracted the general attention of geneticists, since experiments with respiration-deficient yeast[105] and *Neurospora crassa* mutants[239] indicated that mitochondrial formation was in part governed by extrachromosomal genetic determinants. Finally, within the past decade, the problem of mitochondrial biogenesis has also moved within the scope of biochemical investigation, as rapid advances of mitochondrial biochemistry and the discovery of mitochondrial nucleic acids furnished a promising experimental basis from which to explore the assembly of the mitochondrial organelle.

As the study of mitochondrial formation was gradually evolving into a major interdisciplinary effort, it increasingly faced the problem of com-

---

* With a few exceptions, the survey of the literature covered by this article was completed in February 1968.
† Present address: Section of Biochemistry and Molecular Biology, Cornell University, Ithaca, New York.

munication and perspective. When this author decided to write a review on this subject, he was inexperienced enough to conceive a short, yet comprehensive, up-to-date account of all significant developments. He has since become considerably wiser in his aspirations, but still hopes that this survey may fulfill a useful function in drawing together experimental findings from diverse fields, and subjecting them to critical, if admittedly subjective, evaluation.

For additional details, the reader is referred to earlier reviews on this topic[213,215,320,274,131,319a,399a] as well as to the published proceedings of two recent symposia.[351,377] The numerous monographs dealing with chloroplast formation (e.g., see Refs. 136, 138) may also be recommended as valuable sources of information, since the general trends in this area have traditionally foreshadowed those in the field of mitochondrial biogenesis and have often provided stimulating analogies as well as useful ideas for further experiments.

II. MITOCHONDRIAL DNA

Until a few years ago it was generally held that the entire DNA complement of an eucaryotic cell was confined to the nucleus. The well-documented occurrence of extranuclear DNA in the cytoplasm of oocytes[38,149,140,362] as well as in the kinetoplast of flagellates[42,232,164,297,360] was regarded as the exception rather than the rule, since these cells differed in many respects from typical higher cells, especially those of mammalian origin. Even when it was later recognized that the kinetoplasts of trypanosomes and bodonids could, in fact, be interpreted as modified mitochondria,[359,69,286,392] the broader implications of this finding were obscured by the inability to detect DNA in the normal mitochondria of these organisms.

It was not until around 1960 that the possible existence of a mitochondrial DNA was actively explored by discussion as well as experimentation. In retrospect, it appears that this interest was largely triggered by the detection of DNA in chloroplasts[122,309,418,132,14] which had long been known to resemble mitochondria in many of their properties. During the past years, the experimental literature dealing with mitochondrial DNA has increased at an unexpected rate and at present it has reached proportions that preclude a detailed coverage by anything short of a separate full-sized review. The present survey, therefore, limits itself to those aspects of mitochondrial DNA which appear to be directly germane to a discussion of mitochondrial biogenesis. This restriction is justified by the recent appearance of several reviews which exhaustively describe the chemical and physical properties of mitochondrial DNA (e.g., see Refs. 253, 139).

Evidence for the Occurrence of DNA in Mitochondria

The presence of DNA in mitochondrial preparations isolated by differential centrifugation was noted as early as 1951,[160,291] but was generally attributed to contaminating nuclei.[161,223,274] Several years ago, however, evidence from different laboratories indicated that the mitochondria themselves contained a small amount of DNA that could not be ascribed to nuclear contamination. Chèvremont and his colleagues observed that cultured chick embryo fibroblasts treated with DNAase or trihydroxy-N-methylindole incorporated radioactive thymidine into their mitochondria, presumably into a mitochondrial DNA.[65,66,67] In view of the unusual experimental conditions, however, these findings did not lead to the acceptance of DNA as a normal mitochondrial constituent. More convincing cytological evidence for the existence of a mitochondrial DNA was subsequently obtained by Nass and Nass[255,256] who noted that mitochondria of *normal* chick embryo cells exhibited fiber-like inclusions that were specifically removed by treating the ultrathin tissue sections with DNAase and, thus, apparently consisted of DNA (Figure 7-1).

These cytological observations were complemented by the biochemical studies of Schatz *et al.*[332] which indicated that yeast mitochondria purified by flotation in a "Urografin" gradient contained a small and constant amount of DNA. Significantly, the distribution of this DNA in the gradient precisely paralleled that of mitochondrial enzyme activity.

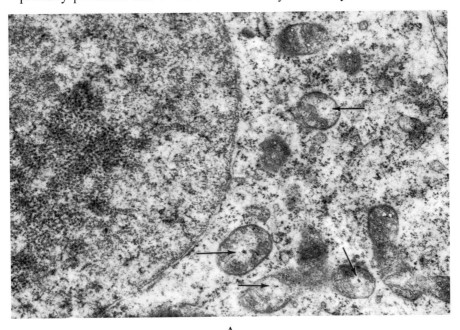

A

FIGURE 7-1. Demonstration of intramitochondrial DNA fibers in chick-embryo tissue fixed with osmium tetroxide (from Nass and Nass[256]). Magnification 29.800X. A control   B treated with RNAase   C treated with DNAase (Reproduced with permission, *J. Cell Biol.* and the authors)

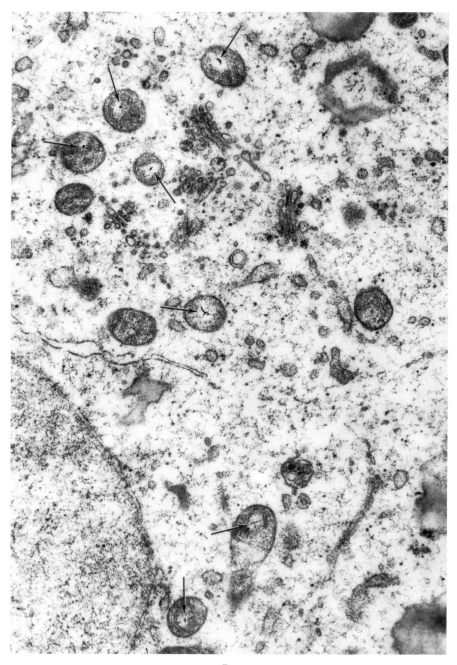

B

FIGURE 7-1. (Continued)

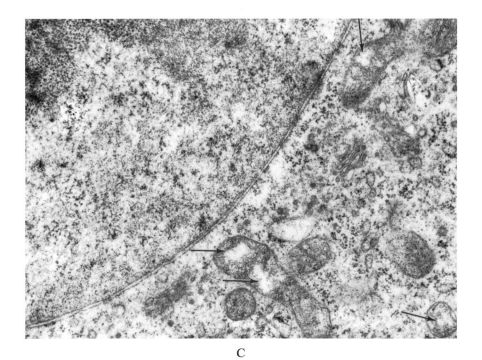

C

FIGURE 7-1. (Continued)

A particularly strong argument for the existence of a DNA indigenous to mitochondria was provided by Luck and Reich[229] who isolated high-molecular weight DNA from purified *N. crassa* mitochondria, and showed that its buoyant density differed from that of the corresponding nuclear DNA.

Luck and Reich also observed that the DNA associated with the *intact* mitochondria, in contrast to isolated mitochondrial DNA, was not degraded by added DNAase. This result suggested an intramitochondrial localization of the DNA and, thus, substantiated the earlier cytological evidence.

These few examples may suffice to illustrate the main experimental approaches which have been taken to demonstrate mitochondrial DNA in molds,[229,145,146,338,116,358a] yeasts,[332,381,382,387,248,427,77,238a] protozoa,[373,364,280,54,148,368,53,273,281,365,258] ferns,[26] amphibian oocytes,[88] mollusks,[285,258] algae,[101,301,100,29] higher plants,[258,187,375,428,63] and animals,[256,77,258,333,257,181,295,259,35,337,347,202,185,55,252,348] as well as in human cells.[73,168,190,77a] Although it is rarely admitted, none of these different approaches furnishes unambiguous evidence for the existence of a mitochondrial DNA. Electron microscopy does not permit a satisfactory chemical identification of DNA, and analytical gradient flotation of isolated mitochondria can not exclude contaminating DNA adsorbed to these organelles. Similarly, as exemplified by a recent controversy regarding mitochondrial DNA from sheep heart,[202 vs 183] the distinct properties of a DNA extracted from mitochondrial preparations do not necessarily prove its mitochondrial origin. Taken together, however, the various lines of evidence carry considerable weight and permit the conclusions that DNA is indeed a genuine, universal mitochondrial constituent.

Amount and Properties

As the presence of DNA in mitochondria became generally accepted, its possible significance received increasing attention. It was obviously attractive to regard this DNA as a mitochondrial gene that codes for mitochondrial proteins and, thereby, controls some essential step(s) of mitochondrial formation. In order to qualify for such a role, however, mitochondrial DNA must meet two minimal requirements: First, there should be enough of it in every mitochondrion to code for at least one reasonably large polypeptide chain; second, it should exhibit the characteristic properties of genetically active DNA from other biological systems. The following discussion shows that these criteria are indeed satisfied.

Although an entirely satisfactory procedure for measuring the mitochondrial DNA content is not yet known, there is ample evidence to suggest that mitochondria from most species contain between 0.2–1.0 μg DNA/mg protein.[35,88,185,229,252,256,259,282,333,337,382] Yeast mitochondria appear to be an exception, since they possess up to 4 μg DNA/mg protein.[332,393] Preliminary evidence suggests, moreover, that the amount of DNA per mitochondrion may be subject to physiological variation, being diminished in glucose-repressed yeast[382] and elevated in rapidly dividing cells.[258]

As different cell types may contain varying amounts of mitochondria which, in turn, may contain varying amounts of DNA, the contribution of mitochondrial DNA to the total cellular DNA content fluctuates considerably. In bovine heart, rat liver, or *N. crassa,* for example, mitochondrial DNA accounts for at most a few percent of the total DNA.[35,337,229] However, this value may rise to 10–20% in certain yeast strains,[248,77,246] and reach 90% or more in oocytes.[88,285] The absolute amount of DNA present in a single bovine heart mitochondrion has been calculated to be in the order of 10^{-17} g.[333] While this amount is certainly small, it can maximally code for some 70 polypeptide chains of molecular weight 17,000 and is, therefore, still compatible with a genetic function. However, as long as the exact redundancy of mitochondrial DNA is unknown, any detailed calculations of its informational capacity are somewhat premature.

While several early attempts to isolate mitochondrial DNA yielded preparations of relatively low molecular weight,[35,199] it is now firmly established that carefully isolated mitochondrial DNA from a variety of cell types exhibits a molecular weight in the order of 10^7 daltons.[382,337,368,49,37,]

[33,89,358a] Although this value is still much lower than that of chromosomal DNA,[56,237a] it is, again, fully consistent with an informational role. Band-sedimentation analyses of freshly isolated mitochondrial DNA from chick liver[33,34] lamb heart,[202] and various oocytes[285,89,90] revealed the presence of two sedimenting species with sedimentation constants of about 39 and 27 S. Treatment with DNAase I converted the 39 S component into the more slowly sedimenting one.[89,33,90,34]

This behavior resembled that of circular DNA isolated from polyoma virus[402,397] and the replicating form of bacteriophage φ 174.[189,61] Indeed, van Bruggen, Borst and colleagues,[49,33] and Nass[252] were able to show that mitochondrial DNA from several cell types exhibited a closed circular structure of rather uniform circumference (approximately 5–6μ). In contrast, the corresponding nuclear DNA was invariably devoid of circular configurations. A circular DNA molecule from chick liver mitochondria is depicted in Figure 7-2.

The circularity of mitochondrial DNA has been confirmed in many different laboratories,[285,347,89,419,348,168,308a] and is of immediate relevance to the problem of mitochondrial biogenesis. On the one hand, it represents one of the most striking features of mitochondrial DNA and may thus help to explain some of the unique features of the mitochondrial genetic system (cf below). On the other hand, it provides a criterion for the intactness of a mitochondrial DNA and, thereby, opens a way for assessing its maximal molecular weight by electron optical methods. Assuming a mass of 1.92×10^6 daltons per μ,[233] a circular DNA molecule 5 μ in circumference would possess a molecular weight of 0.96×10^7 daltons. This value agrees well with that obtained by other methods[382,337,368,89] and suggests that each mitochondrion may contain several circular DNA molecules. Electron micrographs of osmotically ruptured L-cell mitochondria show that each mitochondrion may extrude between two and six DNA circles.[252]

While the detection of circular mitochondrial DNA in a wide variety of cells would suggest that circularity is a general property of mitochondrial DNA, the findings with yeast constitute a puzzling exception. Purified mitochondrial DNA from this organism consists mostly of open-ended molecules which are at least 5 μ long.[349,416,32,34] According to Sinclair *et al.*,[349] circular molecules with a contour length between 0.5 and 7 μ may also be present; however, these appear to be a contaminant since they are more frequent in whole-cell DNA than in purified mitochondrial DNA and, in cesium chloride gradients, band at the density of nuclear DNA, well separated from the bulk of mitochondrial DNA. In contrast, Avers reported [9] that the DNA extruded from osmotically lysed yeast mitochondria consisted predominantly of circles with a circumference ranging from 0.5–10.1 μ. DNA circles were only obtained by lysing relatively intact mitochondrial preparations derived from yeast protoplasts; only open-ended DNA was extruded from damaged mitochondria which had been

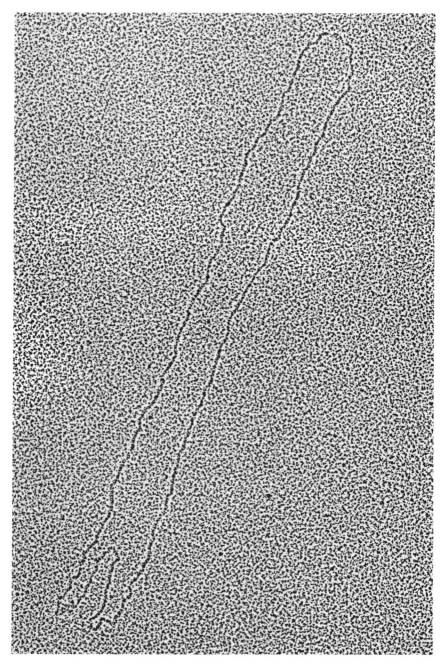

FIGURE 7-2. Circular mitochondrial DNA from chick-liver (open circular configuration) 89,440X (from van Bruggen *et al.*[49]). (Reproduced with permission from *Biochim. et Biophys. Acta,* **119** (1966) 439, Fig. 1d)

isolated by a rather drastic homogenization of whole yeast cells. These findings led Avers to conclude that yeast, like other cell types, contained circular mitochondrial DNA, and that the open-ended strands observed by previous investigators had actually been degradation products. However, the buoyant density of the DNA rings observed by Avers was not determined and it thus remains uncertain whether they were, as claimed, mitochondrial DNA. These reservations are strengthened by the contradictory findings of Sinclair et al.,[349] and by the fact that even "purified" yeast mitochondria may contain large amounts of contaminating nonmitochondrial DNA.[332] Further work will be required to resolve these discrepancies.

In many simpler organisms such as yeast,[382,77,248,349,393,245] molds,[229,116,147] protozoa,[281,54,368,53,373] unicellular algae,[101,301] and mollusks,[285] and in higher plants[370,187] as well as in avian tissues,[295,35,33] the buoyant density of mitochondrial DNA in cesium chloride gradients differs significantly from that of the corresponding nuclear DNA. With these organisms, gradient centrifugation of isolated DNA, or even simply of a cellular lysate,[248] offers a convenient and reliable means of identifying, or purifying, mitochondrial DNA. Upon denaturation, the buoyant density of these mitochondrial DNA's increases by about 0.015 density units, as expected for double-stranded DNA.[295,77,33,100,248,53,368,382,285,394,349] In several cases, (but not all [53]) it was found that the base ratio computed from the buoyant density agreed closely with the ratio determined from the thermal transition temperature, or the actual separation and estimation of the individual bases after acid hydrolysis (e.g., see Refs.[33,382,295,368]). Thus, it appears that the unique buoyant density of a mitochondrial DNA in most cases reflects a unique over-all base composition rather than a presence of some unusual bases or some uncommon linkage between the two complementary DNA strands.

In mammalian cells[337,77,35,185,202,347,36,53] as well as in amphibian oocytes,[88] the over-all base ratio of mitochondrial DNA differs only very little, if at all, from that of the corresponding nuclear DNA. Claims of a buoyant density difference between mitochondrial and nuclear DNA from sheep tissues[77,183] must now be discounted [202] and probably be explained in terms of contamination of the mitochondrial fractions by "satellite" DNA. This enigmatic DNA component has been detected in many cell types[366,188,68,248,295,35,202] and resembles mitochondrial DNA in several of its properties.[121,31] However, it is not a mitochondrial constituent, but appears to be largely, if not exclusively, associated with the nucleus, even though its over-all base composition differs from that of the bulk of nuclear DNA.[35,202,31] On the other hand, Koch and Stockstad [190] suggested that the "satellite" DNA of cultured human liver cells may be associated with the mitochondria rather than the nucleus. This conclusion was based on the observation that the DNA extracted from the human liver mitochondria exhibited a buoyant density that was identical with the reported density value for human "satellite" DNA[76] and considerably different from the density of nuclear DNA. The evidence presented by Koch and Stockstad, however, does not exclude the possibility that the "mitochondrial" DNA was, in fact, derived from nonmitochondrial particles which contaminated the mitochondrial preparations. It is, therefore, still somewhat uncertain whether human cells contain mitochondrial DNA of a unique buoyant density and, thereby, differ from the other mammalian cells which have been investigated until now.

In all cases studied thus far, mitochondrial DNA proved to be more homogeneous than nuclear DNA. Thus, mitochondrial DNA exhibited a narrower band width upon cesium chloride gradient centrifugation as well as a sharper "melting" profile, and renatured under conditions which precluded renaturation of nuclear DNA.[77,337,202,36,285,91,89,37,347,394,349,82]

A further interesting difference between mitochondrial and nuclear DNA was reported by Cummins et al.[82] These authors analyzed the nearest neighbor frequency of RNA that had been synthesized *in vitro* on templates of mitochondrial and nuclear DNA from the slime mold *Physarum polycephalum*. Whereas the frequency of CpG in RNA transcribed from nuclear DNA was found to be much lower than that calculated for random arrangement of nucleotides, it was close to the random value in the RNA transcribed from mitochondrial DNA. Since near-random frequencies are also observed in analogous experiments with DNA from bacteria, but not with that from higher organisms,[374] the experiments of Cummins et al. reveal yet another striking parallel between DNA from mitochondria and from lower forms of life.

In summary, then, there is every indication that the properties of mitochondrial DNA, and its amount per mitochondrion, are consistent with a genetic role.

Replication

Exposure of uni- or multicellular organisms to radioactive DNA precursors results in labeling of both mitochondrial and nuclear DNA.[337,259,338,116,280, 365,26,145,187,264,266,19,254,57,147,263,190] However, the labeling of mitochondrial DNA is usually much more rapid than, and out of phase with, that of nuclear DNA.[337,280,116,254,266,57,147,264,190,356a] These metabolic studies suggest that the synthesis of mitochondrial DNA is controlled by mechanisms which are, in part, independent from the chromosomal genetic system.

Mitochondrial DNA from a variety of cell types has been shown to be metabolically stable,[254,263,266] and to possess physical continuity through at least several cell divisions.[280,77,302] Its pattern of replication in *Saccharomyces cerevisiae* and *Neurospora crassa* cells upon transfer from a ^{15}N to a ^{14}N medium is compatible with, and suggestive of, a classical semiconservative mechanism.[77,302] These transfer experiments also revealed the fact that the precursor pools for the synthesis of mitochondrial and nuclear

DNA are metabolically distinct. Like informational DNA from other sources, mitochondrial DNA thus exhibits precise replication and ordered transmission to the progeny.

Independent evidence from different laboratories suggests that the replication of mitochondrial DNA is mediated by a DNA polymerase localized within the mitochondria themselves. Thus, isolated mitochondria from yeast,[413,414] rat liver,[336,282] and *P. polycephalum*[44] have been shown to incorporate labeled DNA precursors into an acid-soluble product that is degraded by DNAase, and indistinguishable from mitochondrial DNA with respect to buoyant density,[44] chromatographic behavior,[413] and ease of renaturation.[282] The incorporation reaction is inhibited by actinomycin, requires the presence of all four deoxyribonucleotide triphosphates, and, thus, possesses the characteristics of a DNA polymerase. The mitochondrial polymerase from rat liver[270] and yeast[416] can be solubilized by high-speed blending of the isolated mitochondria. The yeast enzyme has recently been considerably purified and found to differ from the corresponding nuclear enzyme with respect to magnesium requirement and chromatographic behavior.[417a] A detailed study of the purified mitochondrial polymerase will be of considerable interest.

Relationship to Nuclear DNA

The available evidence indicates that most, if not all, of mitochondrial- and nuclear-type DNA is indeed localized within the mitochondria and the nucleus, respectively. In living cells, the physical separation between the two DNA species is, therefore, quite clear-cut and probably complete. This does not exclude, however, that genetic messages carried by mitochondrial DNA are also imprinted on the DNA of chromosomes. Unfortunately, attempts to explore this intriguing possibility by DNA–DNA hybridization experiments have yielded conflicting results. According to Dawid and Wolstenholme[90,91] (cf also 87) nd Humm and Humm,[170] mitochondrial DNA from *Rana pipiens* and different mouse tissues hybridizes only to an insignificant extent with the corresponding nuclear DNA. In contrast, duBuy and Riley[55] reported extensive hybridization in analogous experiments with *Leishmania enriettii* and mouse liver. In this latter case, however, contamination of the isolated mitochondrial DNA by nuclear DNA was not rigorously excluded. Moreover, the results of duBuy and Riley may also be questioned on theoretical grounds: In view of the large difference in size between chromosomal DNA and mitochondrial DNA (roughly 10^{10} *vs* 10^7 daltons, cf 237a, 56, 349), hybridization between the two DNA species should be difficult to detect even if complete copies of mitochondrial DNA were integrated into the nuclear genome. Even if such marginal annealing could be reliably measured, it would still be extremely difficult to prove that it reflected a genuine sequence homology between the two DNA components rather than the presence of traces of mitochondrial DNA in the nuclear DNA preparations. It also appears that the specificity, and the pitfalls, of DNA–DNA hybridization techniques involving DNA immobilized on filters[133] or gels[30] are as yet incompletely understood. Although hybridization measurements based on buoyant density shifts are probably more reliable,[90] they seem to be too insensitive to reveal the small homologies between mitochondrial and nuclear DNA that are theoretically plausible. In view of these uncertainties, the existence of sequence homologies between mitochondrial and nuclear DNA is still an open question.

Mitochondrial DNA in the "Petite" Mutant of Yeast

Some twenty years ago, Ephrussi *et al.*,[112,105,109,107] and Mitchell *et al.*[239,240] observed that normal, respiring yeast and *N. crassa* cultures continuously give rise to a small number of spontaneous mutants which are characterized by a defective respiratory system and the lack of cytochromes aa_3 and *b*. These traits are inherited to the daughter cells in a non-Mendelian fashion and, thus, appear to reflect the loss or the inactivation of extrachromosomal hereditary determinants.[244] The nature of these determinants has remained unknown, but it is tempting to speculate that they are identical with mitochondrial DNA. If this is true, then extrachromosomal mutations leading to respiratory deficiency should involve the loss or the modification of mitochondrial DNA. Attempts to test this possibility in experiments with the respiration-deficient, cytoplasmic "petite" mutant of yeast initially led to contradictory conclusions. Corneo *et al.*,[77] Moustacchi and Williamson,[248] and Tewari *et al.*[382] reported that "petite" cells grown in a conventional shaking culture were essentially devoid of mitochondrial DNA, although this DNA was readily detected in the corresponding wild-type strain grown under identical conditions. This suggested that the "petite" mutation involved the permanent loss of mitochondrial DNA. However, the validity of these experimental observations was subsequently contested by Mounolou *et al.*[245] who argued that glucose repression of mitochondrial DNA in the respiration-deficient cells had not been excluded. In support of this view, Mounolou *et al.* found essentially normal amounts of a mitochondrial-type DNA in "petite" cells grown in a chemostat culture which minimized repression by glucose. However, this mitochondrial DNA differed from that present in the wild-type cells by its buoyant density in cesium chloride gradients. No density difference was detected between the nuclear DNA's or between mitochondrial DNA from the wild-type and a chromosomal respiration-deficient mutant. Mounolou *et al.* concluded, therefore, that the "petite" mutation resulted in the modification, rather than the loss, of mitochondrial DNA.

The findings of Mounolou *et al.*, which have been confirmed by Viehhauser *et al.*,[394] and by Mehotra and Mahler[238a] raise several puzzling questions. If the molecular weight of mitochondrial DNA from yeast is taken as 10^7 (Ref. 349), then the buoyant density data[245,238a] indicate that between 20 and 90% of the entire DNA molecule may be altered by the "petite" mutation. It is difficult to visualize the mechanism of such an extensive mutative alteration. An understanding of the "petite" mutation is further complicated by the recent observation[246] that a single wild-type *S. cerevisiae* cell may give rise to several distinct types of "petite" mutants which differ from each other by the buoyant density of their mitochondrial DNA. Although it has not yet been rigorously established that these abnormal mitochondrial DNA's arise by *modification* of the "wild-type" DNA, it appears that mutations of mitochondrial and chromosomal DNA may follow dissimilar mechanisms.

In any case, the presence of a mitochondrial DNA in cytoplasmic "petite" cells is in marked contrast to the situation observed with non-chromosomal, permanently bleached *Euglena* and *Chlamydomonas* mutants which are in other ways quite analogous to the cytoplasmic "petites" of yeast. On screening a large number of these algal mutants, two independent groups of workers found that some mutants contained chloroplast DNA, whereas others did not.[101,301,214] A comparative investigation of different "petite" strains, e.g., of those arising spontaneously or after treatment with different mutagens,[251,52,194] might thus shed further light on the involvement of mitochondrial DNA in extrachromosomal mutations.

In spite of some persisting uncertainties, there appears to be general agreement that the cytoplasmic "petite" mutation profoundly affects mitochondrial DNA. This result is again consistent with the possibility that mitochondrial DNA carries genetic information for the assembly of a functional mitochondrion.

Concluding Remarks

The only known function of DNA is that of carrying genetic information and it is, thus, reasonable to assume that the same holds true for mitochondrial DNA. As discussed in the preceding paragraphs, the physical and chemical properties of this DNA do indeed conform to those expected of an informational DNA. The mode of replication of mitochondrial DNA as well as its pattern of transmission in crosses between different *N. crassa* strains[302] are also fully consistent with the concept that this DNA fulfills a genetic function. Additional, but again circumstantial support for this concept stems from the observation that normal yeast cells can be transformed into cytoplasmic "petites" by exposure to 5-fluorouracil,[247] acridine dyes,[110] or ultraviolet light,[287,323,324,299,407,300] all of which are known to affect nucleic acids. In the case of ultraviolet irradiation, the action spectrum exhibits a maximum around 260 mμ and, thus, further implicates nucleic acid as the target.[300,407] A genetic role of mitochondrial DNA is also suggested by the fact that both amino acid incorporation and RNA synthesis in isolated mitochondria are inhibited by actinomycin, a well-known inhibitor of DNA transcription (cf below). Still more persuasive evidence is provided by experiments, discussed in the following section, which indicate that mitochondrial DNA codes for the RNA moieties of mitochondrial ribosomes. Nevertheless, it has not yet been shown that the DNA of mitochondria codes for a distinct protein and, thus, meets the requirements of a structural gene as defined by Jacob and Monod.[172] Since even less is known about a possible regulatory function of mitochondrial DNA, its role as a mitochondrial gene *sensu stricto* remains to be conclusively established.

III. MITOCHONDRIAL RNA

In the preceding paragraphs we have discussed the possibility that mitochondrial DNA codes for some mitochondrial protein(s) and is, thus, part of a unique mitochondrial genetic system. If we accept this hypothesis, then we might expect that mitochondria contain their own transcription- and translation system mediating the flow of genetic information from the DNA template to the corresponding polypeptide chain(s). In bacteria and eucaryotic cells, DNA–directed polypeptide synthesis requires the concerted action of at least three different classes of RNA: messenger, transfer, and ribosomal RNA. As the following discussion will show, each of these RNA types also seems to be present within the mitochondrial organelles.

RNA Content of Isolated Mitochondria

Numerous investigators have observed that isolated mitochondria contain RNA.[162,250,344,159,151,238,169,259,199,261,318,384,317,410,265] However, the significance of these findings is difficult to evaluate unless the mitochondrial preparations were isolated with special precautions which make it possible to exclude, or at least assess, contamination by cytoplasmic ribosomes. Such precautions include treatment of intact mitochondria with RNAase[308] (cf below), repeated washings of the mitochondria at low centrifugal forces,[275] or radioactivity measurements of mitochondrial fractions that had been isolated from cell homogenates supplemented with labeled microsomes or ribosomes.[215,415] RNA assays carried out in this manner indicate that mitochondria from yeast and *Neurospora crassa* contain 48 and 65 μg RNA/mg protein, respectively,[415,308] whereas rat-liver mitochondria possess only 6.6–8.0 RNA/mg protein.[275] Since both yeast and *N. crassa* lack a pronounced endoplasmic reticulum and, therefore, contain only few membrane-bound cytoplasmic ribosomes, they are especially favorable biological systems for studying mitochondrial RNA. Interestingly, mitochondria from a cytoplasmic "petite" mutant of *S. cerevisiae* contain almost five times less RNA than mitochondria from the corresponding wild-type strain.[415] In contrast, mitochondria from chloramphenicol-repressed wild-type cells (cf below) possess a normal amount of RNA[415] even though they resemble "petite" mitochondria with respect to morphology and enzyme content.[167,71]

Characterization of Mitochondrial RNA

The isolation and partial characterization of a relatively undegraded mitochondrial RNA was first reported by Wintersberger and Tuppy.[417] These authors observed that RNA extracted by phenol from highly purified yeast mitochondria could be separated, by sucrose gradient centrifugation, into three sedimenting species exhibiting sedimentation constants of approximately 4, 16, and 23 S. On the basis of this sedimentation pattern, Wintersberger and Tuppy suggested that mitochondria contained transfer RNA and two RNA species resembling those present in the cytoplasmic ribosomes. These early findings, which have since been confirmed [127,312] and extended to mitochondria from *N. crassa,*[308,206,98] regenerating rat liver,[201] and cultured hamster cells[97] furnished the basis for a rapid succession of more detailed studies which are briefly summarized in the following sections.

Mitochondrial Transfer RNA

More reliable evidence for the existence of mitochondrial transfer RNA was provided by the observation[412] that the low-molecular weight RNA from yeast mitochondria accepted radioactivity upon incubation with ATP, Mg^{++}, a labeled amino acid, and an aminoacyl-RNA synthetase preparation from a yeast 105,000 × g supernatant. The stability characteristics of the labeled product were indistinguishable from those of an authentic sample of aminoacyl-RNA synthesized with cytoplasmic transfer RNA as acceptor. These results suggest that the mitochondrial 4 S RNA, indeed, contains transfer RNA and is thus not merely a breakdown product of the larger RNA species.

Conclusive proof of the existence of distinct mitochondrial transfer RNA's was recently obtained by Barnett and his collaborators.[16–18] These authors were able to show that *N. crassa* cells contained two different sets of transfer RNA's, one localized in the mitochondria, and the other in the soluble cytoplasm. Significantly, the mitochondrial RNA components were not degraded if the intact mitochondria were exposed to venom phosphodiesterase, presumably because of their localization in the interior of the mitochondria. Moreover, the coding properties of mitochondrial and cytoplasmic leucyl-transfer RNA, as measured by ribosomal binding, were clearly different: the mitochondrial leucyl-transfer RNA responded only to polymers containing UC, whereas the cytoplasmic one exhibited coding responses to UC-, UG-, and, ambiguously, U-containing polymers.[114] The genetic code may, thus, be less degenerate in the presumed mitochondrial genetic system than in the chromosomal one.

A further difference between the mitochondrial and the cytoplasmic transfer RNA's was revealed by studying their acceptor specificities.[17] Since these experiments also provided definitive evidence of the existence of distinct mitochondrial aminoacyl-RNA synthetases, they shall be described in some detail. Consolidating and extending previous observations by other investigators,[79,303,318,411,85,196,385] Barnett *et al.*[17] demonstrated that isolated *N. crassa* mitochondria contained aminoacyl-RNA synthetases for at least 15 different amino acids. The mitochondrial leucyl-, phenylalanyl-, and aspartyl-RNA synthetases could be separated from the corresponding cytoplasmic enzymes by chromatography on hydroxylapatite or DEAE-sephadex, and were only active with a mixture of *mitochondrial* transfer RNA's; no activity was observed if a crude preparation of *cytoplasmic* transfer RNA was substituted as an acceptor. Conversely, the three cytoplasmic synthetases were more active with the cytoplasmic RNA's than with the mitochondrial ones, although the pattern of specificity was not as clear-cut as with the corresponding mitochondrial enzymes (Table 7-1).

TABLE 7-1

ACYLATION SPECIFICITIES OF MITOCHONDRIAL AND CYTOPLASMIC ASPARTYL-, LEUCYL-, AND PHENYLALANYL-RNA SYNTHETASES FROM NEUROSPORA CRASSA (ADAPTED FROM BARNETT ET AL.[17]*).

Source of enzyme	Amino acid activated	pmoles of aminoacyl-RNA formed per A_{260} unit per 30 min	
		Mitochondrial transfer RNA	Cytoplasmic transfer RNA
Mitochondria	Aspartate	5.0	~0.1
	Leucine	3.0	~0.1
	Phenylalanine	1.4	0.0
Cytoplasm	Aspartate	0.7	5.9
	Leucine	7.2	9.6
	Phenylalanine	0.6	2.6

* For experimental details cf Ref. 17.

Indeed, the almost absolute acylation specificities of the mitochondrial leucyl-, phenylalanyl-, and aspartyl-RNA synthetases appear to be the exception rather than the rule, since the remaining mitochondrial synthetases exhibited only a partial specificity, or no specificity at all, for the mitochondrial transfer RNA's. In view of the many parallels between nucleic acids from mitochondria and bacteria (cf below), it would be of interest to establish whether those mitochondrial synthetases incapable of acylating cytoplasmic transfer RNA can function with bacterial transfer RNA as acceptor.

Although similar extensive data for other organisms are still lacking, it appears that the existence of distinct mitochondrial transfer RNA's and aminoacyl-RNA synthetases is not limited to *N. crassa*. According to Buck and Nass,[51] rat-liver cells contain at least three different leucyl-transfer RNA's, one of which is exclusively associated with the mitochondria and acylated only by a mitochondrial leucyl-RNA synthetase. On the other hand, phenylalanyl-RNA synthetase isolated from rat-liver mitochondria is unable to acylate one of the two phenylalanyl-transfer RNA's recovered from a rat-liver supernatant. The existence of chromatographically distinct mitochondrial and cytoplasmic leucyl-transfer RNA's in rat liver was also independently demonstrated by Fournier and Simpson.[126] Finally, Suyama and Eyer[371] found *Tetrahymena pyriformis* cells to contain two different sets of leucyl-transfer RNA *plus* the corresponding synthetase, one set being associated with the mitochondria, and the other with the soluble cytoplasm.

These findings make it very likely that all mitochondria are equipped with distinct, and probably complete, sets of transfer RNA's and aminoacyl-RNA synthetases.

Mitochondrial Ribosomal RNA

As mentioned earlier, the bulk of carefully isolated mitochondrial RNA exhibits the sedimentation characteristics of ribosomal RNA. This RNA must also be regarded as a genuine mitochondrial constituent, since it differs from cytoplasmic ribosomal RNA. This important fact was first established by Rogers *et al.*[312] who observed that the sedimentation pattern of ribosomal-type RNA from yeast mitochondria differed slightly, but significantly, from that of RNA extracted from cytoplasmic yeast ribosomes. Interestingly, mitochondrial ribosomal RNA proved to be virtually indistinguishable in sedimentation properties from a bacterial ribosomal RNA (Table 7-2).

TABLE 7-2

SEDIMENTATION PROPERTIES OF RNA EXTRACTED FROM S. CEREVISIAE MITOCHRONDRIA, S. CEREVISIAE CYTOPLASMIC RIBOSOMES, AND ESCHERICHIA COLI RIBOSOMES (ROGERS ET AL.[312]*).

Source of RNA	Sedimentation constants (S)		
S. cerevisiae mitochondria	12.7	17.8	22.4
S. cerevisiae cytoplasmic ribosomes		16.2	24.6
E. coli ribosomes		16.9	22.6

* See Ref. 312 for experimental details.
(Reproduced with permission from *Biochem. Biophys. Res. Comm.*, **27**, 405 (1967) © Academic Press.)

The reproducibility of the sedimentation pattern argued against the possibility that the differences illustrated in Table 7-2 were artifacts reflecting a preferential degradation of the mitochondrial RNA components. On the other hand, the origin and, indeed, the significance of the 12.7 S species observed by Rogers *et al.* remain obscure as this RNA component is not detected in highly purified yeast mitochondria[415,412] or in any other mitochondrial species investigated thus far.[206,98,97,308,201] Apart from this minor uncertainty, however, the essential findings of Rogers *et al.* have been confirmed and extended to *N. crassa*[308,206,98] and cultured hamster cells.

According to experiments with *N. crassa*,[308,206] the ribosomal RNA's from mitochondria and the homologous cytoplasmic ribosomes also differ with respect to nucleotide composition. The cytoplasmic RNA component exhibits a distinctly higher G + C content than the mitochondrial RNA. A similar comparative nucleotide analysis has not yet been reported for other species. However, Dubin and Brown reported that mitochondrial and cytoplasmic ribosomal RNA's from cultured hamster cells differed in their degree of methylation.[97] Although these data are still somewhat preliminary, they might be taken as indirect support for the more complete results obtained with *N. crassa*.

The unique properties of the ribosomal RNA's extracted from isolated mitochondria render it extremely unlikely that these RNA components originate from contaminating cytoplasmic ribosomes. The fact that mitochondria contain their own and distinct ribosomal RNA's immediately raises the question of whether these RNA components, *in situ,* are integrated into ribonucleoprotein structures resembling cytoplasmic ribosomes.

The existence of mitochondrial ribosomes has been under discussion for several years. Numerous workers have observed small granules in electron micrographs of fixed and positively stained mitochondria, and have speculated that they might represent mitochondrial ribosomes.[84,429,401,258,187,375,376,294,226,6] In support of this interpretation, the intramitochondrial particles were shown to possess the following properties: (a) they specifically interacted with histochemical stains considered to be selective for nucleic acids;[401,376,375] (b) they were digested on exposure of the tissue sections to RNAase, DNAase being without effect;[6,375] (c) they were distinctly smaller than the mitochondrial "normal" dense granules.[6] Because of the paucity of the intramitochondrial ribosome-like particles, however, their identification as genuine ribosomes was not generally accepted. Indeed, a search through the pertinent literature reveals the amusing fact that different authors (and even one and the same author at different times) cite the morphological data either as arguments for or against the existence of mitochondrial ribosomes.

Early reports on the isolation of ribosomes from mitochondrial preparations were met with similar reservations. The ribosome-like particles isolated by different workers[304,294,104,166] were either not sufficiently characterized, or exhibited sedimentation constants of around 78 S [294,104] and were, thus, indistinguishable from cytoplasmic ribosomes. Moreover, these "mitochondrial" ribosomes were isolated from preparations of mammalian mitochondria which are notoriously contaminated with membrane-bound cytoplasmic ribosomes. This point is underscored by two recent studies[275,128a] which document the great difficulty of isolating mammalian mitochondria-free of cytoplasmic RNA.

While it would thus appear that the "mitochondrial" ribosomes described in earlier reports were, in fact, of microsomal origin, experiments by Küntzel and Noll [206] and Rifkin *et al.*[308] have now provided strong evidence that *N. crassa* mitochondria do contain ribosomes which, like their constituent RNA components, differ from their counterparts in the cytoplasm. Both groups of workers isolated the mitochondrial ribosomes by high-speed centrifugation of a mitochondrial lysate which had been obtained by exposing the mitochondria to detergent in the presence of magnesium ions. Under the carefully standardized conditions of Küntzel and Noll,[206] the mitochondrial ribosomes exhibited a sedimentation constant of 73 S which was slightly, but significantly lower than the value of 77 S as determined for the cytoplasmic ribosomes.

The sedimentation data of Küntzel and Noll suggest that the ribosomes of mitochondria are somewhat smaller than those present in the surrounding cytoplasm. This notion is supported by the earlier cytological findings of André and Marinozzi[6] which also indicated that the intramitochondrial ribosome-like particles were distinctly smaller than the adjacent cytoplasmic ribosomes.

In contrast to these findings, Rifkin *et al.*[308] were unable to detect a significant difference in sedimentation velocity between mitochondrial and cytoplasmic *N. crassa* ribosomes. The sedimentation constants measured by these workers (89.8 and 89.5 S for the cytoplasmic and mitochondrial ribosomes, respectively) are also much higher than those determined by Küntzel and Noll.[206] Since the two laboratories report virtually identical nucleotide compositions for both types of *Neurospora* ribosomes, these puzzling discrepancies must be attributed to different conditions of sedimentation. This explanation would be compatible with the fact that, under the conditions of Rifkin *et al.*, the sedimentation coefficient of *E. coli* ribosomes was unusually high (81.9 S), whereas Küntzel and Noll measured a coefficient of 70 S, identical with the conventional value. While these discrepancies merit further attention, it appears reasonable to conclude at present that mitochondrial ribosomes of *N. crassa* are somewhat smaller than the corresponding cytoplasmic ribosomes, but that, for some unknown reason, the sedimentation experiments of Rifkin *et al.* were incapable of detecting the rather minor differences.

Sedimentation analysis of *N. crassa* mitochondrial ribosomes failed to reveal discrete ultraviolet-absorbing peaks suggestive of polysomes.[308,206]

However, if the ribosomes were isolated from mitochondria that had been pulse-labeled with radioactive amino acids *in vitro,* most of the ribosome-associated radioactivity sedimented at a velocity typical of a ribosomal dimer.[206] Additional, if less distinct, radioactivity peaks were evident at positions indicative of ribosomal tri- and tetramers. This important experiment thus provides highly suggestive evidence for the existence of mitochondrial polysomes and their participation in mitochondrial protein synthesis.

For the reasons outlined earlier, the isolation and characterization of ribosomes from mammalian mitochondria appears to be an especially challenging task that has not yet been fully accomplished. Nevertheless, an important partial success was achieved by O'Brien and Kalf [276] who isolated a ribosomal particle from carefully purified rat-liver mitochondria and presented preliminary evidence for its involvement in mitochondrial protein synthesis. The isolated particle exhibited a sedimentation constant of 55 S, contained 20–29% RNA, and, upon negative staining, appeared as a globular structure approximately 145 Å in diameter. If the isolated mitochondria were briefly exposed to radioactive leucine, the ribosomal particle acquired radioactivity 50 times faster than the bulk of the mitochondrial proteins. While O'Brien and Kalf regarded the 55 S particle as the monomeric form of a mitochondrial ribosome, the unsuccessful efforts to dissociate it into subunits, and the results obtained with mitochondrial ribosomes from *N. crassa* would suggest that this particle actually represented the larger subunit of a mitochondrial ribosome. A characterization of its RNA would be of considerable interest.

Additional differences between mitochondrial and cytoplasmic ribosomes are suggested by the observation[308] that cytoplasmic *N. crassa* ribosomes dissociate only at magnesium concentrations of around 0.1 mM, whereas the mitochondrial ones already commence to dissociate at magnesium concentrations as high as 4 mM. Moreover, mitochondrial ribosomes appear to be sensitive to antibiotics such as chloramphenicol [305,234] or erythromycin[70,210] which under comparable conditions are without effect on the corresponding cytoplasmic ribosomes.[305,234,357,43,210] Conversely, cycloheximide appears to interact selectively with the cytoplasmic ribosomes of higher cells, leaving the mitochondrial ribosomes completely unaffected.[70,210] These differences are inferred from rather indirect evidence, however, and are, thus, more appropriately discussed in connection with mitochondrial protein synthesis.

Mitochondrial Messenger RNA

While the presence of transfer- and ribosomal RNA's in mitochondria appears well established, the existence of a messenger RNA transcribed from mitochondrial DNA is still uncertain. However, a number of observations may be interpreted as preliminary evidence for a metabolically unstable mitochondrial RNA that possesses certain features of messenger RNA. For example, protein synthesis by isolated mitochondria is partially inhibited by actinomycin,[197,411,181,262] and may, thus, depend on the continuous formation of RNA. This possibility is supported by the finding that isolated yeast mitochondria pulse-labeled with ^{3}H-ATP in the presence of magnesium and a mixture of unlabeled UTP, CTP, and GTP rapidly accumulate a labeled RNA in a process that is almost completely blocked by actinomycin.[412] Upon addition of this inhibitor, the labeled RNA decays with a half-life of approximately 15 minutes and, thus, appears to be metabolically unstable. Moreover, its sedimentation profile in a sucrose gradient differs from that of the stable mitochondrial 4 S and ribosomal RNA.[412] The properties of this short-lived mitochondrial RNA species are, therefore, clearly compatible with, and suggestive of, its function as a mitochondrial messenger. The existence of mitochondrial polysomes[206] also points to the presence of messenger RNA in mitochondria. A more conclusive identification of mitochondrial messenger RNA would be provided by the detection of sequence homologies between mitochondrial DNA and the short-lived RNA species synthesized by isolated mitochondria. Although detailed studies along these lines have not yet been published, Suyama and Eyer have briefly mentioned the fact that isolated *T. pyriformis* mitochondria synthesize a messenger-like RNA that specifically hybridizes with mitochondrial DNA.[372]

A novel and extremely interesting aspect of mitochondrial messenger RNA was recently raised by Attardi and Attardi.[8] These authors identified a rapidly labeled messenger RNA in HeLa cells which was specifically associated with a small fraction ($< 15\%$) of the cellular polysomes. This polysome species appeared to be attached to membranes which were tentatively identified as fragments of the endoplasmic reticulum. The membrane-bound messenger RNA differed from the messenger RNA of the "free" polysomes in its lower content of guanosine *plus* cytosine, its higher turnover rate, and its sedimentation characteristics. Most significantly, it hybridized extensively with a crude preparation of mitochondrial DNA from HeLa cells, whereas hybridization with the total HeLa DNA was at best marginal. Taken at face value, these intriguing observations would suggest that most, if not all, of the messenger RNA associated with the membrane-bound polysomes represents mitochondrial messenger RNA which is synthesized on mitochondrial DNA and exported into the tubuli and cisternae of the endoplasmic reticulum, apparently in order to be translated on cytoplasmic ribosomes. The demonstration of a "migratory" mitochondrial messenger RNA would undoubtedly add a new dimension to our understanding of the mutual interactions between mitochondria and the surrounding cytoplasm. Indeed, it would then be difficult to escape the conclusion that, in a reverse process, messenger RNA's transcribed off nuclear DNA might enter the mitochondria and program the mitochondrial ribosomes for the synthesis of polypeptides coded for by nuclear cistrons. Further work is clearly required, however, before these concepts can be accepted. Neither the polysome-carrying membranes nor the crude "mitochondrial" DNA used in the hybridization experiments by Attardi and Attardi have thus far been adequately characterized. Moreover, the alleged transport of mitochondrial messenger RNA across the mitochondrial membranes raises questions which are, at present, difficult to answer. In any case, it might be argued that the existence of a mitochondrial messenger RNA, migratory or not, is still not conclusively established.

Synthesis of Mitochondrial RNA

As already briefly mentioned above, isolated mitochondria incorporate labeled ribonucleotide triphosphates into an unstable RNA fraction and are, thus, apparently capable of RNA synthesis. However, most of the total mitochondrial RNA is accounted for by ribosomal RNA, compared to which the unstable RNA species is probably quantitatively insignificant. The present section, therefore, summarizes the evidence which indicates that mitochondria are endowed with a DNA-dependent RNA polymerase system which mediates the synthesis of mitochondrial ribosomal RNA and perhaps also that of other mitochondrial RNA components.

As first shown by Wintersberger[410] and, subsequently, by numerous other investigators,[229,181,417,321,205,268,269,265,267,358b] isolated mitochondria from a variety of sources incorporate labeled RNA precursors into a high-molecular weight product. This product can be identified as RNA on the basis of its acid-precipitability, its sensitivity to alkali and RNAase, and its resistance towards DNAase. If the labeled RNA synthesized by isolated *N. crassa* mitochondria with ^{32}P-GTP as precursor is exposed to a mixture of venom phosphodiesterase and alkaline phosphatase, radioactivity and material absorbing at 260 μ are solubilized in a parallel fashion.[229] This indicates that the labeled RNA precursor was not merely attached to the ends of preexisting RNA chains, but was present in internucleotide linkages distributed over sizeable segments of the RNA molecule.

In most cases studied so far, the mitochondrial incorporation reaction is magnesium-dependent, requires the presence of all four ribonucleotide triphosphates, and is inhibited by a biologically active member of the actinomycin family. However, the inhibition by actinomycin appears to be inversely related to the intactness of the isolated mitochondria[267] apparently because the mitochondrial membranes are relatively impermeable to the antibiotic. Other inhibitors include pyrophosphate,[267] acriflavin,[417] or a glucose-hexokinase trap.[267] No inhibition is observed with P_i,[229] a typical inhibitor of polynucleotide phosphorylase. All these results suggest that the mitochondrial RNA synthesizing system reflects the activity of a DNA-directed RNA polymerase which is quite similar to the more thoroughly studied enzymes from other sources.

The following lines of evidence indicate that the RNA polymerase activity associated with mitochondrial preparations is not merely due to contaminating bacteria or cell nuclei:

(a) Yeast mitochondria purified by flotation in a "Urografin" gradient[332] exhibit an active RNA polymerase,[417,412] but no detectable nuclear DNA.[394]

(b) In intact rat-liver mitochondria (but not in mitochondria rendered more permeable by disruptive treatments), the incorporation reaction is unaffected by actinomycin C_1, DNAase, or RNAase, and differs in these respects from the RNA polymerase associated with isolated rat-liver nuclei.[267]

(c) Sterile preparations of rat-liver mitochondria exhibit the same RNA polymerase activity as nonsterile mitochondrial preparations containing up to 10^5 bacteria per mg mitochondrial protein.[205]

(d) The incorporation of ^{14}C-ATP into RNA by intact rat-liver mitochondria is strongly inhibited by atractyloside, whereas the corresponding activity of the isolated nuclei is completely unaffected.[321] Since atractyloside specifically inhibits the entry of adenine nucleotides into the interior of mitochondria,[62,155,186] these experiments can be regarded as a particularly compelling argument for the existence of a unique mitochondrial-RNA polymerase system.

The RNA species synthesized by isolated mitochondria has not yet been definitely identified, but it appears to be determined, in part, by the duration of the labeling experiment. As mentioned earlier, short pulses (10 min or less) with a radioactive precursor lead to the formation of an unstable RNA species that may represent messenger RNA.[412] In contrast, the RNA synthesized in long-term labeling experiments with yeast mitochondria (60 min) appears to be quite stable and, upon sucrose gradient centrifugation, exhibits three major peaks corresponding to 4 S RNA and the two ribosomal RNA's.[412,417] These experiments would, thus, suggest that isolated yeast mitochondria can synthesize all of their major RNA components. For some unexplained reason, however, the long-term incorporation process catalyzed by yeast mitochondria is only some 60% inhibited by actinomycin which renders the above conclusion somewhat uncertain. Unfortunately, the RNA products of long-term labeling experiments *in vitro* have not yet been studied with other types of mitochondria.

While labeling experiments with isolated mitochondria have thus furnished only rather tentative information about the RNA species made by the mitochondrial RNA synthesizing system, recent hybridization experiments strongly suggest that this system is indeed involved in the formation of mitochondrial ribosomal RNA. Suyama reported [369] that *T. pyriformis* mitochondria contain an RNA that hybridizes extensively (up to

6.8%) with mitochondrial DNA from *Tetrahymena,* but not with the corresponding nuclear DNA, or with DNA from *Bacillus subtilis.* This RNA was tentatively identified as ribosomal RNA, since it was present in the pellet upon high-speed centrifugation of a mitochondrial lysate and resembled ribosomal RNA in its chromatographic behavior on methylated albumin adsorbed to kieselguhr. Its atypical sedimentation pattern (14 and 18 S) was suggested to reflect partial degradation by the very active nucleases of the *Tetrahymena* cells.

Fukuhara[127] and Wintersberger[415] reported analogous experiments with *S. cerevisiae* that yielded results quite similar to those obtained with *Tetrahymena.* However, the data furnished by the yeast system are somewhat more clear-cut, since the mitochondrial RNA used for the hybridization studies exhibited the typical sedimentation pattern of a ribosomal RNA. In addition, it was demonstrated that the RNA of the cytoplasmic ribosomes hybridized very poorly, if at all, with mitochondrial DNA; on the other hand, it exhibited considerable annealing (some 2%) with nuclear DNA, in essential agreement with the results observed with bacterial systems.

It should be pointed out that, in all studies reported thus far, no saturation plateau was reached if a fixed amount of mitochondrial DNA was hybridized with increasing amounts of the corresponding mitochondrial ribosomal RNA. The true extent of sequence homology between these two nucleic acid species is, therefore, still unknown. The highest hybridization value observed for the yeast system (7%)[415] is quite similar to that obtained with *Tetrahymena* (6.8%, cf Ref. 369). It would, thus, appear that the maximal attainable hybridization is in the order of 10%.

Since mitochondrial DNA possesses a molecular weight of roughly 10^7 daltons and is, thus, relatively small, its extremely efficient hybridization with mitochondrial ribosomal RNA is not unexpected. If it is assumed that each molecule of mitochondrial DNA carries one cistron for each of the two mitochondrial ribosomal RNA's (combined molecular weight approximately 1.7×10^6),[3] hybridization values as high as 17% would be theoretically possible. These preliminary calculations suggest that the genetic messages directing the synthesis of mitochondrial ribosomal RNA occupy about one third of the total mitochondrial genome. At present, it is not known which type of genetic information, if indeed any at all, is carried by the remaining segments of mitochondrial DNA. Conceivably, these segments could include cistrons for mitochondrial messenger RNA and, hence, for mitochondrial proteins.

It is still not settled whether only one, or both, of the mitochondrial ribosomal RNA's are coded for by mitochondrial DNA. In this context, it is of interest that exposure of cultured hamster cells to actinomycin has relatively little effect on the *in vivo* synthesis of the smaller mitochondrial ribosomal RNA component, while the synthesis of the larger one, together with that of all cytoplasmic RNA's, is completely blocked.[96] In view of the well-documented impermeability of intact mammalian mitochondria to actinomycin,[267,404] these observations suggested to Dubin[96] that only the smaller one of the two mitochondrial ribosomal RNA's was synthesized in the interior of the mitochondria, whereas the larger one was made outside the mitochondria, presumably by the chromosomal system. However, the complexity of this experimental system and the possibility of side effects of actinomycin[163,1,306] make it difficult to accept this conclusion. Hybridization experiments with the separated ribosomal RNA components will undoubtedly soon furnish a more definitive answer to this important question.

A priori, it seems reasonable to assume that mitochondrial DNA also codes for the mitochondrial transfer RNA's. Unexpectedly, however, Suyama[369] was unable to detect significant hybridization between the DNA and the low-molecular weight RNA from *Tetrahymena* mitochondria. Even though attempts to hybridize this RNA fraction with nuclear DNA yielded equivocal results, Suyama concluded that most, if not all, of the mitochondrial transfer RNA's are transcribed off nuclear cistrons, and are only subsequently incorporated into the mitochondria. While this notion seems to offer the best explanation for the experimental results, it appears, in many ways, rather unsatisfying. For example, it is difficult to conceive transport mechanisms discriminating between cytoplasmic and mitochondrial transfer RNA. Moreover, the fact that "petite" mitochondria lack all typical RNA components, including 4 S RNA[415] favors the possibility that this latter RNA species, too, is coded for by mitochondrial DNA. It might also be pointed out that the low-molecular weight RNA extracted from mitochondria, including those of *Tetrahymena,* is probably a very inhomogeneous fraction only a small portion of which might be accounted for by transfer RNA. Finally, the small size of transfer RNA's (approximately 25,000 daltons, cf Ref. 58) might impose a critical limit to the stability of the resulting hybrids, especially if hybridization is carried out at elevated temperatures. While this author is reluctant to pitch personal views against experimental facts, he, nevertheless, feels that the identity of the DNA templates for mitochondrial transfer RNA's should be left open until it has been shown that these RNA's can hybridize with either nuclear or mitochondrial DNA.

Concluding Remarks

The present evidence allows the general conclusion that mitochondria contain complete sets of transfer RNA's and aminoacyl-RNA synthetases as well as ribosomal RNA's integrated into ribosomal structures. All these components differ from their respective counterparts in the cytoplasm. At least the ribosomal RNA's are coded for by mitochondrial DNA, and are, thus, apparently synthesized within the mitochondria by a mitochondrial

DNA-dependent RNA polymerase system. Together with the preliminary evidence for the existence of mitochondrial messenger RNA, these facts strongly suggest that mitochondria are endowed with a distinct transcription and translation system that may provide the basis for a limited genetic autonomy of the mitochondrial organelle.

IV. MITOCHONDRIAL PROTEIN SYNTHESIS

As first described in detail by McLean *et al.*,[238] isolated mitochondria are capable of incorporating radioactive amino acids into acid-insoluble polypeptides. The evidence summarized in the preceding sections makes it tempting to regard this incorporation process as the manifestation of a mitochondrial genetic system involving transcription and translation of mitochondrial DNA. However, the mechanism of mitochondrial amino acid incorporation, and the relationship of this process to mitochondrial biogenesis, are still under discussion. To a large measure, these uncertainties appear to be explained by the fact that the *in vitro* incorporation process is very slow, and responds in a complex manner to a great number of seemingly minor experimental variables (cf below).

These variables, as well as the possible interrelationships between mitochondrial protein synthesis and the reactions of energy coupling, have been summarized in several earlier reviews.[345,215,293] In the present survey, mitochondrial protein synthesis will be mainly discussed in terms of its possible bearing on the biogenesis of the mitochondrial organelle.

Evidence for a Mitochondrial Amino Acid Incorporating System

During the past decade, a considerable body of literature has accumulated which describes the incorporation of amino acids by isolated mitochondria from yeast,[411,412,143,210,222] molds,[206] protozoa,[234] sea urchins,[133a] insects,[48,198,271] higher plants,[86,350a,63,20,86a] various tissues of frogs,[314] guinea pigs,[11] rabbits,[11] cattle,[383a,199,195,196,198–200,386,346] sheep,[181,182,350] rats,[238,303–305,191,79,318,319,313,384,180,197–200,46,385,315,188a,404,425,177,40,142,184,367] as well as various tumors.[137,314] Thus, there is little doubt that the capacity for amino-acid incorporation is a general feature of mitochondrial preparations. Moreover, a number of observations suggest that the incorporating activity is not due to contaminating bacteria, microsomes, or other nonmitochondrial subcellular particles, and that it is, thus, apparently catalyzed by the mitochondria themselves. It is important to remember, however, that this latter view derives mainly from a process of exclusion and is, therefore, still somewhat uncertain. Moreover, the arguments against a contribution by each of the various potential contaminants are not equally persuasive.

The most compelling evidence exists against a significant contribution by microsomes, since the mitochondrial incorporating system exhibits many properties which differentiate it from the protein synthesizing system involving cytoplasmic ribosomes. Thus, amino-acid incorporation by isolated mitochondria does not require the addition of pH 5 enzymes from a high-speed supernatant,[196,238] and is not inhibited by RNAase[184,142,196, 411,234,318,180] or cycloheximide,[70,416,210] both of which completely abolish the activity of the microsomal system. Conversely, the mitochondrial incorporation process is sensitive to chloramphenicol [305,234,196,180] and several macrolide antibiotics,[70,210] whereas protein synthesis by microsomes is not affected by these inhibitors.[305,234,357,43,102,210]

Thus, it is generally conceded that contaminating microsomes do not significantly participate in the incorporation process catalyzed by mitochondrial preparations. On the other hand, a contribution by bacteria is much more difficult to exclude, since most of the major distinguishing features of the mitochondrial amino-acid incorporating process appear to be shared by the corresponding bacterial systems. Moreover, the search for specific inhibitors distinguishing between these two types of systems has, thus far, been unsuccessful.[293] Indeed, it has been claimed that completely sterile mitochondrial preparations are incapable of incorporating amino acids, and that the incorporating activity observed in different laboratories mainly reflected the presence of contaminating bacteria.[92,322] However, reinvestigations by other authors have failed to substantiate these claims.[205,203,21,425,207,143,320a] In one laboratory it was even observed that mitochondria containing up to 10^5 bacteria per mg of mitochondrial protein incorporated amino acids at essentially the same rate as completely sterile mitochondrial preparations.[205,203] Experiments in other laboratories, however, indicate that contamination by bacteria may be a serious interfering factor unless virtually sterile mitochondrial preparations are used.[21,320a] A significant bacterial contribution is also rendered unlikely by the fact that the incorporating activity of mitochondrial preparations is, in some instances, related to the physiological status of the tissue from which they were isolated. Thus, the *in vitro* incorporating activity of locust flight-muscle mitochondria decreases almost tenfold upon molting of the insects.[48,271] Similarly, partial hepatectomy of normal rats,[40] or administration of triiodo-thyronine to thyroidectomized rats[47,379,317] significantly enhances the activity of the isolated mitochondria. Finally, mitochondria from liver[315] and brain[188a] of new-born rats incorporate amino acids considerably more rapidly than the corresponding mitochondrial preparations from adult rats (cf Ref. 271). Thus, it would appear that, under carefully controlled conditions, contaminating bacteria account at most for a very minor fraction of the amino acid incorporation by isolated mitochondria.

Since mitochondrial preparations frequently contain other subcellular particles such as lysosomes[7] or peroxisomes,[99,15] a possible contribution by these contaminants to amino-acid incorporation must also be considered. This possibility is often ignored, although at present it can not be directly excluded. However, yeast mitochondria actively incorporate amino acids[411,210] even though yeast cells appear to lack subcellular structures resembling lysosomes or peroxisomes.[335,329a] Moreover, by their very nature such particles should not be expected to be capable of amino acid incorporation, whereas mitochondria contain many components diagnostic for the presence of a protein synthesizing system.

Although a number of uncertainties remain, the present evidence would appear to justify the conclusion that the ability of incorporating amino acids is an inherent property of the mitochondrion.

The Nature of the Energy Donor

There seems to be general agreement that mitochondrial amino-acid incorporation is energy-dependent (cf Ref. 293). Although conflicting experimental findings have been reported concerning the relative efficiency of various respiratory substrates and/or adenine nucleotides in sustaining the incorporation process, the data of several authors clearly indicate that this process is stimulated by respiratory substrates and depressed by electron transport inhibitors and agents interfering with oxidative phosphorylation (cf Ref. 293). The fact that these inhibitions and stimulations are often not as clear-cut as with other energy-linked processes, such as Ca^{++}-transport or reversed electron transfer, should not cause unreasonable concern. Indeed, it has to be anticipated that the presence of endogenous substrates, or the relative insensitivity of some mitochondrial preparations to uncouplers of oxidative phosphorylation (e.g., 284), would tend to obscure the energy-dependence of a mitochondrial process as slow as amino-acid incorporation.

Several workers have reported that respiration-supported amino-acid incorporation by isolated mitochondria is not inhibited by oligomycin, and may, thus, proceed at the expense of a nonphosphorylated high-energy intermediate of oxidative phosphorylation.[187,249,266] However, this conclusion is now generally questioned since the rate of amino-acid incorporation by isolated mitochondria is generally between 0.1 and 4 pmoles amino acid incorporated per minute per mg protein[411,48] and is, thus, at least four to five orders of magnitude lower than the maximal rate of oxidative phosphorylation. Colli and Pullman have shown directly that the residual ATP synthesis in oligomycin-inhibited mitochondrial particles is still several times faster than amino-acid incorporation.[74] Moreover, a convincing study by Wheeldon and Lehninger indicates that respiration-supported protein synthesis in rat-liver mitochondria may, under certain conditions, be strongly inhibited by oligomycin, whereas no inhibition is observed if added ATP is the energy source.[404] Thus, it appears that ATP is the immediate major energy donor and that the obligate involvement of high-energy intermediates, if it indeed occurs, is quantitatively insignificant.

It should be remembered, however, that these conclusions refer only to the *over-all* incorporation process, and do not specify the energy requirements of its various steps. Amino-acid incorporation by isolated mito-chondria is undoubtedly an extremely complex process which appears to involve several reactions not shared by cell-free bacterial and micro-somal protein synthesizing systems. For example, the data of Wheeldon and Lehninger[404] and Monroy and Pullman[241] indicate that the labeled amino acid, prior to its actual incorporation into acid-insoluble peptides, undergoes an energy-dependent transport into the interior of the mito-chondria. Since the rate-limiting steps of mitochondrial amino-acid incorpo-ration are unknown, the energy requirements of the over-all process may reflect those of any given energy-linked reaction preceding the actual peptide-bond formation. Mitochondrial transport processes appear to be generally capable of utilizing high-energy intermediates of oxidative phos-phorylation, and it is, thus, still reasonable to assume that these inter-mediates may also energize certain steps of the mitochondrial amino-acid in-corporation process. A definitive conclusion concerning the immediate energy donor(s) for mitochondrial protein synthesis will, therefore, have to await the preparation of soluble mitochondrial incorporating systems.

The Mechanism of the Incorporation Reaction

Although the rate of amino-acid incorporation by isolated mitochondria is extremely slow compared to mitochondrial electron transport and oxi-dative phosphorylation, it may, in certain favorable instances, reach 10% of the estimated *in vivo* rate.[48,271] Moreover, the apparent incorporation rates must be regarded as minimal values since the added labeled amino acid is undoubtedly diluted by an intramitochondrial amino-acid pool, the size of which is difficult to assess. A further underestimation of the "true" incorporation rate appears to be introduced by a rapid decay of the labeled product to acid-soluble peptides (cf below). The rate of amino-acid in-corporation by isolated mitochondria is, thus, not inconsistent with the notion that this process is of some biological significance.

The mechanism of mitochondrial amino-acid incorporation is still un-certain, partly because attempts to solubilize the over-all process from the mitochondria have thus far been unsuccessful. Early reports[184,181] of sol-uble mitochondrial incorporating systems have not been confirmed,[196,271] and must be viewed with considerable reservation. Claims of incorporating submitochondrial particles prepared by sonic irradiation[195,196] could like-wise not be consistently substantiated.[271,416] Indeed, the fact that in-corporation by these particles proceeded for as long as three hours in the absence of a respiratory substrate or ATP would suggest that a major portion of the incorporation was due to contaminating bacteria. A limited success in obtaining a simplified mitochondrial incorporation system has apparently been achieved with the preparation of active submitochondrial "digitonin" particles.[23,404,199] However, these particles are still rather com-plex structures, as evidenced by the atractyloside-sensitivity of their oxidative phosphorylation system[396,50] and their capacity for respiratory control.[216,50]

Nevertheless, a number of observations are consistent with the notion that the mechanism of mitochondrial amino-acid incorporation resembles that of the more familiar bacterial and microsomal incorporating systems. For example, it has been shown that the labeled amino acids are not merely attached to the ends of preformed protein molecules, but are actually incorporated into the interior of polypeptide chains.[184,318,180,86,367, 411,350,320a] The sensitivity of the mitochondrial incorporating process to puromycin[345,180,411,314,271,404] represents a further criterion in common with typical protein synthesizing systems. Additional parallels to these latter systems are revealed by the observation that amino-acid incorporation in partially damaged mitochondrial preparations is sensitive to actinomycin,[181,411,198,271,350] indicative of an involvement of a short-lived messenger RNA. Finally, the fact that mitochondria contain full complements of transfer RNA's and amino-acyl-RNA synthetases as well as ribosomes and polysome-like aggregates (cf above) strongly suggests that mitochondrial amino-acid incorporation proceeds essentially via the same steps as protein synthesis in other systems described thus far. While direct evidence for an involvement of mitochondrial transfer RNA's and aminoacyl-RNA synthetases is still lacking, a participation of the mitochondrial ribosomes is indicated by the results of the pulse-labeling experiments mentioned in the preceding section, as well as by the sensitivity of the mitochondrial incorporation process to chloramphenicol.[305,234] This antibiotic appears to inhibit protein synthesis specifically in bacterial and chloroplast systems containing 70 S type ribosomes, and does not seem to affect the 80 S cytoplasmic ribosomal system of higher cells (cf Ref. 391). The selective inhibition of mitochondrial amino acid incorporation by chloramphenicol is, therefore, not only suggestive of the involvement of ribosomes, but also accords with the observation that mitochondrial ribosomes from *N. crassa* resemble bacterial ribosomes with respect to their sedimentation coefficient.[206]

While conclusive information is still lacking, the present evidence is fully consistent with the notion that mitochondrial protein synthesis follows the same mechanism as other DNA-directed protein synthesizing systems, especially those of bacterial origin.

The Identity of the Protein(s) Synthesized by Mitochondria

Several workers have noted that isolated mitochondria incorporate added labeled amino acids predominantly into an insoluble membrane fraction, whereas negligible incorporation occurs into the readily solubilized mitochondrial protein fraction,[318,313,384,48,411,271] or into malate dehydrogenase, or cytochrome *c*.[319,346] Recent experiments by Neupert *et al.*,[272] Beattie *et al.,*[23] and Work[422,423] have shown that the labeled membrane fraction is identical with the mitochondrial inner membrane, and that little, if any, label is incorporated into the outer membrane (Table 7-3).

TABLE 7-3

SPECIFIC RADIOACTIVITY AND ENZYME CONTENT OF INNER AND OUTER MEMBRANE FRACTIONS DERIVED FROM RAT-LIVER MITOCHONDRIA LABELED WITH ^{14}C-LEUCINE IN VITRO (ADAPTED FROM NEUPERT ET AL.[272]).

	Mitochondria	Outer membrane	Inner membrane
Succinate dehydrogenase[+]	14.5	11.0	40.4
Monoamine oxidase[+]	224	3180	962
Specific radioactivity[++]	44	66	238
Specific radioactivity calculated for pure fraction[+++]	—	2	314

[+] μmole per hour per mg protein.
[++] cpm per mg protein.
[+++] calculated on the assumption that succinate dehydrogenase and monoamine oxidase are specific for the inner and outer membrane, respectively.

In line with other investigations to be mentioned later, these important findings suggest that the mitochondrial protein synthesizing machinery is mainly concerned with the formation of the mitochondrial inner membrane. However, this interpretation presupposes that isolated mitochondria are indeed capable of synthesizing at least one discrete inner membrane protein—an assumption which has not been verified thus far. It is generally agreed [411,177,271,404] that the majority of the label incorporated into the membranes is present in an insoluble protein fraction resembling the "structural protein" of Criddle *et al.*[80] However, "structural protein" can not be regarded as discrete protein, since it can be electrophoretically resolved into several components.[150,389,78,339,334a] Indeed, a recent study shows conclusively that "structural protein" from beef-heart mitochondria contains large amounts of denatured mitochondrial adenosine triphosphatase.[334a] If "structural protein" is isolated from rat-liver mitochondria previously incubated with labeled amino acids, and subjected to acrylamide gel electrophoresis, some protein bands appear to be more labeled than others.[150,339] However, the published data[150] are not sufficiently clear-cut to permit the conclusion that one, or several, distinct protein species in the mitochondrial "structural protein" preparation are selectively labeled.

Mitochondrial "contractile protein" [277,260] has also been suggested [182] as a product of mitochondrial protein synthesis. This claim was based on the observation that isolated lamb-heart mitochondria incorporated labeled amino acids into a protein fraction which resembled the mitochondrial "contractile protein" of Ohnishi and Ohnishi[277] in that it catalyzed an ATPase reaction and could be extracted from the mitochondria with 0.6 M KCl. These observations, however, appear to be of doubtful significance, since according to Vignais *et al.*[395] and Conover and Bárány[75] the existence of a mitochondrial "contractile protein" with the properties described by Ohnishi and Ohnishi must in all probability be discounted.

The search for the protein synthesized by isolated mitochondria may be complicated by the fact that, according to experiments with rat liver mitochondria, the labeled product is unstable and appears to be continuously degraded to acid-soluble neutral peptides.[404] This decay is greatly stimulated by puromycin and seems to be energy-dependent. At present, it is unexplained but could be interpreted as an abortive protein synthesizing process which involves the discharge of incomplete polypeptides from their site of synthesis (presumably the mitochondrial ribosome) and their subsequent degradation by mitochondrial endopeptidases (cf Ref. 390). If this interpretation is correct, then most, or all, of the amino-acid incorporation catalyzed by isolated mitochondria may well represent an aberrant activity which leads to the assembly of biologically inactive polypeptides. These might be liberated in a largely unfolded state, and aggregate to insoluble polymers which are subsequently recovered in the "structural protein" fraction. While this notion may be overly pessimistic, it can not be excluded at present, and may, thus, serve as a remainder that the biological significance of amino-acid incorporation by isolated mitochondria is still an open question.

Another approach for identifying the protein(s) synthesized by mitochondria makes use of the selective inhibition of mitochondrial protein synthesis in intact cells. *A priori*, such an inhibition might be assumed to specifically lower the intracellular concentration of the proteins made by the mitochondrial system. Inhibitors like chloramphenicol or erythromycin lend themselves especially well to such studies, since they are relatively nontoxic to higher cells and, as mentioned earlier, specifically block amino acid incorporation by isolated mitochondria without affecting the corresponding cytoplasmic systems. In a series of papers, Linnane and his colleagues reported that *S. cerevisiae* cells grown on a fermentable carbon source in the presence of chloramphenicol or erythromycin were respiration-deficient and lacked the cytochromes aa_3, b, *and* c_1.[70,167,71,220,408] On the other hand, their content of cytochrome *c* and other easily extractable mitochondrial enzymes such as fumarase and malate dehydrogenase was roughly the same as, or even higher than, that of the nonrepressed cells. Electron micrographs of the chloramphenicol-grown cells revealed the presence of atypical mitochondrial profiles exhibiting an apparently normal outer membrane, but only very few and poorly defined cristae.[71] Similar, if less drastic, effects of chloramphenicol were subsequently observed with the flagellate *Polytomella caeca*[117] and with cultured mammalian cells.[204,120a]

The results obtained with yeast cells led Linnane *et al.* to conclude[71,220] that the normally insoluble inner membrane proteins such as cytochrome aa_3, b, and c_1, as well as succinate dehydrogenase, are synthesized by the mitochondrial system, whereas the entire outer membrane, as well as the more easily solubilized mitochondrial enzymes, are formed in the cytoplasm. While the latter conclusion is supported by the results of the incorporation experiments with isolated mitochondria, the former is not, since neither cytochrome aa_3 nor succinate dehydrogenase are labeled by isolated mitochondria.[177]

This discrepancy could be rationalized by the previously mentioned assumption that amino-acid incorporation by isolated mitochondria represents an atypical process which does not reflect the biosynthetic capacity of the mitochondria *in vivo*. It is equally possible, however, that the synthesis of most insoluble inner membrane proteins is controlled by nuclear genes and is only indirectly affected by chloramphenicol and erythromycin. For example, these antibiotics could conceivably inhibit the synthesis of some protein (e.g., a "structural protein") that is essential for the integration of the various respiratory carriers into the membrane structure. Through some type of feedback mechanism, loss of this organizational protein could then prevent the formation of significant quantities of respiratory carriers, even though these carriers are synthesized on cytoplasmic ribosomes. Alternatively, the results of the experiments with isolated mitochondria and with chloramphenicol-repressed cells could be reconciled by the postulate that the normally insoluble respiratory carriers such as cytochrome aa_3 or succinate dehydrogenase are synthesized on mitochondrial ribosomes under the direction of messenger RNA's transcribed from nuclear cistron.

In addition to chloramphenicol-repressed wild-type yeast cells, the cytoplasmic "petite" mutant of *S. cereviasiae*[112,105,107] has also been employed as an experimental system for identifying the product(s) of mitochondrial protein synthesis. As mentioned earlier, the nonrespiring mitochondria of this mutant lack ribosomal RNA and normal mitochondrial DNA; apparently, they contain a defective protein synthesizing system permanently blocked at the transcription level. In many of their other properties, however, the mitochondria of the mutant resemble those of chloramphenicol-repressed wild-type cells. Thus, they exhibit few, if any, distinct cristae,[426a] lack the cytochromes aa_3, b, and c_1, and contain almost normal or elevated levels of cytochrome c, fumarase, aconitase, and other easily extractable mitochondrial enzymes.[354] According to Kováč and Weissová[193] and Schatz,[329] they also contain near-normal amounts of ATPase (F_1). However, the enzyme associated with the mutant mitochondria is cold-labile[329] and oligomycin-insensitive,[193,329] and differs in these respects from the enzyme bound to the wild-type mitochondria which is cold-stable and strongly inhibited by oligomycin. F_1 purified from the mutant mitochondria exhibits the same enzymic and immunological properties and the same sedimentation coefficient as wild-type F_1[334,329] and can be rendered cold-stable and oligomycin-sensitive upon binding to F_1-deficient submitochondrial bovine heart particles.[329] These results indicate that the "petite" mutant can still synthesize unaltered F_1 and that this important inner membrane protein is therefore not formed by the mitochondrial system. On the other hand, it appears from these studies that the mutant has lost the ability to synthesize a component that is involved in the specific binding of F_1 to the mitochondrial inner membrane and that may be functionally analogous to CF_0 from bovine heart mitochondria.[179] Therefore, it appears fruitful to explore the possibility that CF_0, or a constituent thereof, is formed by the mitochondrial protein synthesizing system. This possibility would also be in line with recent experiments of Tuppy *et al.*[389] These authors examined a preparation of mitochondrial "structural protein" from the "petite" mutant by acrylamide gel electrophoresis and immunological methods and found that it lacked at least one component present in the corresponding preparation from the wild-type strain. Closely similar results were also briefly reported by Work.[424,422] Since experiments with bovine heart mitochondria suggest that "structural protein," ore one of its subfractions, is an essential component of CF_0,[178] it is tempting to speculate that one and, perhaps, the major product of mitochondrial protein synthesis is an insoluble inner membrane protein that is essential for CF_0-function and that, upon fractionation of the wild-type yeast mitochondria, is recovered in the "structural protein" fraction. On the other hand, the recent identification of denatured mitochondrial ATPase in mitochondrial "structural protein"[334a] suggests an alternate explanation of the results of Tuppy *et al.*[389] and Work.[424,422] Since the "petite" mutation labilizes the linkage between ATPase and the mitochondrial inner membrane,[329] the difference between the "structural proteins" from the wild-type and the mutant could merely reflect a loss of ATPase protein from the mutant mitochondria during their isolation.

A different conclusion about the product of mitochondrial protein synthesis was reached by Woodward and Munkres[420] on the basis of experiments with *N. crassa* mutants. These experiments indicated that mitochondrial "structural protein" from the respiration-deficient, cytoplasmic "poky" mutant of *N. crassa*[239,240] differed from the corresponding protein of the wild-type strain by the substitution of a single tryptophane residue by cysteine. In contrast, mitochondrial "structural protein" from a phenotypically identical, but chromosomal *N. crassa* mutant was indistinguishable from the wild-type protein. Woodward and Munkres suggested, therefore, that "structural protein" was synthesized by the mitochondrial system under the direction of mitochondrial DNA, and that a point mutation of this DNA in the course of the "poky" mutation resulted in the synthesis of an abnormal mitochondrial "structural protein." The implications of this interesting hypothesis were subsequently considerably extended by the claim that "structural proteins" from nuclei, mitochondria, "microsomes," and even a high-speed supernatant from *N. crassa* were all identical and altered in the same way by the "poky" mutation.[421] This puzzling finding might mean that *N. crassa* cells synthesize a universal cellular structural protein which is coded for by mitochondrial DNA and assembled in different cellular compartments via a migratory messenger RNA transcribed from mitochondrial DNA (cf Ref. 8). However, Tuppy and Swetly found that "structural protein" from normal yeast mitochondria differed from the corresponding insoluble proteins isolated from other subcellular yeast fractions by its ability to form complexes with ATP that were dissociable by atractyloside.[388] In view of this contradictory evidence and the serious objections that have been raised concerning the homogeneity of mitochondrial "structural protein," [150,389,78,334a] any final decision regarding the data reported by Woodward and Munkres appears to be unwarranted.

The Participation of the Cytoplasmic Protein Synthesizing System in Mitochondrial Formation

As mentioned before, the outer mitochondrial membrane and the easily extractable mitochondrial proteins are neither labeled by isolated mitochondria nor diminished in chloramphenicol-repressed or "petite" mutant yeast cells. Thus, it is reasonable to assume that these components are not products of the mitochondrial-protein synthesizing system, but are made in the cytoplasm under the direction of nuclear genes. A participation of nuclear cistrons in mitochondrial formation is also implicated by the existence of respiratory-deficient yeast[64,106,341,288,298,287] and *N. crassa*[399] mutants whose characteristic traits are inherited in a strictly classical, Mendelian pattern. Additional evidence for an involvement of the chromosomal genetic system is furnished by the labeling pattern of the various mitochondrial protein fractions *in vivo*. Both Truman[383] and Beattie *et al.*[20] observed that a few minutes after administration of ^{14}C-leucine to rats, the soluble proteins of liver and kidney mitochondria were distinctly less labeled than the insoluble ones. With increasing time, however, the specific radioactivity of the soluble proteins rose sharply until, after several hours, all mitochondrial protein fractions were equally labeled.[20] This time course of labeling is to be expected if the soluble mitochondrial proteins are made in the cytoplasm and only subsequently transferred into the mitochondria.

More direct evidence for such a transfer was obtained by Kadenbach[177] who reported that isolated rat-liver microsomes labeled with radioactive leucine in a cell-free system were capable of releasing labeled polypeptides into isolated, admixed, rat-liver mitochondria. This transfer reaction was time- and energy-dependent, and resulted in essentially equal labeling of all mitochondrial protein fractions. Transfer of a distinct mitochondrial protein was not demonstrated, however.

Conclusive proof for the extramitochondrial synthesis of a typical mitochondrial protein stems from investigations on the biosynthesis of cytochrome c in yeast and mammalian cells. Wild-type *S. cerevisiae* cells contain two different cytochromes c, designated as iso-1- and iso-2-cytochrome c, respectively, which differ in their primary structure.[343,355] In most aerobically-grown, wild-type strains, the major component is iso-1-cytochrome c, constituting some 95% of the total cellular cytochrome c complement. Genetic analyses of yeast mutants specifically deficient in cytochrome c indicated that the synthesis of iso-1-cytochrome c is controlled by at least one chromosomal gene CY_1.[343,340] This gene could be identified as the *structural* gene for iso-1-cytochrome c, since a reversion of the non-functional mutant allele cy_{1-2} resulted in the synthesis of an altered iso-1-cytochrome c containing one residue less of glutamic acid, and one residue more of tyrosine than the wild-type hemoprotein.[342] The revertant gene segregated from the wild-type gene in a characteristic Mendelian fashion and occupied the same genetic locus as the mutant gene cy_{1-2}. These important results, which are summarized in Figure 7-3, constitute unambiguous proof that, in baker's yeast, iso-1-cytochrome c is coded for by a chromsomal cistron (cf Refs. 361, 278). In addition, the data obtained

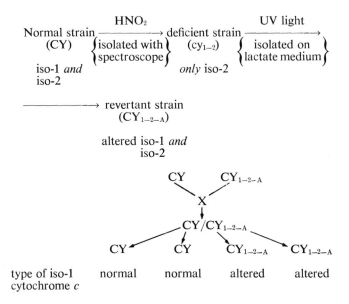

FIGURE 7-3. Synopsis of the experiments demonstrating chromosomal coding of yeast iso-1-cytochrome c (adapted from Sherman *et al.*,[342]).

with yeast suggest that the cytochrome *c* of higher cells may likewise be a product of the chromosomal genetic system.

Recent studies on the biosynthesis of cytochrome *c* in mammalian cells fully support this contention. With the aid of a newly developed micro-method for the isolation of cytochrome *c*, Gonzalez-Cadavid and Campbell showed that, in rat-liver cells, roughly one tenth of the total cellular cytochrome *c* was associated with the microsomes.[134] When rats were injected with ^{14}C-lysine, this microsomal cytochrome *c* was labeled much more rapidly than the mitochondrial one. Moreover, the time course of labeling of the two cytochrome *c* fractions indicated that cytochrome *c* was synthesized *in toto* on the microsomes and subsequently transferred to the mitochondria.[135] Preliminary experiments by Freeman *et al.*[125] on the labeling of cytochrome *c* in intact Krebs II ascites tumor cells furnished comparable results. The combined evidence obtained by widely differing methods and from different biological systems leaves little doubt that cytochrome *c* is a product of the chromosomal genetic system.

Even if the outer mitochondrial membrane is permeable to cytochrome *c,* the transfer of this hemoprotein from the cytoplasmic ribosomes into the mitochondria poses a number of difficult problems. Does this transfer process involve simple diffusion, the required selectivity being imparted by specific binding sites for cytochrome *c* on the mitochondrial inner membrane (cf Ref. 173), or does it proceed via special transport systems which recognize those microsomal protein products destined to be incorporated into the mitochondria? We do not know. Even tentative evidence concerning this vital point is unavailable at present. However, a promising experimental advance was recently achieved by the demonstration[176] that radioactive cytochrome *c* can be recovered from isolated rat-liver mitochondria that had been incubated with cell sap as well as with microsomes labeled with ^{14}C-lysine. It proved to be essential that the microsomes had been labeled within intact liver slices, i.e., under conditions approaching those existing *in vivo*. Microsomes labeled with ^{14}C-lysine in a cell-free system were incapable of labeling the cytochrome *c* pool of the admixed isolated mitochondria, although they still transferred some unidentified radioactive polypeptides into the mitochondria (cf above). These findings may be viewed as yet another powerful argument for the extramitochondrial synthesis of cytochrome *c*, and may open the way for a better understanding of the interactions between mitochondria and the cytoplasm. A more detailed investigation of this interesting transfer system should be most rewarding.

Concluding Remarks

In concluding this section, it must be emphasized that the numerous studies on mitochondrial protein synthesis have, thus far, mainly produced negative answers. We can state with reasonable confidence that certain mitochondrial constituents, such as the outer membrane or the easily solubilized proteins, are *not* made within the mitochondria, but we are as yet unable to specify a single distinct product of the mitochondrial protein synthesizing system.

There is mounting evidence, however, that such a system participates in mitochondrial biogenesis, and that it does so by acting in concert with the cytoplasmic protein synthesizing system. The participation of this latter system is, at present, perhaps the most persuasive biochemical argument for the view that both chromosomal and nonchromosomal genetic messages direct the assembly of a mitochondrion.

There is also evidence to indicate that the contribution of the chromosomal cistrons is not limited to the synthesis of mitochondrial protein. Wilgram and Kennedy[406] have shown some time ago that, in rat-liver cells, the final step in the formation of lecithine, the phosphorylcholine-triglyceride transferase reaction, is almost exclusively localized in the microsomes. It follows, therefore, that lecithin, which represents a major mitochondrial constituent, is not formed within the mitochondria (cf Ref. 175). This result reemphasizes the conclusions derived from studies on mitochondrial protein synthesis, since it shows that the task of assembling a mitochondrion significantly exceeds the biosynthetic capacity of this organelle.

V. THE MECHANISM OF MITOCHONDRIAL FORMATION

The foregoing sections have dealt in considerable detail with the biochemical correlates of mitochondrial semiautonomy, but have told us little about the pathways of mitochondrial formation in living cells. If we now focus our attention on this question, we leave the relatively secure area of biochemical investigation and enter a rather emotional field that has traditionally abounded with hypotheses and speculations. However, the past few years have witnessed several significant developments that afford some distinction between the various proposals advanced thus far. In addition, the recent evidence for a mitochondrial genetic system imposes a number of important restrictions on the mechanism of mitochondrial formation and thereby further limits the number of plausible possibilities. Finally, the present discussion is greatly aided by the fact that the bewildering array of different hypotheses can be reduced to relatively few recurring concepts. These concepts are: (a) mitochondrial formation from other intracellular structures; (b) *de novo* formation; and (c) formation by growth and division of preexisting mitochondria. The following survey directs itself to these three basic hypotheses and attempts to evaluate the experimental basis on which they rest.

Mitochondrial Formation from Nonmitochondrial Precursor Structures

There is hardly an intracellular structure that has not, at one time or another, been implicated in mitochondrial formation. Mitochondria have been suggested to arise from the nucleus,[159,249,25,39] the nucleolus,[103] the Golgi apparatus,[218,219] the "microsomes," [310] the cell membrane,[28,27,311] or from pinocytosis vesicles.[129] However, the evidence in support of these hypotheses is nearly always limited to morphological observations suggesting similarities, or close contacts, between mitochondria and other cellular structures. These observations refer to an artificially static situation which lacks the dimensions of time and direction and is, thus, inherently unsuited for the analysis of vectorial events. Moreover, many of the earlier electron micrographs cited as evidence for precursor relationships are quite unsatisfactory by present standards, and can no longer be accepted as compelling arguments for a genuine contact, much less a continuity, between mitochondria and other membrane systems. A mitochondrial origin from nonmitochondrial precursor structures is also difficult to reconcile with the structural and biochemical individuality of the different cellular membranes. Many of the above mentioned hypotheses were advanced at a time when electron microscopy of biological membranes was largely restricted to positively stained and embedded specimens in which the various cytomembranes exhibited a closely similar fine structure and seemed to represent minor variations of a universal biological "unit membrane." However, the introduction of negative staining techniques[41] subsequently exposed profound structural differences between the various cellular membrane systems.[120,279,363,83] According to a large body of recent evidence, equally great differences exist with respect to the chemical and enzymic properties[215,316,217] so that there appears to be little in common between a mitochondrion and, for example, a microbody or the cell membrane. Even more important, it is difficult to visualize how the existence of a specific mitochondrial DNA and a mitochondrial genetic system can be plausibly fitted into any hypothesis postulating a mitochondrial origin from other structured elements. The experiments on mitochondrial formation in *N. crassa* also militate against such a mechanism (cf below). At present, it appears quite unlikely that mitochondria arise by transformation of other cellular structures. The operation of such a pathway under certain conditions is, of course, in no way excluded, nor is it ruled out that nonmitochondrial structures participate in mitochondrial formation. On the contrary, as outlined in the preceding section, it is now generally held that the endoplasmic reticulum synthesizes mitochondrial proteins and, possibly also, mitochondrial lipids. However, this interaction merely reflects the integration of the mitochondrial organelle into the cellular environment and is quite distinct from the relationship between a product and its precursor.

Mitochondrial Formation *de novo*

The term "*de novo* formation" is largely operational and generally implies an origin from precursors whose size is below the limit of resolution of the electron microscope. At present, this amounts to formation exclusively from individual molecules.

This consideration immediately illuminates the problem of proving, or disproving, *de novo* formation of mitochondria. Not only is it intrinsically difficult to establish the absence of an intracellular structure, but the validity of an experimental observation will also diminish with time as the methods of observation improve. Most of the early claims of mitochondrial *de novo* formation carry little weight today, since they were based on observations with the light microscope which can no longer be regarded as a reliable instrument for mitochondrial investigations. This fact is perhaps most clearly illustrated by the history relating to the experiments of Beckwith[24] and Harvey[152,153] which were long viewed as particularly strong arguments for mitochondrial *de novo* formation. In these experiments, fertilized egg cells of *Hydractinia echinata* and other organisms were centrifuged so that their intracellular contents separated into a distinct pellet of mitochondria and yolk particles, a clear intermediate phase, and a floating lipid layer which appeared to be devoid of mitochondria. Several of these centrifuged eggs spontaneously divided in such a fashion that the entire mitochondrial pellet was transmitted to only one of the two resulting blastomeres. Although the other seemed to be devoid of mitochondria, it sometimes remained viable and, upon development into a larva, ultimately restored its mitochondrial complement. Obviously, the interpretation of this experiment rested upon the assumption that the centrifugation procedure had succeeded in concentrating all of the cell's mitochondria into the particulate pellet. While this assumption appeared to be supported by conventional histochemical tests, it did not withstand the resolving power of the electron microscope. This instrument clearly revealed the presence of mitochondria in the floating lipid layer, and thus, invalidated the conclusions drawn from these brilliantly conceived experiments.[212]

In spite of this result, the discussion on mitochondrial *de novo* formation has continued unabated, and has recently received a strong impetus from experiments with *S. cerevisiae*. If grown aerobically, this organism contains an active, cyanide-sensitive respiratory system involving the cytochromes aa_3, b, c_1, and c and, as already discussed earlier, possesses numerous well-defined mitochondria which closely resemble those of higher cells.[354,2,398,426] In addition, however, *S. cerevisiae* is equipped with a powerful glycolytic system and can, therefore, also grow in the absence of oxygen if a fermentable substrate is present.[354] The anaerobically-grown cells lack a cyanide-sensitive respiration, and the entire classical cytochrome complement, but adaptively regain these characteristics upon aeration.[354,426,353,356,358]

At first sight, it would seem that this adaptive process represents an especially simple and clear-cut system for investigating the assembly of yeast mitochondria. In spite of considerable effort, however, the fate of these mitochondria during anaerobic growth and respiratory adaptation has remained uncertain. In 1959, Heyman-Blanchet and her colleagues briefly reported the isolation of mitochondria-like particles from anaerobically-grown yeast cells, but presented little evidence to support this claim.[156] Several years later, Wallace and Linnane were unable to discern mitochondrial profiles in electron micrographs of the anaerobic cells, and proposed that the appearance of typical mitochondria during respiratory adaptation involved *de novo* formation of these organelles from structureless precursors.[400] An examination of yeast cells during various stages of adaptation suggested to Wallace and Linnane that mitochondrial formation was initiated by the appearance of numerous electron-transparent vesicles which later filled with amorphous, electron-dense material, and finally evolved into primitive mitochondria.

A subsequent report by Morpurgo *et al.*[243] indicated, however, that the effect of anaerobic growth on the cytology of *S. cerevisiae* was more complex, being decisively influenced by the composition of the growth medium. According to this study, yeast cells grown anaerobically in a simple yeast extract–glucose–salt medium, similar to that used by Wallace and Linnane,[400] seemed to be completely devoid of mitochondria. However, they also appeared to be considerably damaged since they failed to adapt to oxygen and exhibited a survival rate of only 50%. On the other hand, cells grown in a medium supplemented with ergosterol and unsaturated fatty acids were healthy and capable of respiratory adaptation, but at the same time contained numerous mitochondrial profiles even prior to adaptation.

Still another view was advanced by Polakis *et al.*[289] who found that electron micrographs of anaerobically grown yeast cells did not reveal any mitochondrial structures even when the cells had been grown in the presence of a lipid supplement. However, the partially adapted cells contained numerous electron-transparent vacuoles similar to those described by Wallace and Linnane.[400]

These puzzling discrepancies suggest that the anaerobic propagation of yeast cells and their adaptation to oxygen involve many unknown variables and, thus, call for a rigorous control of experimental conditions. Many of the above-mentioned inconsistencies may perhaps be explained by the fact that the anaerobic yeast cultures studied by different workers were grown in the presence of different concentrations of glucose (cf below) and isolated at different points in their growth cycle. Moreover, the cells were often harvested and homogenized under conditions that would be expected to allow for considerable respiratory adaptation. In the opinion of this author, much of the blame for the confusion must also be attributed to the present limitations of electron microscopy. Because of their thick cell wall and possibly also because of their high RNA content, yeast cells are notoriously difficult objects for electron microscopical studies. Additional uncertainties arise from the fact that anaerobic growth profoundly alters the lipid composition of the yeast cells[339a,192,174a,277a] and thus may modify the staining characteristics of the intracellular membrane systems. Electron microscopy of stained and embedded anaerobic yeast cells would, thus, appear to be of limited value for ascertaining the absence of mitochondrial structures.

In an alternate approach to this question, we have attempted to isolate the different subcellular particles of anaerobically grown *S. cerevisiae* and to study their biochemical properties.[328,331,81,277a] Particular emphasis was placed on the maintenance of strict anaerobiosis during cell growth and the prevention of respiratory adaptation during the subsequent handling of the anaerobic cells. Regardless of whether these had been grown in the presence or the absence of a lipid supplement, they invariably contained subcellular particles that exhibited an oligomycin-sensitive, magnesium-stimulated ATPase activity. This ATPase was conclusively identified as mitochondrial ATPase (F_1) since it was inhibited by the naturally-occurring F_1 inhibitor of Pullman and Monroy[292] and a specific antiserum against purified F_1 from aerobic, respiring yeast mitochondria. The anaerobic particles resembled mitochondria from aerobically grown yeast cells also with respect to their buoyant density in "Urografin" gradients and their contents of succinic- and NADH dehydrogenase, flavin, and ferrochelatase. A further analogy to aerobic yeast mitochondria was discovered in experiments with anaerobically-grown cells of the cytoplasmic "petite" mutant: The mitochondria-like particles from these cells contained oligomycin-*in*-sensitive F_1 and, thus, exhibited the characteristic mitochondrial lesion that had earlier been noted with the aerobic "petite" cells (cf the section on mitochondrial protein synthesis). The properties of F_1 associated with the mitochondria-like particles are summarized in Table 7-4.

Perhaps the most persuasive argument for the mitochondrial nature of these nonrespiring particles was provided by the observation that they contained mitochondrial DNA (density = 1.685 g/ml). This finding, together with the above-mentioned effect of the cytoplasmic "petite" mutation, raised the possibility that the mitochondria-like particles still contain a functional mitochondrial genetic system (cf Refs. 59a, 296, 127).

Direct evidence for this view was obtained by exposing the anaerobically-grown cells to radioactive leucine in the presence of cycloheximide.[329a] Under these conditions, at least 90% of the total radioactivity incorporated into the cellular proteins was found to be associated with the mitochondria-like particles. This labeling process was obviously mediated by a mitochondrial protein synthesizing system since it was inhibited by chloramphenicol and was not observed with cells of the cytoplasmic "petite" mutant.

When the selectively labeled anaerobic cells were washed free of cycloheximide, suspended in an excess of nonlabeled leucine, and adapted to oxygen, the respiring mitochondria formed during adaptation proved to be strongly radioactive.[329a] Since *de novo* formation of mitochondria should have resulted in essentially nonlabeled organelles, this result indicated that respiratory adaptation in yeast induces differentiation of the mitochondria-like particles.

TABLE 7-4

PROPERTIES OF ATPase* ASSOCIATED WITH YEAST PROMITOCHONDRIA (CRIDDLE and SCHATZ [81])

Source of Promitochondria	ATPase Activity (μmole of ATP/min/mg protein)			
	no additions	+F_1-inhibitor	+oligomycin	+F_1-antiserum
Wild type grown with lipid supplement	1.35	0.12	0.18	0.21
Wild type grown without lipid supplement	0.71	0.069	0.11	0.11
"Petite" mutant grown with lipid supplement	0.52	0.061	0.49	0.041

* ATPase was assayed as described.[292] The effect of inhibitors was determined by measuring the ATPase activity of particles which had been incubated for 10 min at 30° with 0.5 mg of F_1-inhibitor,[292] 130 μg of oligomycin or 2 mg of F_1-antiserum per mg of particle protein, respectively.

In summary, then, it appears that anaerobically grown yeast cells contain undifferentiated "promitochondria" which evolve into fully developed mitochondria upon exposure to oxygen and which are thus in many ways analogous to the "proplastids" of etiolated plant cells.[130,242,115] Electron micrographs of frozen-etched anaerobic yeast cells have clearly revealed the presence of mitochondrial structures which are morphologically quite similar to those observed in aerobically grown yeast cells.[288a] (Figure 7-4) Anaerobic growth of *S. cerevisiae* is, thus, associated with a de-differentiation, rather than a complete loss, of the mitochondrial organelles.

In any case, the rapidly consolidating evidence for the existence of a specific mitochondrial genetic system makes it increasingly difficult to sustain the thesis that mitochondria can arise solely from nonstructured precursors. Cogent arguments against this mechanism of formation also stem from experiments with *N. crassa* and *T. pyriformis* which are discussed in the following section. Unless additional evidence is forthcoming to support the possibility of mitochondrial *de novo* formation, the operation of this pathway must be regarded as rather unlikely.

FIGURE 7-4. Electron micrograph of a frozen-etched yeast cell grown anaerobically in the absence of a lipid supplement. Numerous promitochondria are clearly visible (arrows). Magnification 20,400×.

Mitochondrial Formation by Growth and Division

The idea that mitochondria grow by the acquisition of cytoplasmic constituents and multiply by fission was first advanced by Altmann[4] and has ever since remained especially intriguing to cytologists.[213,409,235,236,108,5,]

[378,118,131,274] Numerous electron microscopists have noted dumbbell-shaped mitochondrial profiles in various tissues and have interpreted them as mitochondria undergoing division,[118,378,72,208,94,12,154] However, many of these electron micrographs are of doubtful significance, since they were not taken from serially sectioned specimens. Even when genuine mitochondrial constrictions were adequately documented, it could not be excluded that they were merely reversible events which mirrored mitochondrial plasticity rather than mitochondrial division. It is equally difficult to interpret the mitochondrial images that appear to be divided by a median double membrane and that are, therefore, often thought to represent mitochondria in the terminal stages of their fission cycle. The cinematographic studies of Frederic[124] suggest that mitochondria are extremely motile structures which readily pinch off parts and coalesce with one another; and, thus, it is entirely possible that many of the "dividing" mitochondria were actually in the process of fusing with another mitochondrion.

It is rather doubtful whether electron micrographs are inherently capable of proving the occurrence of mitochondrial division. They are even less suited to answer the question of whether such an event, if it could indeed be verified, was of any significance for the biogenesis of the mitochondrial organelle. Nevertheless, it is important that dumbbell-shaped and bisected mitochondria do occur in living cells and that the morphological evidence is fully consistent with a mechanism of mitochondrial formation by growth and division.

Conclusive proof for this mechanism was obtained by Luck in biochemical experiments with exponentially growing *N. crassa* cells.[224,225] These experiments made use of the fact that a choline-less *N. crassa* mutant rapidly incorporated externally added choline into its phospholipids so that radioactive choline served as an efficient and stable marker for the mitochondrial structures. Thus, if the mutant was first grown in the presence of labeled choline and then transferred to a nonlabeled growth medium, the origin of the mitochondria formed after the transfer could be deduced by quantitative radioautography of the isolated mitochondrial fractions. Specifically, three distinct labeling patterns would have been possible, each of them indicative of a different mechanism of mitochondrial formation:

(a) A random distribution of label over all of the mitochondria, the average grain count per mitochondrion diminishing as a function of growth in the nonlabeled medium. This pattern of labeling would be consistent with mitochondrial formation by division.

(b) A nonrandom distribution in which the percentage of labeled mitochondria decreased in proportion to growth in the nonlabeled medium and in which the average number of grains associated with each of the labeled mitochondria remained unchanged. This pattern would reflect *de novo* formation from nonlabeled precursors.

(c) An intermediate pattern in which the label was, again, nonrandomly distributed, but in which the percentage of labeled mitochondria decreased less rapidly than would be required by a *de novo* mechanism. This would be compatible with mitochondrial formation from labeled precursors such as nonmitochondrial membrane structures.

The experimentally determined labeling pattern precisely conformed to the first possibility. Luck concluded, therefore, that most, if not all, of the mitochondria formed during exponential growth in the nonlabeled medium had arisen by growth and division of preexisting mitochondria.

As a further test of this conclusion, Luck studied the density of the mitochondria which were formed by the cholineless mutant after a sudden increase of the external choline concentration.[227,228] This experimental approach was based on the observation[226] that an increase in the choline content of the growth medium from 1 to 10 μg/ml did not measurably affect the growth rate of the *Neurospora* cells but significantly decreased the buoyant density of their mitochondria. Mitochondria isolated from cells grown at different choline levels could thus readily be identified, and separated from each other, by isopycnic centrifugation in sucrose density gradients.

Therefore, if the exponentially growing mutant was shifted from a low to a high choline medium, the density of the newly formed mitochondria should identify the mechanism of mitochondrial formation. If mitochondria arose by the random addition of new material and subsequent fission, their density should gradually and uniformly decrease until it finally reaches the density value of the "high choline" mitochondria. On the other hand, if the mitochondria were formed *de novo,* they should immediately exhibit this low density. Since the density of the preexisting mitochondria would remain unchanged, the transient emergence of two distinct mitochondrial populations should be observed. Finally, if the mitochondria originated from other membrane structures, a more complex situation should prevail: while the density of the "old" mitochondria would not change, that of the "new" ones would initially exhibit an intermediate value. A distinct low-density population should appear only after a significant lag which would be proportional to the size of the precursor membrane pool.

Again, the first model proved to be correct. As illustrated in Figure 7-5, the transfer to the "high choline" medium induced a gradual and uniform decrease of mitochondrial density without the emergence of a light mitochondrial subfraction. It is also evident that the gradient method used by Luck would have been capable of detecting such a subfraction had it occurred in significant amounts. This impressive concordance of results obtained by quite different experimental procedures leaves little doubt that mitochondria arise by growth and division, at least in exponentially growing *N. crassa* cells.

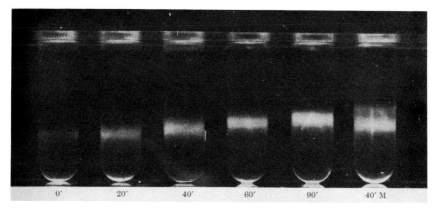

FIGURE 7-5. Buoyant density of mitochondria isolated from *N. crassa* cells at various times after transfer from a "low choline" to a "high choline" medium. The mitochondrial preparation shown in the last centrifuge tube (marked 40′ M) had been isolated from an artificial mixture (1:4 by wet weight) of cells grown for 15 hours in "high choline" and cells first grown in "low choline," and subsequently exposed for 40 min to "high choline" (from Luck[227]). (Reproduced with permission *J. Cell Biol.* and the authors)

Although it has not yet been rigorously established that these conclusions can be extended to other biological systems, a wealth of suggestive evidence indicates that this is indeed the case. In an experiment analogous to that described earlier by Luck, Parsons labeled the intramitochondrial DNA in living *T. pyriformis* cells with radioactive thymidine and followed the fate of this mitochondrial marker by phase-contrast autoradiography.[280] He found that the label remained randomly distributed over the entire mitochondrial population during four cellular division cycles, thus strongly suggesting mitochondrial growth and division. A further argument for this mechanism was obtained by Broesemer *et al.*[45] who studied the development of the indirect flight muscle of *Locusta migratoria*. During imaginal molting of this insect, the mitochondrial mass in the muscle increased sixtyfold whereas the composition of the individual mitochondria remained essentially unchanged. Moreover, the mitochondrial increase was not accompanied by the appearance of intracellular structures that could have been interpreted as incomplete mitochondria or as mitochondrial precursors.

Little evidence is available bearing on the mechanism of mitochondrial formation in mammalian cells. However, the fact that a Poisson distribution of liver mitochondrial grain counts is observed after injecting rats with tritiated thymidine[95] is consistent with, and suggestive of, mitochondrial growth by random accretion of cytoplasmic material. If a *de novo* mechanism were operative, one would expect to find *two* distinct populations of mitochondria, one highly labeled and the other relatively free of radio-activity.

For many years, the experiments of Fletcher and Sanadi (123) on the turnover of rat-liver mitochondria were also considered as persuasive arguments for mitochondrial formation via growth and division. In these interesting experiments, the mitochondrial proteins and lipids were labeled *in vivo* by injecting the rats with ^{35}S-methionine and ^{14}C-acetate, respectively. At specified times later, the animals were killed and their radio-active liver mitochondria fractionated into lipids, water-soluble proteins, cytochrome *c,* and insoluble proteins. The decline in specific radioactivity of these four fractions indicated that they all exhibited an identical biological half-life of 10.3 days. Several years later, Neubert *et al.*[263] carried out a similar experiment by labeling the mitochondrial DNA of rat liver with radioactive thymidine and again obtained a half-life of about 9 days. From these results, it looked as if mitochondria were turning over as a unit, a concept consistent with, but not required by, a mechanism of mitochondrial formation by growth and division.

Subsequent investigations by other authors have shown, however, that this concept represents an oversimplification and can no longer be maintained.[22,13,231,380] A more detailed kinetic analysis of the turnover of various mitochondrial constituents revealed that mitochondria from liver, kidney, and brain of the rat possessed components of widely different half-lives. For example, the water-soluble proteins of rat-kidney mitochondria turned over much more rapidly than the insoluble proteins.[22] Other data indicated that both the soluble and the insoluble mitochondrial proteins[13,231,380] as well as the mitochondrial lipids[13,380,283,144] each comprised a number of distinct components with different biological half-lives. These results are strengthened by the experiments of Luck with *N. crassa* which demonstrated that mitochondrial proteins and phospholipids are, within certain limits, metabolically independent.[226]

While the characteristics of mitochondrial turnover in higher cells do not prove mitochondrial formation by growth and division, neither do they exclude such a mechanism. For the time being, they may be taken as yet another indication for the complexity of the events that culminate in the assembly of a mitochondrion. This complexity is further borne out by the enormous variability of the mitochondrial organelle in response to environmental conditions. An extensive literature has accumulated on the abnormal mitochondria that are formed in various organisms upon malnutrition, acute poisoning, or disease (cf Refs. 157, 158, 307, 59, 405, 211). The influence of the external choline concentration on the density of *N. crassa* mitochondria is another example of a reversible mitochondrial alteration.[226] Even more striking mitochondrial changes occur in maize roots during maturation[230] and in *S. cerevisiae* cells upon growth in the presence of high concentrations of glucose.[113,426,327,174,290] Perhaps the most drastic case, however, is the de-differentiation of yeast mitochondria during anaerobic growth which has been discussed in one of the preceding paragraphs. At present, it is not known whether the different external conditions primarily affect the mitochondrial genetic system or the chromosomal one. It is evident, however, that the phenomenon of mitochondrial variability offers promising experimental possibilities for studying the regulatory mechanisms governing mitochondrial formation.

It was mentioned earlier that every mechanism of mitochondrial formation must be compatible with the existence for a specific mitochondrial genetic system. Obviously, a mechanism postulating growth and division of preexisting mitochondria meets this essential requirement and is thereby distinctly superior to alternate mechanisms proposed thus far. By the same token, it can easily account for the phenomenon of mitochondrial mutation which has already been briefly touched upon in an earlier section of this review. Mitochondrial mutants were first discovered by Ephrussi et al.[112,105,109,107] and Mitchell et al.[239,240] in their now classical studies on respiration- and cytochrome-deficient, extrachromosomally inherited yeast and *N. crassa* mutants. While the extrachromosomal genetic determinants affected by these mutations remain to be conclusively identified, the evidence discussed in an earlier section suggests that they are, in fact, mitochondrial DNA, and that the primary mutational event occurs within the mitochondria themselves.

Since the loss of respiratory cytochromes represents a rather drastic impairment of mitochondrial function, it is doubtful whether similar mutations are tolerated by higher organisms. More recently, however, other nonchromosomal yeast mutants have been discovered that have suffered a more subtle modification of their mitochondrial function. Linnane et al.[221,222] isolated several yeast mutants which differed from the normal parent strains in that mitochondrial formation was no longer repressed by growth in the presence of erythromycin. This resistance was traced to a decreased erythromycin-sensitivity of the mitochondrial protein-synthesizing system and was inherited in a nonchromosomal fashion. Present evidence would be consistent with the possibility that the mitochondrial DNA of these strains contains altered cistrons for mitochondrial ribosomal RNA and, thus, causes the synthesis of atypical mitochondrial ribosomes, which are insensitive to erythromycin.

A more detailed discussion of mitochondrial mutants and their possible relationship to phenotypically similar chromosomal mutants clearly exceeds the scope of the present review. It may suffice, therefore, to emphasize that growth and division of the mutated mitochondria represents an attractively simple mechanism for explaining the transmission of the abnormal mitochondrial phenotype to the daughter cells.

Diacumakos et al.[93] have now presented direct evidence that mitochondria may indeed transmit genetic information. This important fact was established by injecting normal *N. crassa* cells with nuclei and mitochondria that had been isolated from a cytoplasmic *N. crassa* mutant. This mutant, termed abnormal-1, was characterized by a lack of cytochrome aa_3, the inability to form conidia, and slow and irregular growth.[128] Whereas none of the normal cells that had received the isolated nuclei were noticeably altered, many of those injected with the abnormal mitochondria permanently acquired the characteristics of the mutant donor strain. The details of this "mitochondrial transformation" are not at all clear at present, but it is conceivable that the injected mutant mitochondria simply outgrew their normal counterparts. Again, mitochondrial formation by division offers the most logical explanation for these experimental results.

The findings of Diacumakos *et al.* not only underscore the reality of a mitochondrial genetic system, but also indicate that a single *N. crassa* cell may, at least temporarily, contain different types of mitochondria. A similar transient coexistence of different mitochondrial populations in a common cytoplasm may also occur in yeast cells shortly after exposure to acriflavin.[111] According to some authors,[9,10,325,326] even a *permanent* coexistence of different mitochondrial "clones" may be possible in certain cells. Although this possibility is still far from established, it would once again be most easily accounted for by mitochondrial growth and division.

VI. CONCLUSIONS

In view of the present evidence, it can no longer be questioned that mitochondria are capable of growth and division and that these events represent the major route of mitochondrial formation in exponentially growing *N. crassa* cells. It is equally clear that such a mechanism is consistent with a large number of additional observations that are difficult to reconcile with alternate pathways of mitochondrial formation. For the time being, we may conclude that each mitochondrion arises by growth and fission of a preexisting mitochondrion.

This hypothesis, however, merely provides a conceptual framework for investigating the biochemistry of mitochondrial formation. The different turnover rates of the various mitochondrial constituents and the phenomenon of mitochondrial variability bear ample witness to the fact that the mechanism of mitochondrial formation is considerably more intricate and subject to numerous regulations which we are just beginning to understand. The various interactions between mitochondria and the surrounding cytoplasm undoubtedly hold many answers to these still unsolved and important questions.

The fact that mitochondria may grow and divide and are capable of mutation appears to support the long-held view (cf Ref. 4) that mitochondria originated from bacterial endosymbionts which were gradually integrated into their host. Indeed, mitochondria resemble bacteria with respect to size,[215] the distribution of respiratory enzymes,[237] and their nucleic acid components, and it is safe to predict that additional resemblances will be found as our knowledge expands. Impressive as these similarities are, they do not prove an evolutionary origin of mitochondria from free-living organisms, since actually the reverse may be the case. Nevertheless, the possible phylogenetic relationship between mitochondria and bacteria is too fascinating a topic to be entirely disregarded, and it will undoubtedly continue to enliven lunch table discussions, scientific meetings and, last but not least, reviews (cf Ref. 330).

ACKNOWLEDGMENTS

I am greatly indebted to my colleagues Dr. E. Wintersberger and Dr. R. S. Criddle for many extensive and stimulating discussions concerning the various views expressed in this chapter and for valuable criticism during the preparation of the manuscript. My thanks are also due to Dr. M. M. K. Nass and Dr. E. F. J. van Bruggen who kindly provided me with prints of their electron micrographs, and to Dr. A. W. Linnane and Dr. W. E. Barnett who sent me some of their papers in advance of publication. I also wish to acknowledge the generous financial assistance of the Austrian Biochemical Society.

REFERENCES

1. Acs, G., Reich, E., and Valanju, S. (1963), *Biochim. Biophys. Acta* **76**, 68.
2. Agar, H. D., and Douglas, H. C. (1957), *J. Bacteriol.* **73**, 365.
3. Allen, F. W. (1962), "Ribonucleoproteins and Ribonucleic Acids," Elsevier Publ. Co., Amsterdam.
4. Altmann, R. (1890), "Die Elementarorganismen und ihre Beziehungen zu den Zellen," Veit und Co., Leipzig.
5. André, J. (1962), *J. Ultrastruct. Res.* Suppl. 3.
6. André, J., and Marinozzi, V. (1965), *J. Microsc.* **4**, 615.
7. Appelmans, F., Wattiaux, R., and de Duve, C. (1955), *Biochem. J.* **59**, 438.
8. Attardi, B., and Attardi, G. (1967), *Proc. Natl. Acad. Sci.* **58**, 1051.
9. Avers, C. J. (1967), *Proc. Natl. Acad. Sci.* **58**, 620.
10. Avers, C. J., Rancourt, M. W., and Lin, F. H. (1965), *Proc. Natl. Acad. Sci.* **54**, 527.
11. Bachelard, H. S. (1966), *Biochem. J.* **100**, 131.
12. Bahr, G. F., and Zeitler, E. (1962), *J. Cell Biol.* **15**, 489.
13. Bailey, E., Taylor, C. B., and Bartley, W. (1967), *Biochem. J.* **104**, 1026.
14. Baltus, E., and Brachet, J. (1963), *Biochim. Biophys. Acta* **76**, 490.
15. Baudhuin, P., Müller, M., Poole, B., and de Duve, C. (1965), *Biochem. Biophys. Res. Commun.* **20**, 53.
16. Barnett, W. E., and Braun, D. H. (1967), *Proc. Natl. Acad. Sci.* **57**, 452.
17. Barnett, W. E., Braun, D. H., and Epler, J. (1967), *Proc. Natl. Acad. Sci.* **57**, 1775.

18. Barnett, W. E., and Epler, J. L. (1966), *Proc. Natl. Acad. Sci.* **55**, 184.
19. Bass, R., Neubert, D., and Morris, H. P. (1966), *Arch. Exptl. Pathol. Pharmakol.* **255**, 2.
20. Beattie, D. S., Basford, R. E., and Koritz, S. B. (1966), *Biochemistry* **5**, 926.
21. Beattie, D. S., Basford, R. E., and Koritz, S. B. (1967), *J. Biol. Chem.* **242**, 3366.
22. Beattie, D. S., Basford, R. E., and Koritz, S. B. (1967), *J. Biol. Chem.* **242**, 4584.
23. Beattie, D. S., Basford, R. E., and Koritz, S. B. (1967), *Biochemistry* **6**, 3099.
24. Beckwith, C. J. (1914), *J. Morphol.* **25**, 189.
25. Bell, P. R., and Mühlethaler, K. (1964), *J. Cell Biol.* **20**, 235.
26. Bell, P. R., and Mühlethaler, K. (1964), *J. Mol. Biol.* **8**, 853.
27. Ben Geren, B., and Schmitt, F. O. (1954), *Proc. Natl. Acad. Sci.* **40**, 863.
28. Ben Geren, B., and Schmitt, F. O. (1956), "Symposium on the Fine Structure of Cells," p. 251, Interscience Publishers, New York.
29. Bisalputra, T., and Bisalputra, A. A. (1967), *J. Cell Biol.* **33**, 511.
30. Bolton, E. T., and McCarthy, B. J. (1962), *Proc. Natl. Acad. Sci.* **48**, 1390.
31. Bond, H. E., Flamm, W. G., Burr, H. E., and Bond, S. B. (1967), *J. Mol. Biol.* **27**, 289.
32. Borst, P. (1967), *Biochem. J.* **105**, 37 P.
33. Borst, P., van Bruggen, E. F. J., Ruttenberg, G. J. C. M., and Kroon, A. M. (1967), *Biochim. Biophys. Acta* **149**, 156.
34. Borst, P., van Bruggen, E. F. J., and Ruttenberg, G. J. C. M., cf Ref. 377, p. 51.
35. Borst, P., and Ruttenberg, G. J. C. M. (1965), in "Regulation of Metabolic Processes in Mitochondria," BBA Library Vol. 7, p. 454 (J. M. Tager, S. Papa, E. Quagliariello, and E. C. Slater, eds.), Elsevier Publ. Co., Amsterdam.
36. Borst, P., and Ruttenberg, G. J. C. M. (1966), *Biochim. Biophys. Acta* **114**, 645.
37. Borst, P., Ruttenberg, G. J. C. M., and Kroon, A. M. (1967), *Biochim. Biophys. Acta* **149**, 140.
38. Brachet, J. (1962), *J. Cell. Comp. Physiol.* Suppl. 1, **60**, 1.
39. Brandt, P. W., and Pappas, G. D. (1959), *J. Biophys. Biochem. Cytol.* **6**, 91.
40. Braun, G. A., Marsh, J. B., and Drabkin, D. L. (1963), *Biochim. Biophys. Acta* **72**, 645.
41. Brenner, S., and Horne, R. W. (1959), *Biochim. Biophys. Acta* **34**, 103.
42. Bresslau, E., and Scremin, L. (1924), *Arch. Protistenkunde* **48**, 509.
43. Bretthauer, R. K., Marcus, L., Chaloupka, J., Halvorson, H. O., and Bock, R. M. (1963), *Biochemistry* **2**, 1079.
44. Brewer, E. N., DeVries, A., and Rusch, H. P. (1967), *Biochim. Biophys. Acta* **145**, 686.
45. Brosemer, R. W., Vogell, W., and Bücher, Th. (1963), *Biochem. Z.* **338**, 854.
46. Bronk, J. R. (1963), *Proc. Natl. Acad. Sci.* **50**, 524.

47. Bronk, J. R. (1963), *Science* **141,** 816.
48. Bronsert, U. and Neupert, W., cf. Ref. 35. p. 426.
49. van Bruggen, E. F. J., Borst, P., Ruttenberg, G. J .C .M., Gruber, M., and Kroon, A. M. (1966), *Biochim. Biophys. Acta* **119,** 437.
50. Bruni, A., cf Ref. 35, p. 275.
51. Buck, C., and Nass, M. M. K. (1968), *Fed. Proc.* **27,** 342.
52. Bulder, C. J. E. A. (1964), *Antonie van Leeuwenhoek* **30,** 1.
53. du Buy, H. G., Mattern, C. F. T., and Riley, F. L. (1966), *Biochim. Biophys. Acta* **123,** 298.
54. du Buy, H. G., Mattern, C. F. T., and Riley, F. L. (1965), *Science* **147,** 754.
55. du Buy, H. G., and Riley, F. L. (1967), *Proc. Natl. Acad. Sci.* **57,** 790.
56. Cairns, J. (1963), *Cold Spring Harbor Sympos. Quant. Biol.* **28,** 43.
57. Cameron, I. L. (1966), *Nature* **209,** 630.
58. Cantoni, G. L. (1966), "Molecular Architecture in Cell Physiology," p. 147 (T. Hayashi, and A. G. Szent-Györgyi, eds.), Prentice-Hall, Inc., Englewood, N.J.
59. Carafoli, E., Margreth, A., and Buffa, P. (1962), *Nature* **196,** 1101.
59a. Carnevali, F., Piperno, G., and Tecce, G. (1966), *Atti Accad. Naz. Lincei, Rend., Cl. Sci. Fis. Mat. Nat.* **41,** 194.
60. Chamberlin, M. and Berg, P. (1962), *Proc. Natl. Acad. Sci.* **48,** 81.
61. Chandler, B., Hayashi, M., Hayashi, M. N., and Spiegelman, S. (1964), *Science* **143,** 47.
62. Chappell, J. B., and Crofts, A. R. (1965), *Biochem. J.* **95,** 707.
63. Chatterjee, S. K., Das, H. K., and Roy, S. C. (1966), *Biochim. Biophys. Acta* **114,** 349.
63a. Chatterjee, S. K., Mukherjee, T., Das, H. K., Nath, K., and Roy, S. C. (1966), *Ind. J. Biochem.* **3,** 239.
64. Chen, S. Y., Ephrussi, B., and Hottinguer, H. (1950), *Heredity* **4,** 337.
65. Chèvremont, M. (1962), *Biochem. J.* **85,** 25 P.
66. Chèvremont, M. (1963), "Cell Growth and Cell Division," p. 323 (R. J. C. Harris, ed.), Academic Press, Inc., New York.
67. Chèvremont, M., Chèvremont-Comhaire, S., and Baeckeland, E. (1959), *Arch. Biol. (Liège)* **70,** 833.
68. Chun, E. H., and Littlefield, J. W. (1963), *J. Mol. Biol.* **7,** 245.
69. Clark, T. B., and Wallace, F. G. (1960), *J. Protozool.* **7,** 115.
70. Clark-Walker, G. D., and Linnane, A. W. (1966), *Biochem. Biophys. Res. Commun.* **25,** 8.
71. Clark-Walker, G. D., and Linnane, A. W. (1967), *J. Cell Biol.* **34,** 1.
72. Claude, A. (1965), *J. Cell Biol.* **27,** 146 A.
73. Clayton, D. A., and Vinograd, J. (1967), *Nature,* **216,** 652.
74. Colli, W., and Pullman, M. E. (1964), *Ciencia Cult. (Sao Paolo)* **16,** 187.
75. Conover, T. E., and Bárány, M. (1966), *Biochim. Biophys. Acta* **127,** 235.
76. Corneo, G., Ginelli, E., and Polli, E. (1967), *J. Mol. Biol.* **23,** 619.
77. Corneo, G., Moore, C., Sanadi, D. R., Grossman, L. I., and Marmur, J. (1966), *Science* **151,** 687.
77a. Corneo, G., Zardi, L., and Polli, E. (1968), *J. Mol. Biol.* **26,** 419.

78. Cotman, C. W., and Mahler, H. R. (1967), *Arch. Biochem. Biophys.* **120,** 384.
79. Craddock, V. M., and Simpson, M. V. (1961), *Biochem. J.* **80,** 348.
80. Criddle, R. S., Bock, R. M., Green, D. E., and Tisdale, H. (1962), *Biochemistry* **1,** 827.
81. Criddle, R. S., and Schatz, G. (1969), *Biochemistry* **8,** 322.
82. Cummins, J. E., Rusch, H. P., and Evans, T. E. (1967), *J. Mol. Biol.* **23,** 281.
83. Cunningham, W. P., and Crane, F. L. (1965), *Plant Physiol.* **40,** 1041.
84. Dalton, A. J., Potter, M., and Merwin, R. M. (1961), *J. Natl. Cancer Inst.* **26,** 1221.
85. Das, H. K., Chatterjee, S. K., and Roy, S. C., (1964), *Biochim. Biophys. Acta* **87,** 478.
86. Das, H. K., Chatterjee, S. K., and Roy, S. C. (1964), *J. Biol. Chem.* **239,** 1126.
86a. Das, H. K., and Roy, S. C. (1961), *Biochim. Biophys. Acta* **53,** 445.
87. Dawid, I. B. (1965), *J. Mol. Biol.* **12,** 581.
88. Dawid, I. B. (1966), *Proc. Natl. Acad. Sci.* **56,** 269.
89. Dawid, I. B., and Wolstenholme, D. R. (1967), *J. Mol. Biol.* **28,** 233.
90. Dawid, I. B., and Wolstenholme, D. R., cf Ref. 377, p. 83.
91. Dawid, I. B., and Wolstenholme, D. R. (1968), *Biophys. J.* **8,** 65.
92. von der Decken, A., Löw, H., and Sandell, S., cf Ref. 35, p. 415.
93. Diacumakos, E. G., Garnjobst, L., and Tatum, E. L. (1965), *J. Cell Biol.* **26,** 427.
94. Diers, L. (1966), *J. Cell Biol.* **28,** 527.
95. Droz, B., and Bergeron, M. (1965), *Compt. rend.* **261,** 2757.
96. Dubin, D. T. (1967), *Biochem. Biophys. Res. Commun.* **29,** 655.
97. Dubin, D. T., and Brown, R. E. (1967), *Biochim. Biophys. Acta* **145,** 538.
98. Dure, L. S., Epler, J. L., and Barnett, W. E. (1967), *Proc. Natl. Acad. Sci.* **58,** 1883.
99. de Duve, C., Beaufay, H., Jacques, P., Rahman-Li, Y., Sellinger, O. Z., Wattiaux, R., and de Coninck, S. (1960), *Biochim. Biophys. Acta* **40,** 186.
100. Edelman, M., Epstein, H. T., and Schiff, J. A. (1966), *J. Mol. Biol.* **17,** 463.
101. Edelman, M., Schiff, J. A., and Epstein, H. T. (1965), *J. Mol. Biol.* **11,** 769.
102. von Ehrenstein, G., and Lipmann, F. (1961), *Proc. Natl. Acad. Sci.* **47,** 941.
103. Ehret, C. F., and Powers, E. L. (1955), *Exptl. Cell Res.* **9,** 241.
104. Elaev, N. R. (1964), *Biokhimiya* **29,** 413.
105. Ephrussi, B. (1950), *The Harvey Lectures,* Series **46,** 45.
106. Ephrussi, B. (1953), "Nucleo-Cytoplasmic Relations in Microorganisms," Clarendon Press, Oxford.
107. Ephrussi, B. (1956), *Naturwiss.* **43,** 505.
108. Ephrussi, B. (1958), *J. Cell. Comp. Physiol.* **52** (Suppl.) 35.
109. Ephrussi, B., L'Heritier, P., and Hottinguer, H. (1949), *Ann. Inst. Pasteur* **77,** 64.
110. Ephrussi, B., and Hottinguer, H. (1950), *Nature* **166,** 956.

111. Ephrussi, B., and Hottinguer, H. (1951), *Cold Spring Harbor Sympos. Quant. Biol.* **16**, 75.

112. Ephrussi, B., Hottinguer, H., and Chimenes, A. M. (1949), *Ann. Inst. Pasteur* **76**, 351.

113. Ephrussi, B., Slonimski, P. P., Yotsuyanagi, Y., and Tavlitzki, J. (1956), *Compt. Rend. Lab. Carlsberg, Ser. physiol.* **26**, 87.

114. Epler, J. L., and Barnett, W. E. (1967), *Biochem. Biophys. Res. Commun.* **28**, 328.

115. Epstein, H. T., and Schiff, J. A. (1961), *J. Protozool.* **8**, 427.

116. Evans, T. E. (1966), *Biochem. Biophys. Res. Commun.* **22**, 678.

117. Evans, D. A., and Lloyd, D. (1967), *Biochem. J.* **103**, 22P.

118. Fawcett, D. W. (1955), *J. Natl. Cancer Inst.* **15**, 1475.

119. Federman, M., and Avers, C. J. (1967), *J. Bacteriol.* **94**, 1236.

120. Fernández-Morán, H. (1962), *Circulation* **26**, 1039.

120a. Firkin, F. C., and Linnane, A. W. (1968), *Biochem. Biophys. Res. Commun.* **32**, 398.

121. Flamm, W. G., Bond, H. E., Burr, H. E., and Bond, S. B. (1966), *Biochim. Biophys. Acta* **123**, 652.

122. Flaumenhaft, E., Conrads, S. M., and Katz, J. J. (1960), *Science* **132**, 892.

123. Fletcher, M. J., and Sanadi, D. R. (1961), *Biochim. Biophys. Acta* **51**, 356.

124. Frederic, J. (1958), *Arch. Biol. (Liège)* **69**, 167.

125. Freeman, K. B., Haldar, D., and Work, T. S. (1967), *Biochem. J.* **105**, 947.

126. Fournier, M., and Simpson, M. V. (1968), cf Ref. 377, p. 227.

127. Fukuhara, H. (1967), *Proc. Natl. Acad. Sci.* **58**, 1065.

128. Garnjobst, L., Wilson, J. F., and Tatum, E. L. (1965), *J. Cell Biol.* **26**, 413.

128a. Georgatsos, J. G., and Papasarantopoulou, N. (1968), *Arch. Biochem. Biophys.* **126**, 771.

129. Gey, G. (1956), *The Harvey Lectures,* Series **50**, 154.

130. Gibor, A., and Granick, S. (1962), *J. Protozool.* **9**, 327.

131. Gibor, A., and Granick, S. (1964), *Science* **145**, 890.

132. Gibor, A., and Izawa, M. (1963), *Proc. Natl. Acad. Sci.* **50**, 1164.

133. Gillespie, D., and Spiegelman, S. (1965), *J. Mol. Biol.* **12**, 829.

133a. Giudice, G. (1960), *Exp. Cell Res.* **21**, 222.

134. González-Cadavid, N. F., and Campbell, P. N. (1967), *Biochem. J.* **105**, 427.

135. González-Cadavid, N. F., and Campbell, P. N. (1967), *Biochem. J.* **105**, 443.

136. Goodwin, T. W. ed. (1966), "Biochemistry of Chloroplasts," Academic Press, Inc., New York.

137. Graffi, A., Butschak, G., and Schneider, E. J. (1965), *Biochem. Biophys. Res. Commun.* **21**, 418.

138. Granick, S. (1963), "Cytodifferentiation and Macromolecular Synthesis," p. 144 (M. Locke, ed.), Academic Press, Inc., New York.

139. Granick, S., and Gibor, A. (1967), *Prog. Nucl. Acid Res. Mol. Biol.* **6,** 143.
140. Grant, P. (1965), "The Biochemistry of Animal Development," p. 488 (R. Weber, ed.), Academic Press, Inc., New York.
141. Greengard, O. (1959), *Biochim. Biophys. Acta* **32,** 270.
142. Greengard, O., and Campbell, P. N. (1959), *Biochem. J.* **72,** 305.
143. Grivell, L. A. (1967), *Biochem. J.* **105,** 44 C.
144. Gurr, M. I., Prottey, C., and Hawthorne, J. N. (1965), *Biochim. Biophys. Acta* **106,** 357.
145. Guttes, E., and Guttes, S. (1964), *Science* **145,** 1057.
146. Guttes, S., Guttes, E., and Hadek, R. (1966), *Experientia* **22,** 452.
147. Guttes, E. W., Hanawalt, P. C., and Guttes, S. (1967), *Biochim. Biophys. Acta* **142,** 181.
148. Guttman, H. N., and Eisenman, R. N. (1965), *Nature* **207,** 1280.
149. Haggis, A. J. (1964), *Develop. Biol.* 10, 358.
150. Haldar, D., Freeman, K., and Work, T. S. (1966), *Nature* **211,** 9.
151. Harel, L., Jacob, A., and Moulé, Y. (1957), *Exp. Cell Res.* **13,** 181.
152. Harvey, E. B. (1946), *J. Exptl. Zool.* **102,** 253.
153. Harvey, E. B. (1953), *J. Histochem. Cytochem.* **1,** 265.
154. Hawley, E. S., and Wagner, R. P. (1967), *J. Cell Biol.* **35,** 489.
155. Heldt, H. W., Jacobs, H., and Klingenberg, M. (1965), *Biochem. Biophys. Res. Commun.* **18,** 174.
156. Heyman-Blanchet, T., Zajdela, F., and Chaix, P. (1959), *Biochim. Biophys. Acta* **36,** 569.
157. Hoch, F. L. (1962), *New Engl. J. Med.* **266,** 446.
158. Hoch, F. L. (1962), *New Engl. J. Med.* **266,** 498.
159. Hoffman, H., and Grigg, G. W. (1958), *Exp. Cell Res.* **15,** 118.
160. Hogeboom, G. H., and Schneider, W. C. (1952), *J. Biol. Chem.* **197,** 611.
161. Hogeboom, G. H., and Schneider, W. C. (1955), "The Nucleic Acids," Vol. 2, p. 199 (Chargaff, E., and Davidson, J. N., eds.), Academic Press Inc., New York.
162. Hogeboom, G. H., Schneider, W. C., and Palade, G. E. (1948), *J. Biol. Chem.* **172,** 619.
163. Honig, G. R., and Rabinovitz, M. (1965), *Science* **149,** 1504.
164. Horne, R. W., and Newton, B. A. (1958), *Exp. Cell Res.* **15,** 103.
165. Horowitz, N. H., Bonner, D., and Houlahan, M. B. (1945), *J. Biol. Chem.* **159,** 145.
166. Howell, R. R., Loeb, J. N., and Tomkins, G. M. (1964), *Proc. Natl. Acad. Sci.* **52,** 1241.
167. Huang, M., Biggs, D. R., Clark-Walker, G. D., and Linnane, A. W. (1966), *Biochim. Biophys. Acta* **114,** 434.
168. Hudson, B., and Vinograd, J. (1967), *Nature* **216,** 647.
169. Hulsmans, H. A. M. (1961), *Biochim. Biophys. Acta* **54,** 1.
170. Humm, D. G., and Humm, J. H. (1966), *Proc. Natl. Acad. Sci.* **55,** 114.
171. Hurwitz, J., Bresler, A., and Diringer, R. (1960), *Biochem. Biophys. Res. Commun.* **3,** 15.
172. Jacob, F., and Monod, J. (1961), *Cold Spring Harbor Sympos. Quant. Biol.* **26,** 193.

173. Jacobs, E. E., and Sanadi, D. R. (1960), *J. Biol. Chem.* **235**, 531.
174. Jayaraman, J., Cotman, C., Mahler, H. R., and Sharp, C. W. (1966), *Arch. Biochem. Biophys.* **116**, 224.
174a. Jollow, D., Kellerman, G. M., and Linnane, A. W. (1968), *J. Cell Biol.* **37**, 221.
175. Kadenbach, B., *4th Meeting Fed. Europ. Biochem. Socs.*, Oslo 1967, Abstract 75.
176. Kadenbach, B. (1967), *Biochim. Biophys. Acta* **138**, 651.
177. Kadenbach, B. (1966), *Biochim. Biophys. Acta* **134**, 430.
178. Kagawa, Y., and Racker, E. (1966), *J. Biol. Chem.* **241**, 2461.
179. Kagawa, Y., and Racker, E. (1966), *J. Biol. Chem.* **241**, 2467.
180. Kalf, G. F. (1963), *Arch. Biochem. Biophys.* **101**, 350.
181. Kalf, G. F. (1964), *Biochemistry* **3**, 1702.
182. Kalf, G. F., and Gréce, M. A. (1964), *Biochem. Biophys. Res. Commun.* **17**, 674.
183. Kalf, G. F., and Gréce, M. A. (1966), *J. Biol. Chem.* **241**, 1019.
184. Kalf, G. F., and Simpson, M. V. (1959), *J. Biol. Chem.* **234**, 2943.
185. Kazakova, T. B., and Markosian, K. A. (1966), *Nature* **211**, 79.
186. Kemp, A., Jr., and Slater, E. C. (1964), *Biochim. Biophys. Acta* **92**, 178.
187. Kislev, N., Swift, H., and Bogorad, L. (1965), *J. Cell Biol.* **25**, 327.
188. Kit, S. (1961), *J. Mol. Biol.* **3**, 711.
188a. Klee, C. B., and Sokoloff, L. (1965), *Proc. Natl. Acad. Sci.* **53**, 1014.
189. Kleinschmidt, A. K., Burton, A., and Sinsheimer, R. L. (1963), *Science* **142**, 961.
190. Koch, J., and Stockstad, E. L. R. (1967), *Europ. J. Biochem.* **3**, 1.
191. Korner, A. (1959), *Exp. Cell Res.* **18**, 594.
192. Kováč, L., Šubik, J., Russ, G., and Kollár, K. (1967), *Biochim. Biophys. Acta* **144**, 94.
193. Kováč, L., and Weissová, K. (1968), *Biochim. Biophys. Acta* **153**, 55.
194. Kraepelin, G. (1967), *Z. Allg. Mikrobiol.* **7**, 287.
195. Kroon, A. M. (1963), *Biochim. Biophys. Acta* **69**, 184.
196. Kroon, A. M. (1963), *Biochim. Biophys. Acta* **72**, 391.
197. Kroon, A. M. (1963), *Biochim. Biophys. Acta* **76**, 165.
198. Kroon, A. M. (1964), *Biochim. Biophys. Acta* **91**, 145.
199. Kroon, A. M., cf Ref. 35, p. 397.
200. Kroon, A. M. (1965), *Biochim. Biophys. Acta* **108**, 275.
201. Kroon, A. M., cf Ref. 377, p. 207.
202. Kroon, A. M., Borst, P., van Bruggen, E. F. J., and Ruttenberg, G. J. C. M. (1966), *Proc. Natl. Acad. Sci.* **56**, 1836.
203. Kroon, A. M., Botman, M. J., and Saccone, C., *4th Meeting Fed. Europ. Biochem. Socs.*, Oslo 1967, Abstract 19.
204. Kroon, A. M., and Jansen, R. J. (1968), *Biochim. Biophys. Acta* **155**, 629.
205. Kroon, A. M., Saccone, C., and Botman, M. J. (1967), *Biochim. Biophys. Acta* **142**, 552.
206. Küntzel, H., and Noll, H. (1967), *Nature* **215**, 1340.
207. Lado, P., and Schwendimann, M. (1967), *Ital. J. Biochem.* **15**, 279.
208. Lafontaine, J. G., and Allard, C. (1964), *J. Cell Biol.* **22**, 143.

209. Laird, A. K., Nygaard, Q., Ris, H., and Barton, A. D. (1953), *Exp. Cell Res.* **5**, 147.
210. Lamb, A. J., Clark-Walker, G. D., and Linnane, A. W. (1968), *Biochim. Biophys. Acta,* **161**, 415.
211. Lance, C. (1961), *Compt. Rend.* **252**, 933.
212. Lansing, A. I., Hillier, J., and Rosenthal, T. B. (1952), *Biol. Bull.* **103**, 294.
213. Lederberg, J. (1952), *Physiol. Rev.* **32**, 403.
214. Leff, J., Mandel, M., Epstein, H. T., and Schiff, J. A. (1963), *Biochem. Biophys. Res. Commun.* **13**, 126.
215. Lehninger, A. L. (1964), "The Mitochondrion," John Wiley & Sons, Inc. New York.
216. Lehninger, A. L., and Gregg, C. T. (1963), *Biochim. Biophys. Acta* **78**, 12.
217. Lenaz, G. (1968), *Ital. J. Biochem.* **17**, 129.
218. Lever, J. D. (1956), *Endocrinol.* **58**, 163.
219. Lever, J. D. (1956), *J. Biophys. Biochem. Cytol.* **2**, Suppl. 313.
220. Linnane, A. W., cf Ref. 377, p. 333.
221. Linnane, A. W., Lamb, A. W., Christodoulou, C., and Lukins, H. B. (1968), *Proc. Natl. Acad. Sci.,* **59**, 1288.
222. Linnane, A. W., Saunders, G. W., Gingold, E. B., and Lukins, H. B. (1968), *Proc. Natl. Acad. Sci.,* **59**, 903.
223. Lowther, D. A., Green, N. M., and Chapman, J. A. (1961), *J. Biophys. Biochem. Cytol.* **10**, 373.
224. Luck, D. J. L. (1963), *J. Cell Biol.* **16**, 483.
225. Luck, D. J. L. (1963), *Proc. Natl. Acad. Sci.* **49**, 233.
226. Luck, D. J. L. (1965), *J. Cell Biol.* **24**, 445.
227. Luck, D. J. L. (1965), *J. Cell Biol.* **24**, 461.
228. Luck, D. J. L. (1966). "Probleme der biologischen Reduplikation" (P. Sitte, ed.), Springer Verlag, Berlin.
229. Luck, D. J. L., and Reich, E. (1964), *Proc. Natl. Acad. Sci.* **52**, 931.
230. Lund, H. A., Vatter, A. E., and Hanson, J. B. (1958), *J. Biophys. Biochem. Cytol.* **4**, 87.
231. Lusena, C. V., and Depocas, F. (1966), *Can. J. Biochem.* **44**, 497.
232. Lwoff, A., and Lwoff, M. (1931), *Bull. Biol. Franc. Belg.* **65**, 170.
233. MacHattie, L. A., Berns, K. I., and Thomas, C. A., Jr. (1965), *J. Mol. Biol.* **11**, 648.
234. Mager, J. (1960), *Biochim. Biophys. Acta* **38**, 150.
235. Manton, I. (1959), *J. Mar. Biol.* **38**, 319.
236. Manton, I. (1961), *J. Exp. Bot.,* **12**, 421.
237. Marr, A. G. (1960), *Ann. Rev. Microbiol.* **14**, 241.
237a. McCarthy, B. J. (1965), *Prog. Nucl. Acid Res. Mol. Biol.* **4**, 129.
238. McLean, J. R., Cohn, G. L., Brandt, I. K., and Simpson, M. V. (1958), *J. Biol. Chem.* **233**, 657.
238a. Mehotra, B. D., and Mahler, H. R. (1968), *Arch. Biochem. Biophys.* **128**, 685.
239. Mitchell, M. B., and Mitchell, H. K. (1952), *Proc. Natl. Acad. Sci.* **38**, 442.

240. Mitchell, H. B., Mitchell, H. K., and Tissières, A. (1953), *Proc. Natl. Acad. Sci.* **39**, 606.
241. Monroy, G. C., and Pullman, M. E., cited in Ref. 293.
242. Moriber, L. G., Hershenov, B., Aaronson, S., and Bensky, B. (1963), *J. Protozool.* **10**, 80.
243. Morpurgo, G., Serlupi-Creszenzi, G., Tecce, G., Valente, F., and Venettacci, D. (1964), *Nature* **201**, 897.
244. Mortimer, R. K., and Hawthorne, D. C. (1966), *Ann. Rev. Microbiol.* **20**, 151.
245. Mounolou, J. C., Jakob, H., and Slonimski, P. P. (1966), *Biochem. Biophys. Res. Commun.* **24**, 218.
246. Mounolou, J. C., Jakob, H., and Slonimski, P. P., cf Ref. 377, p. 473.
247. Moustacchi, E., and Marcovich, H. (1963), *Compt. Rend.* **256**, 5646.
248. Moustacchi, E., and Williamson, D. H. (1966), *Biochem. Biophys. Res. Commun.* **23**, 56.
249. Mühlethaler, K., and Bell, P. R. (1962), *Naturwiss.* **49**, 63.
250. Muntwyler, E., Seifter, S., and Harkness, D. M. (1950), *J. Biol. Chem.* **184**, 181.
251. Nagai, S., Yanagishima, N., and Nagai, H. (1961), *Bacteriol. Rev.* **25**, 404.
252. Nass, M. M. K. (1966), *Proc. Natl. Acad. Sci.* **56**, 1215.
253. Nass, M. M. K. (1967), "Organizational Biosynthesis," p. 503 (H. J. Vogel, J. O. Lampen, and V. Bryson, eds.), Academic Press, Inc., New York.
254. Nass, S. (1967), *Biochim. Biophys. Acta* **145**, 60.
255. Nass, M. M. K., and Nass, S. (1963), *J. Cell Biol.* **19**, 593.
256. Nass, S., and Nass, M. M. K. (1963), *J. Cell Biol.* **19**, 613.
257. Nass, S., and Nass, M. M. K. (1964), *J. Natl. Cancer Inst.* **33**, 777.
258. Nass, M. M. K., Nass, S., and Afzelius, B. A. (1965), *Exp. Cell Res.* **37**, 516.
259. Nass, S., Nass, M. M. K., and Hennix, U. (1965), *Biochim. Biophys. Acta* **95**, 426.
260. Neifakh, S. A., and Kazakova, T. B. (1963), *Nature* **197**, 1106.
261. Neubert, D., Ref. 35, p. 451.
262. Neubert, D., and Helge, H., Second Meeting Fed. Europ. Biochem. Socs., Vienna 1965, e 84.
263. Neubert, D., Bass, R., and Helge, H. (1966), *Naturwiss.* **53**, 23.
264. Neubert, D., Oberdisse, E., and Bass, R., cf Ref. 377, p. 103.
265. Neubert, D., and Helge, H. (1965), *Biochem. Biophys. Res. Commun.* **18**, 600.
266. Neubert, D., Helge, H., and Bass, R. (1965), *Arch. Exptl. Path. Pharmakol..* **252**, 258.
267. Neubert, D., Helge, H., and Merker, H.-J. (1965), *Biochem. Z.* **343**, 44.
268. Neubert, D., Helge, H., and Teske, S. (1966), *Arch. Exptl. Path. Pharmakol.* **252**, 452.
269. Neubert, D., and Morris, H. P. (1966), *Arch. Exptl. Path. Pharmakol.* **255**, 51.

270. Neubert, D., Oberdisse, E., Schmieder, M., and Reinisch, I. (1967), *Z. physiol. Chem.* **348,** 1709.
271. Neupert, W. (1967), Thesis, University of Munich (Germany).
272. Neupert, W., Brdiczka, D., and Bücher, Th. (1967), *Biochem. Biophys. Res. Commun.* **27,** 488.
273. Nilova, V. K., and Sukhanova, K. M. (1966), *Doklady Akademij Nauk SSR.* **168,** 921.
274. Novikoff, A. B. (1961), "The Cell," Vol. 2, p. 229 (J. Brachet, and A. E. Mirsky, eds.), Academic Press, Inc., New York.
275. O'Brien, T. W., and Kalf, G. F. (1967), *J. Biol. Chem.* **242,** 2172.
276. OBrien, T. W., and Kalf, G. F. (1967), *J. Biol. Chem.* **242,** 2180.
277. Ohnishi, T., and Ohnishi, T. (1962), *J. Biochem.* **51,** 380.
277a. Paltauf, F., and Schatz, G. (1969), *Biochemistry* **8,** 335.
278. Parker, J. H., and Sherman, F., 12th Int. Congr. Genetics, 1968, in press.
279. Parsons, D. F. (1963), *Science* **140,** 985.
280. Parsons, J. A., (1965), *J. Cell Biol.* **25,** 641.
281. Parsons, J. A., and Dickson, R. C. (1965), *J. Cell Biol.* **27,** 77 A.
282. Parsons, P., and Simpson, M. V. (1967), *Science* **155,** 91.
283. Pascaud, M. (1964), *Biochim. Biophys. Acta* **84,** 528.
284. Penefsky, H. S., Pullman, M. E., Datta, A., and Racker, E. (1960), *J. Biol. Chem.* **235,** 3330.
285. Pikó, L., Tyler, A., and Vinograd, J. (1967), *Biol. Bull.* **132,** 68.
286. Pitelka, D. R. (1961), *Exp. Cell Res.* **25,** 87.
287. Pittman, D. (1959), *Cytologia* **24,** 315.
288. Pittman, D., Webb, J. M., Roshanmanesh, A., and Coker, L. E. (1960), *Genetics* **45,** 1023.
288a. Plattner, H., and Schatz, G. (1969), *Biochemistry* **8,** 339.
289. Polakis, E. S., Bartley, W., and Meek, G. A. (1964), *Biochem. J.* **90,** 369.
290. Polakis, E. S., Bartley, W., and Meek, G. A. (1965), *Biochem. J.* **97,** 298.
291. Potter, V. R., Recknagel, R. O., and Hurlbert, R. B. (1951), *Federat. Proc.* **10,** 646.
292. Pullman, M. E., and Monroy, G. C. (1963), *J. Biol. Chem.* **238,** 3762.
293. Pullman, M. E., and Schatz, G. (1967), *Ann. Rev. Biochem.* **36,** 539.
294. Rabinowitz, M., DeSalle, L., Sinclair, J., Stirewalt, R., and Swift, H. (1966), *Fed. Proc.* **25,** 581.
295. Rabinowitz, M., Sinclair, J., DeSalle, L., Haselkorn, R., and Swift, H. H. (1965), *Proc. Natl. Acad. Sci.* **53,** 1126.
296. Rabinowitz, M., Snoble, E., Sanghavi, P., Getz, G. S., and Heywood, J., 4th Meeting Europ. Fed. Biochem. Socs. Oslo 1967, Abstract 181.
297. Randall, J. T. (1959), *Protozool.* **6,** 30.
298. Raut, C. (1953), *Exp. Cell Res.* **4,** 295.
299. Raut, C. (1954), *J. Cell. Comp. Physiol.* **44,** 463.
300. Raut, C., and Simpson, W. L. (1955), *Arch. Biochem. Biophys.* **57,** 218.
301. Ray, D. S., and Hanawalt, P. C. (1965), *J. Mol. Biol.* **11,** 760.
302. Reich, E., and Luck, D. J. L. (1966), *Proc. Natl. Acad. Sci.* **55,** 1600.
303. Reis, P. J., Coote, J. L., and Work, T. S. (1959), *Nature* **184,** 165.
304. Rendi, R. (1959), *Exp. Cell Res.* **17,** 585.
305. Rendi, R. (1959), *Exp. Cell Res.* **18,** 187.

306. Revel, M., and Hiatt, H. H. (1964), *Proc. Natl. Acad. Sci.* **51**, 810.
307. Reynolds, E. S., Thiers, R. E., and Vallee, B. L. (1962), *J. Biol. Chem.* **237**, 3546.
308. Rifkin, M. R., Wood, D. D., and Luck, D. J. L. (1967), *Proc. Natl. Acad. Sci.* **58**, 1025.
308a. Rion, G., and Delain, E. (1969), *Proc. Natl. Acad. Sci.* **62**, 210.
309. Ris, H., and Plaut, W. (1962), *J. Cell Biol.* **13**, 383.
310. Robertis, E. D. P., Nowinski, W. W., and Saez, F. S. (1954), General Cytology, Saunders, Philadelphia.
311. Robertson, J. D. (1961), "Regional Neurochemistry," p. 497, Pergamon Press Inc., Oxford, England.
312. Rogers, P. J., Preston, B. N., Titchener, E. B., and Linnane, A. W. (1967), *Biochem. Biophys. Res. Commun.* **27**, 405.
313. Roodyn, D. B. (1962), *Biochem. J.* **85**, 177.
314. Roodyn, D. B., cf. Ref. 35. p. 383.
315. Roodyn, D. B. (1965), *Biochem. J.* **97**, 782.
316. Roodyn, D. B. (ed.) (1967), "Enzyme Cytology," London, Academic Press.
317. Roodyn, D. B., Freeman, K. B., and Tata, J. R. (1965), *Biochem. J.* **94**, 628.
318. Roodyn, D. B., Reis, P. J., and Work, T. S. (1961), *Biochem. J.* **80**, 9.
319. Roodyn, D. B., Suttie, J. W., and Work, T. S. (1962), *Biochem. J.* **83**, 29.
319a. Roodyn, D. B., and Wilkie, D. (1968), "The Biogenesis of Mitochondria," Methuen, Ltd., London.
320. Rouiller, Ch. (1960), *Intern. Rev. Cytol.* **9**, 227.
320a. Rouslin, W., and Schatz, G., unpublished experiments.
321. Saccone, C., Gadaleta, M. N., and Quagliariello, E. (1967), *Biochim. Biophys. Acta* **138**, 474.
322. Sandell, S., Löw, H., and von der Decken, A. (1967), *Biochem. J.* **104**, 575.
323. Sarachek, A. (1958), *Cytologia* **23**, 143.
324. Sarachek, A. (1959), *Cytologia* **24**, 507.
325. Sarkissian, I. V., and McDaniel, R. G. (1967), *Proc. Natl. Acad. Sci.* **57**, 1262.
326. Sarkissian, I. V., and Srivastana, H. K. (1967), *Genetics* **57**, 843.
327. Schatz, G. (1963), *Biochem. Biophys. Res. Commun.* **12**, 448.
328. Schatz, G. (1965), *Biochim. Biophys. Acta* **96**, 342.
329. Schatz, G. (1968), *J. Biol. Chem.* **243**, 2192.
329a. Schatz, G., unpublished experiments.
330. Schatz, G. (1969), this volume, p. 251.
331. Schatz, G., and Criddle, R. S. (1968), "Symposium Mitochondria—Structure and Function," (L. Ernster and Z. Drahota, eds.), 5th Meeting Fed. Europ. Biochem. Socs., Prague 1968, in press.
332. Schatz, G., Haslbrunner, E., and Tuppy, H. (1964), *Biochem. Biophys. Res. Commun.* **15**, 127.
333. Schatz, G., Haslbrunner, E., and Tuppy, H. (1964), *Mh. Chemie* **95**, 1135.

334. Schatz, G., Penefsky, H. S., and Racker, E. (1967), *J. Biol. Chem.* **242,** 2552.

334a. Schatz, G., and Saltzgaber, J. (1969), *Biochim. Biophys. Acta* **180,** 186.

335. Schatz, G., Tuppy, H., and Klima, J. (1963), *Z. Naturforsch.* **18b,** 145.

336. Schmieder, M., and Neubert, D. (1966), *Arch. Exptl. Path. Pharmakol.* **255,** 68.

337. Schneider, W. C., and Kuff, E. L. (1965), *Proc. Natl. Acad. Sci.* **54,** 1650.

338. Schuster, F. L. (1965), *Exp. Cell Res.* **39,** 329.

339. Sebald, W. (1968), 5th Meeting Fed. Europ. Biochem. Socs., Prague 1968, Abstract 374.

339a. Serlupi-Crescenzi, G., and Barcellona, S. (1966), Ann. Ist. Super. Sanita **2,** 431.

340. Sherman, F. (1964), *Genetics* **49,** 39.

341. Sherman, F., and Slonimski, P. P. (1964), *Biochim. Biophys. Acta* **90,** 1.

342. Sherman, F., Stewart, J. W., Margoliash, E., Parker, J., and Campbell, W. (1966), *Proc. Natl. Acad. Sci.* **55,** 1498.

343. Sherman, F., Taber, H., and Campbell, W. (1965), *J. Mol. Biol.* **13,** 21.

344. Siekevitz, P., and Watson, M. L. (1956), *J. Biophys. Biochem. Cytol.* **2,** 653.

345. Simpson, M. V. (1962), *Ann. Rev. Biochem.* **31,** 333.

346. Simpson, M. V., Skinner, D. M., and Lucas, J. M. (1961), *J. Biol. Chem.* **236,** PC 81.

347. Sinclair, J. H., and Stevens, B. J. (1966), *Proc. Natl. Acad. Sci.* **56,** 508.

348. Sinclair, J. H., Stevens, B. J., Gross, N., and Rabinowitz, M. (1967), *Biochim. Biophys. Acta* **145,** 528.

349. Sinclair, J. H., Stevens, B. J., Sanghavi, P., and Rabinowitz, M. (1967), *Science* **156,** 1234.

350. Singh, V. N., Raghupathy, E., and Chaikoff, I. L. (1964), *Biochem. Biophys. Res. Commun.* **16,** 12.

350a. Sissakian, N. M., and Filippovich, I. I. (1957), *Biokhimiya* **22,** 375.

351. Sitte, P. (ed.) (1966), "Probleme der biologischen Reduplikation," Springer Verlag, Berlin.

352. Skinner, D. M., and Triplett, L. L. (1967), *Biochem. Biophys. Res. Commun.* **28,** 892.

353. Slonimski, P. P. (1950), *Compt. Rend.* **231,** 375.

354. Slonimski, P. P. (1953), "La Formation des Enzymes Respiratoires chez la Levure," Masson, Paris.

355. Slonimski, P. P., Acher, R., Péré, G., Sels, A., and Somlo, M. (1965), "Mécanismes de régulation des activités cellulaires chez les micro-organismes," p. 435, Ed. du CNRS, Paris.

356. Slonimski, P. P., and Hirsch, H. M. (1952), *Compt. Rend.* **235,** 914.

356a. Smith, D., Tauro, P., Schweizer, E., and Halvorson, H. O. (1968), *Proc. Natl. Acad. Sci.* **60,** 936.

357. So, A. G., and Davie, E. W. (1963), *Biochemistry* **2,** 132.

358. Somlo, M., and Fukuhara, H. (1965), *Biochem. Biophys. Res. Commun.* **19,** 587.

358a. Sonenshein, G. E., and Holt, C. E. (1968), *Biochem. Biophys. Res. Commun.* **33,** 361.

358b. South, D. J., and Mahler, H. R. (1968), *Nature* **218**, 1226.
359. Steinert, M. (1960), *J. Biophys. Biochem. Cytol.* **8**, 542.
360. Steinert, G., Firket, H., and Steinert, M. (1958), *Exp. Cell Res.* **15**, 632.
361. Stewart, J. W., and Sherman, F., 12. Intern. Congr. Genetics, 1968, in press.
362. Stich, H. F. (1962), *Exp. Cell Res.* **26**, 136.
363. Stoeckenius, W. (1963), *J. Cell Biol.* **17**, 443.
364. Stone, G. E., and Miller, O. L., Jr. (1964), *J. Cell Biol.* **23**, 89A.
365. Stone, G. E., Miller, O. L., Jr., and Prescott, D. M. (1964), *J. Protozool.* **11**, Suppl. 24.
366. Sueoka, N. (1961), *J. Mol. Biol.* **3**, 31.
367. Suttie, J. W. (1962), *Biochem. J.* **84**, 382.
368. Suyama, Y. (1966), *Biochemistry* **5**, 2214.
369. Suyama, Y. (1967), *Biochemistry* **6**, 2829.
370. Suyama, Y., and Bonner, W. D., Jr. (1966), *Plant Physiol.* **41**, 383.
371. Suyama, Y., and Eyer, J. (1967), *Biochem. Biophys. Res. Commun.* **28**, 746.
372. Suyama, Y., and Eyer, J., cited in Ref. 369.
373. Suyama, Y., and Preer, J. R., Jr. (1965), *Genetics* **52**, 1051.
374. Swartz, M. N., Trautner, T. A., and Kornberg, A. (1962), *J. Biol. Chem.* **237**, 1961.
375. Swift, H. (1965), *Amer. Naturalist* **99**, 201.
376. Swift, H., Adams, B. J., and Larsen, K. (1964), *J. Roy. Microsc. Soc.* **83**, 161.
377. Tager, J. M., Papa, S., Quagliariello, E., and Slater, E. C. (eds.), (1968), "Round Table Discussion on Biochemical Aspects of the Biogenesis of Mitochondria," Adriatica Editrice, Bari.
378. Tamishian, T. N., Powers, E. L., and Devine, R. L. (1956), *J. Biophys. Biochem. Cytol.* **2**, Suppl. 325.
379. Tata, J. R., cf. Ref. 35, p. 489.
380. Taylor, C. B., Bailey, E., and Bartley, W. (1967), *Biochem. J.* **105**, 605.
381. Tewari, K. K., Jayaraman, J., and Mahler, H. R. (1965), *Biochem. Biophys. Res. Commun.* **21**, 141.
382. Tewari, K. K., Vötsch, W., Mahler, H. R., and Mackler, B. (1966), *J. Mol. Biol.* **20**, 453.
383. Truman, D. E. S. (1963), *Exp. Cell Res.* **31**, 313.
383a. Truman, D. E. S. (1964), *Biochem. J.* **91**, 59.
384. Truman, D. E. S., and Korner, A. (1962), *Biochem. J.* **83**, 588.
385. Truman, D. E. S., and Korner, A. (1962), *Biochem. J.* **85**, 154.
386. Truman, D. E. S., and Löw, H. (1963), *Exp. Cell Res.* **31**, 330.
387. Tuppy, H., Haslbrunner, E., and Schatz, G. (1965), *Mh. Chemie* **96**, 1831.
388. Tuppy, H., and Swetly, P. (1968), *Biochim. Biophys. Acta* **153**, 293.
389. Tuppy, H., Swetly, P., and Wolff, I. (1968), *Europ. J. Biochem.* **5**, 339.
390. Umaña, R., and Dounce, A. L. (1964), *Exp. Cell Res.* **35**, 277.
391. Vasquez, D. (1966), *Sympos. Soc. Gen. Microbiol.* **16**, 169.
392. Vickermann, K. (1962), *Trans. Roy. Soc. Trop. Med. Hyg.* **56**, 270.
393. Viehhauser, G., Wintersberger, E., and Tuppy, H., 2nd Intern. Biophysics Congr., Vienna 1966, Abstract 113.

394. Viehhauser, G., Wintersberger, E., and Tuppy, H., in preparation.
395. Vignais, P. V., Vignais, P. M., Rossi, C. S., and Lehninger, A. L. (1963), *Biochem. Biophys. Res. Commun.* **11**, 307.
396. Vignais, P. V., Vignais, P. M., and Stanislas, E. (1962), *Biochim. Biophys. Acta* **60**, 284.
397. Vinograd, J., Lebowitz, J., Radloff, R., Watson, R., and Laipis, P. (1965), *Proc. Natl. Acad. Sci.* **53**, 1104.
398. Vitols, E., North, R. J., and Linnane, A. W. (1961), *J. Biophys. Biochem. Cytol.* **9**, 689.
399. Wagner, R. P., and Mitchell, H. K. (1955), "Genetics and Metabolism," p. 186, John Wiley & Sons, Inc., New York.
399a. Wagner, R. P. (1969), *Science* **163**, 1026.
400. Wallace, P. G., and Linnane, A. W. (1964), *Nature* **201**, 1191.
401. Watson, M. L., and Aldridge, W. G. (1964), *J. Histochem. Cytochem.* **12**, 96.
402. Weil, R., and Vinograd, J. (1963), *Proc. Natl. Acad. Sci.* **50**, 730.
403. Weiss, S. B. (1960), *Proc. Natl. Acad. Sci.* **46**, 1020.
404. Wheeldon, L. W., and Lehninger, A. L. (1966), *Biochemistry* **5**, 3533.
405. Wheeler, H., and Hanchey, P. J. (1966), *Science* **154**, 1569.
406. Wilgram, G. F., and Kennedy, E. P. (1963), *J. Biol. Chem.* **238**, 2615.
407. Wilkie, D. (1963), *J. Mol. Biol.* **7**, 527.
408. Wilkie, D., Saunders, G., and Linnane, A. W. (1967), *Genet. Res.* **10**, 199.
409. Wilson, E. B., "The Cell in Heredity and Development," New York, Macmillan, 1928.
410. Wintersberger, E. (1964), *Z. Physiol. Chemie* **336**, 285.
411. Wintersberger, E. (1965), *Biochem. Z.* **341**, 409.
412. Wintersberger, E., cf. Ref. 35, p. 439.
413. Wintersberger, E. (1966), *Biochem. Biophys. Res. Commun.* **25**, 1.
414. Wintersberger, E., cf. Ref. 377, p. 189.
415. Wintersberger, E. (1967), *Z. Physiol. Chemie* **348**, 1701.
416. Winterberger, E., personal communication.
417. Wintersberger, E., and Tuppy, H. (1965), *Biochem. Z.* **341**, 399.
417a. Wintersberger, U., and Wintersberger, E. (1968), 5th Meeting Fed. Europ. Biochem. Socs., Prague 1968, Abstract 86. (1969), 6th Meeting Fed. Europ. Biochem. Socs., Madrid 1969, Abstract 774.
418. Wollgiehn, R., and Mothes, K. (1963), *Naturwiss.* **50**, 95.
419. Wolstenholme, D. R., and Dawid, I. B. (1967), *Chromosoma* **20**, 445.
420. Woodward, D. O., and Munkres, K. D. (1966), *Proc. Natl. Acad. Sci.* **55**, 872.
421. Woodward, D. O., and Munkres, K. D., *Proc. Rutgers Sympos.* 1966.
422. Work, T. S. (1967), *Biochem. J.* **105**, 38 P.
423. Work, T. S., cf. Ref. 377, p. 367.
424. Work, T. S., Intern. Conf. Biol. Membranes, Frascati, Italy, June 1967.
425. Yellin, T. O., Butler, B. J., and Stein, H. H. (1967), *Fed. Proc.* **26**, 833.
426. Yotsuyanagi, Y. (1962), *J. Ultrastruct. Res.* **7**, 121.
426a. Yotsuyanagi, Y. (1962), *J. Ultrastruct. Res.* **7**, 141.
427. Yotsuyanagi, Y. (1966), *Compt. Rend.* **262**, 1348.
428. Yotsuyanagi, Y., and Guerrier, C. (1965), *Compt. Rend.* **260**, 2344.
429. Zeigel, R. F. (1962), *J. Natl. Cancer Inst.* **28**, 269.

INDEX